AF710075

Sally Faulkner

Finite Mathematics

Under the Editorship of Carl B. Allendoerfer

Finite Mathematics

Hugh G. Campbell & Robert E. Spencer

Virginia Polytechnic Institute and State University

Macmillan Publishing Co., Inc.

New York

Collier Macmillan Publishers

London

Copyright © 1974, Hugh G. Campbell and Robert E. Spencer

PRINTED IN THE UNITED STATES OF AMERICA

All rights reserved. No part of this book may be reproduced or transmitted in any form or by any means, electronic or mechanical, including photocopying, recording, or any information storage and retrieval system, without permission in writing from the Publisher.

Macmillan Publishing Co., Inc.
866 Third Avenue, New York, New York 10022

Collier-Macmillan Canada, Ltd.

LIBRARY OF CONGRESS CATALOGING IN PUBLICATION DATA

Campbell, Hugh G
Finite mathematics.

Bibliography: p.
1. Mathematics—1961– I. Spencer, Robert E., joint author. II. Title.
QA39.2.C33 510 73-2761
ISBN 0-02-318570-8

Printing: 1 2 3 4 5 6 7 8 Year: 4 5 6 7 8 9 80

To Linnie, Dorothy, and Dorothy Lynn

Preface

One distinguishing feature of this book is its preoccupation with motivation—a vital part of the learning process. The authors have devoted considerable energy toward the goal of furnishing the reader with reasons for studying the mathematics included in this book. Almost every section in the book includes a subsection entitled "Applications"; these subsections, however, are not designed as comprehensive lists of real-world applications, nor is any claim made concerning the realism of many examples. Rather, these "indications of usefulness" are designed to point out in a limited and elementary way that the material is relevant to a broad spectrum of disciplines and also that it has considerable potential for future applications. Many of the applications are documented to permit further investigation by the reader.

A second distinguishing feature of this book is its organizational form. A conscious effort was made to organize the book in such a way that the instructor can alter the emphasis or length of a course to suit his needs. Many of the sections are independent of other sections and are so advertised at the outset of each chapter; all the applications are optional; and the format of the text is such that the theory can be emphasized or avoided as desired. The chart shown in Table A is given to summarize the flexibility of topic selection. Using Table A, instructors may construct alternative courses by varying the emphasis placed on each of the seven chapters; for example, one course can be constructed that gives major emphasis to Chapters 4, 5, 6, and 7 with as few as 6 lessons on Chapters 1, 2, and 3, whereas another course might give major emphasis to Chapters 1, 2, 3, and 4 with minimum or moderate emphasis on Chapters 5, 6, and 7.

Ordinarily, the material found in Chapter 1 is relegated to a less prominent position in the study of finite mathematics. The authors have found, however, that this material is ideal for an introduction to finite mathematics and certainly facilitates the study of logic. This is probably due in part to the fact that a great many diagrams are involved which allow the student to obtain a visual conception of some of the basic ideas. The material in the first chapter also offers the opportunity to present, very early in the course, diverse applications of interest.

Above all, the guiding principle in the construction of this book has been consideration for the needs of the student. The sincerity of the authors in this regard is illustrated by the attempt to motivate through the subsections on applications, the list of new vocabulary at the end of each chapter, the numerous exercises, and the many examples illustrating the concepts.

The authors express sincere appreciation to the many people who have assisted us in the production of this book. Special gratitude is expressed

Table A

	SUGGESTED SECTIONS FOR MINIMUM EMPHASIS	SUGGESTED SECTIONS FOR MODERATE EMPHASIS	SUGGESTED SECTIONS FOR MAXIMUM EMPHASIS
CHAPTER 1 (Systems)	1.1, 1.2 (1 or 2 lessons)	1.1–1.3 (3 lessons)	Entire chapter (4 lessons)
CHAPTER 2 (Logic)	2.1, 2.2 (2 lessons)	2.1–2.3 (3 lessons)	Entire chapter (5 lessons)
CHAPTER 3 (Sets)	3.1, 3.2, 3.5 (3 lessons)	3.1–3.3, 3.5, 3.6 (5 lessons)	Entire chapter (6 lessons)
CHAPTER 4 (Probability)	4.1–4.4, 4.10 (5 lessons)	4.1–4.6, 4.8, 4.10 (8 lessons)	Entire chapter (10 lessons)
CHAPTER 5 (Matrices)	5.1–5.3, 5.5 (4 lessons)	5.1–5.8 (8 lessons)	Entire chapter (10 lessons)
CHAPTER 6 (Linear Programming)	6.1, 6.2 (2 lessons)	6.1–6.4 (4 lessons)	Entire chapter (6 lessons)
CHAPTER 7 (Game Theory)	7.1, 7.2 (2 lessons)	7.1–7.3 (3 lessons)	Entire chapter (4 lessons)
TOTAL	19–20 lessons	34 lessons	45 lessons

to Dr. Carl Allendoerfer, Mr. Everett Smethurst, and Dr. H. Earl Spencer for their time and efforts in behalf of this project. Our respective wives, Allen and Carole, also deserve considerable thanks for their encouragement during the development of the book.

HUGH G. CAMPBELL
ROBERT E. SPENCER

Blacksburg, Virginia

Contents

0 An Orientation

Over the years the development of mathematics has closely paralleled the need for quantification of natural and other phenomena. From the time man invented symbols that correspond to our numerals 1, 2, . . . , there has been a relentless march toward the use of mathematics in ever more diverse ways. Certainly, the rate of this advance has not been constant and often has been related to the intellectual and political climates of the age as well as the maturity and influence of sister disciplines. Early interest in astronomy, for example, provided considerable impetus for the development of related mathematical concepts and techniques. Over a period of years, many of these concepts and techniques have led to a comparatively sophisticated development of the engineering sciences; the development of the computer and space travel give ample evidence of this fact. The computer and other engineering developments, in turn, are now having a tremendous effect upon modern mathematics, as well as having the effect of opening vast new areas for quantification in the social, management, and life sciences.

In this book we have chosen a few mathematical topics that have shown considerable promise in the social, management, and life sciences. Extensive quantification in these areas is a relatively new thing, and there is not yet universal acceptance of the usefulness of the mathematical topics in these disciplines. Although many criticisms are probably valid and useful, we should not lose sight of the fact that similar battles were fought over the usefulness of calculus in the natural sciences back in the 17th and 18th centuries. It should be emphasized that the art of application of the topics of this book to the behavioral, management, and life sciences is in its infancy, and while there is considerable reason for hope, much work remains to be done.

One of the predominant characteristics of modern mathematics is the stress placed on the structure of mathematics; this approach is very important as far as applications are concerned because it lends itself readily to the applied concept of quantitative models. Although the development of structure will be emphasized, it is not the purpose of this book to give an airtight, logical development of the topics selected. The purpose is to give the reader an idea of the nature of mathematical structure and modern mathematical thought, along with material that may be useful in quantitative applications in his discipline. In constructing a mathematical structure we begin with a foundation consisting of certain undefined and defined terms, together with a set of assumptions. Beyond this foundation we make conjectures and more definitions. Those conjectures that are justified according to certain rules of logic are called theorems and become part of the superstructure, so to speak. Applications then follow naturally from the structure. In other words, mathematical techniques can be thought of not as a collection of unrelated magical practices, but as natural operations performed within a unifying structure.

To illustrate these and other important ideas, a treatment of the customary topics of sets and logic is given by constructing in each of the first three chapters a specific mathematical structure which has elements or building blocks consisting of entities other than numbers. In the third chapter the material is further unified by showing that each of the first three structures is really a special case of a single abstract structure known as a Boolean Algebra.

1 Systems and the Algebra of Components

This chapter has a twofold purpose. (1) The reader is introduced to the idea of using a mathematical structure in which elements other than numbers are combined by various operations. Ideally, this will considerably broaden the reader's perspective of both the structure and uses of mathematics. (2) The study of logic, which is to be studied in Chapter 2, is made easier by first studying the analogous and easier material of this chapter. The diagrams of this chapter provide a visual conception of some of the basic ideas. Furthermore, this material is fundamental in the design of various electrical and electronic devices such as the digital computer. Although it is beyond the scope of this text to investigate the construction of these devices, some of the applications are intended to make clearer their nature, design, and use. Also, for those readers who have access to a computer terminal that is capable of accepting the APL computer language, an appendix to this chapter is provided for instruction in its use in making evaluations.

Prerequisites: high school algebra
Suggested Sections for Moderate Emphasis: 1.1–1.3
Suggested Sections for Minimum Emphasis: 1.1, 1.2 (For those who prefer to reserve more time for Chapters 4, 5, 6, and 7, a one- or two-lesson survey of 1.1 and 1.2 will satisfy the minimum prerequisites for Chapters 2 and 3.)

1.1 Components and Connectives

The word "system" is quite common in our current language. We frequently hear about such diverse and important systems as computer systems, digestive systems, business systems, spacecraft systems, political systems, transportation systems, and many others. It is important in various disciplines that participants have some idea of what is meant by a system and the various ways that mathematics can serve in analyzing, modifying, or improving identifiable systems; often this service can be achieved by means of constructing mathematical models of physical, natural, or social systems. This procedure will be illustrated later.

Webster defines the word system to be "a set or arrangement of things (components) so related or connected as to form a unity or organic whole." In this chapter we shall be concerned with those systems whose components are connected with parallel and series connections. In order to give the reader specific illustrations of the abstract concepts that we are trying to teach, we shall make our presentation of the algebra of components by giving the word **component** a specific interpretation. Consider a set of "on–off" switches similar to a common light switch found near the door of most rooms or a set of switches found in the interior of a digital computer. We shall use the letters p, q, r, s, t, and so on, to represent these switches. Switches can be connected in parallel as shown in Figure 1.1, or they can be connected in series as shown in Figure 1.2.

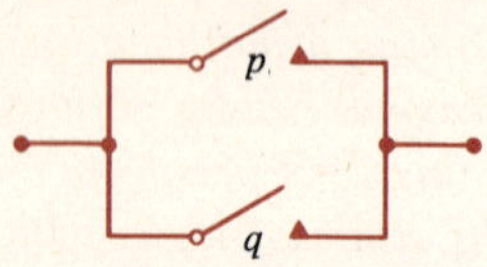

Figure 1.1. *Parallel connection of switches.*

Figure 1.2. *Series connection of switches.*

We shall designate a parallel connection of switches by the symbol $+$, and hence the arrangement of switches in **parallel** as shown in Figure 1.1 can be expressed symbolically as $p + q$. The symbol "$+$" can be read "is connected in parallel with." A connection of switches in **series** as shown in Figure 1.2 can be expressed as $r \times s$. The symbol "$\times$" can be read "is connected in series with." When two or more single switches are connected in parallel or series (or both), we have a system of switches. For reasons that will become apparent later, we choose to call such a system a **compound switch.** Large and complicated compound switches can be built, and, in fact, one of the purposes of this study is to point the way toward the analysis of compound switches with the view of simplifying them. Such simplification can be very useful in certain applications. Any single switch or compound switch that makes up part or all of a compound switch can be thought of as a **component** of a **system of switches.** The compound switch consisting of components p, q, and r and arranged according to Figure 1.3 can be expressed as $(p \times q) + (p \times r)$. We shall prove later that the compound switch $(p \times q) + (p \times r)$ will function in an identical manner as the simpler compound switch $p \times (q + r)$, which is shown in Figure 1.4.

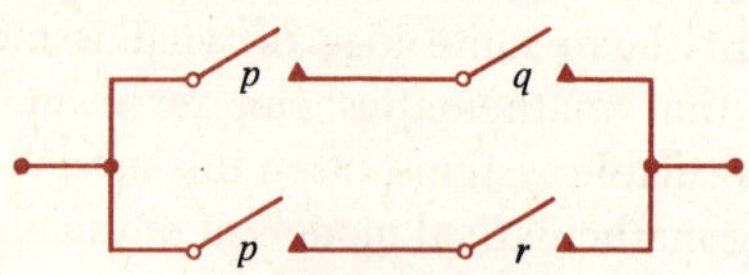

Figure 1.3

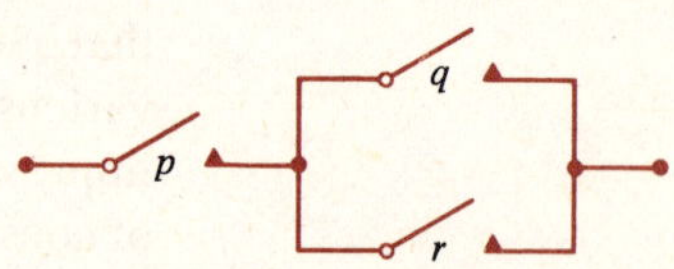

Figure 1.4

In Figure 1.3, notice that switch p appears twice. Two or more switches can be constructed in such a way that they are "on" at the same time and "off" at the same time; when this is the case, the same letter is used to represent the switches that behave in an identical manner.

Example 1

Sketch the diagram of the compound switch $[q + (r \times q)] \times p$. Start from within, namely $(r \times q)$, and sketch that as shown in Figure 1.5. Then connect compound switch $(r \times q)$ with q in parallel as shown in Figure 1.6. Finally, connect compound switch $[q + (r \times q)]$ in series with p as shown in Figure 1.7. ▮

Figure 1.5

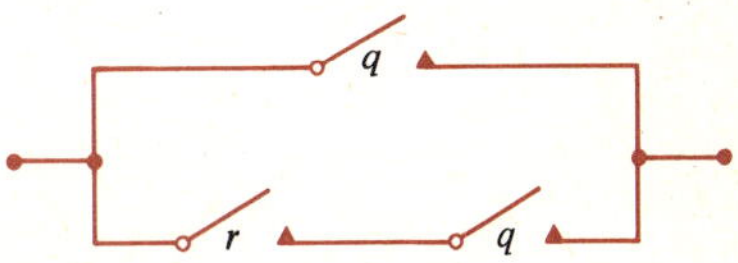

Figure 1.6

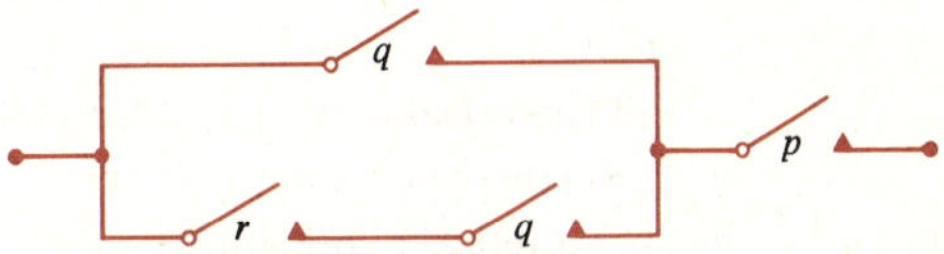

Figure 1.7

In this section we have begun to lay the foundation for a new mathematical structure which we will call the Algebra of Switches (components). The elements or building blocks of this structure are not numbers but are single or compound switches (components) p, q, r, s, t, and so on. Over these elements, the operations presented so far are "connection in parallel," denoted by the symbol $+$, and "connection in series," denoted by the symbol $\times$. Our construction is continued in the remaining sections of the chapter. Hereafter when the word "switch" ("component") is used, it means a single or compound switch (component).

APPLICATIONS

The applications given at the end of each section are certainly not meant to be an exhaustive listing of the applications of that particular section, nor is any claim made concerning the realism of many of the examples. Rather, these "evidences of usefulness" are supplied primarily in the interest of motivating the reader and with the hope that occasionally one will provide a spark that may arouse a much deeper, productive study in some areas of interest. All applications are optional in that they are not prerequisite to later material in the main part of the text.

Example 2

Consider a specific system having n parts or components that are interrelated in a certain way. A "system" may be given a variety of interpretations, such as a spacecraft, a football team, a management team, a computer, a living organism, and so forth. If the components of a system are interrelated in such a way that the entire system fails when any one of the components fails, then the system is called a **series system.** On the other hand a **parallel system** is one that will fail only if all of its components fail. Naturally we can have systems with both parallel and series systems as components. For example, consider a space flight in which we consider as the primary components the spacecraft, the pilot, and the copilot. Such a system can be diagrammed as shown in Figure 1.8.

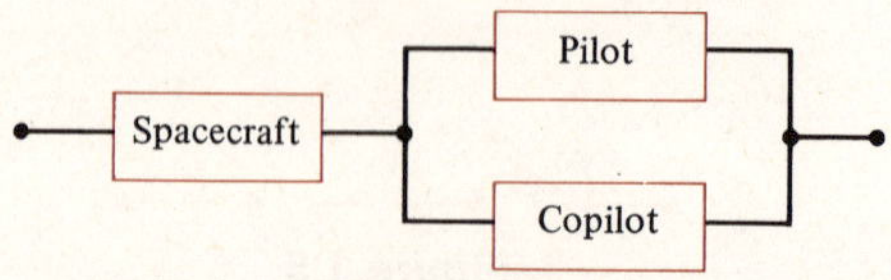

Figure 1.8

Let the **reliability** of a system be defined as the probability that the system will perform satisfactorily for a specified time interval under specified environmental conditions. In Chapter 4 after a study of probability is made, we can verify that the reliability R_s of a series system consisting of n independent components is the product of the reliabilities $R_1, R_2, \ldots, R_n$ of the respective components. That is, $R_s = (R_1)(R_2)\cdots(R_n)$. For example, if a series system had 3 components with reliabilities of $\frac{1}{3}$, $\frac{3}{4}$, and $\frac{5}{6}$, then the reliability R_s of the system would be

$$R_s = (\tfrac{1}{3})(\tfrac{3}{4})(\tfrac{5}{6}) = \tfrac{15}{72} \approx 21 \text{ per cent.}$$

We can also verify later that the reliability R_p of a parallel system with n independent components is given by the formula

$$R_p = 1 - [(1 - R_1)(1 - R_2)\cdots(1 - R_n)].$$

For example, if a parallel system had components with reliabilities of $\frac{1}{3}$, $\frac{3}{4}$, and $\frac{5}{6}$, then the reliability of the system would be

$$\begin{aligned} R_p &= 1 - [(1 - \tfrac{1}{3})(1 - \tfrac{3}{4})(1 - \tfrac{5}{6})] \\ &= 1 - [(\tfrac{2}{3})(\tfrac{1}{4})(\tfrac{1}{6})] \\ &= 1 - \tfrac{1}{36} \\ &= \tfrac{35}{36} \approx 97 \text{ per cent.} \end{aligned}$$

The reliability of a system can be improved by the appropriate inclusion of parallel components. In space travel such parallel subsystem components are often referred to as "backup systems." Further discussion of reliability may be found in Example 6 of Section 4.4, and in Miller [35], pp. 362–364. ∎

EXERCISES

Suggested minimum assignment: Exercises 1, 4, 5, 8, and 9.

In Exercises 1–4 sketch the diagrams of the following switches:

1. $(p + q) \times (r + s)$.
2. $(p \times q) + (r \times s)$.
3. $[p + (p \times q)] + (p \times r)$.
4. $[p \times (p + q)] \times (p + r)$.

5–8. For Exercises 5–8 write the symbolic expressions for the compound switches shown in Figures 1.9–1.12.

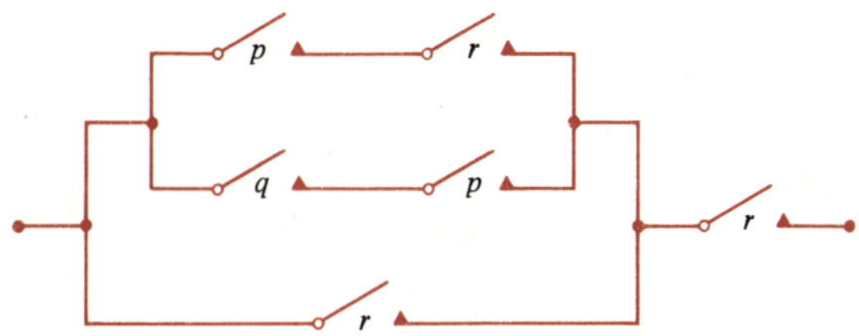

Figure 1.9

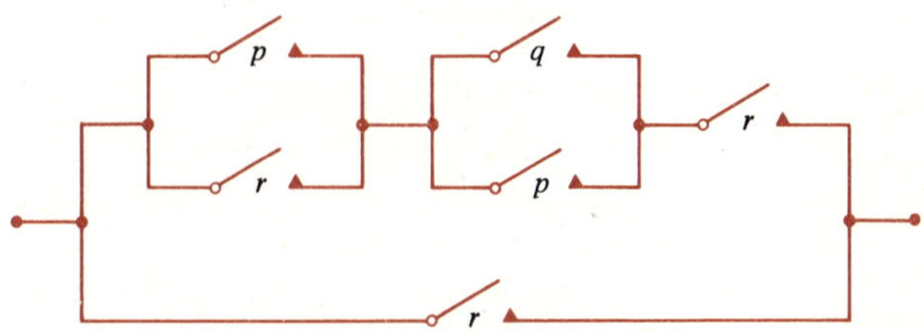

Figure 1.10

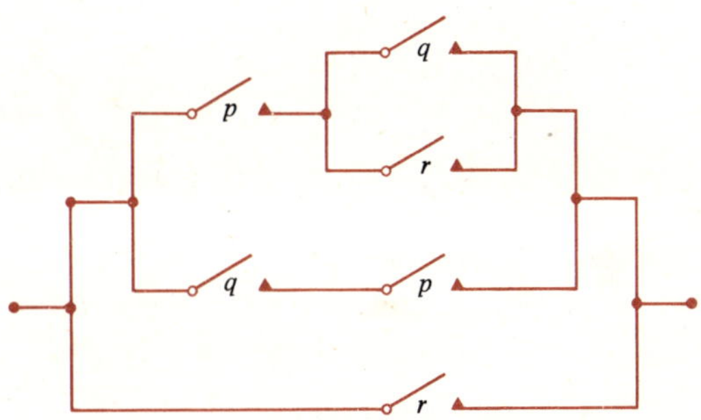

Figure 1.11

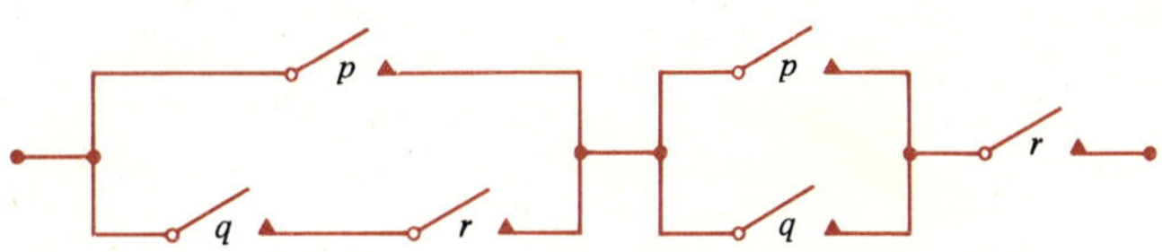

Figure 1.12

9. List all the components of $p \times (q + r)$.
10. List all the components of $(p \times q) + (r \times s)$.

1.2 Evaluation of Components

An ordinary light switch is either closed or open (that is, on or off); to reflect these two states, a switch p is assigned the value 1 if closed, and 0 if open. By a closed switch we mean that electrical current *will* flow through the switch from terminal A to terminal B (Figure 1.13), and by an open switch we mean that electrical current will *not* flow through the switch from terminal A to terminal B (Figure 1.14). Of course, one need not be confined to the flow of electricity; one can just as well consider the flow of something else, such as information, money, or influence.

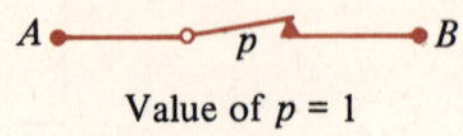

Figure 1.13. *Closed switch: Electricity will flow from A to B.*

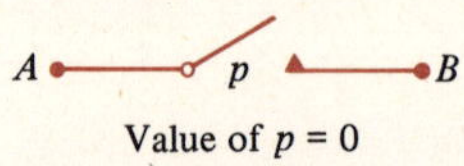

Figure 1.14. *Open switch: Electricity will not flow from A to B.*

Values of the compound switches $p + q$ and $p \times q$ are defined according to whether electrical current will flow through the compound switch from one terminal to another terminal. For example, $p + q$ has value 1 if p has value 1, regardless of the value of q, because current will flow from terminal A to terminal B through the switch p as shown in Figure 1.15.

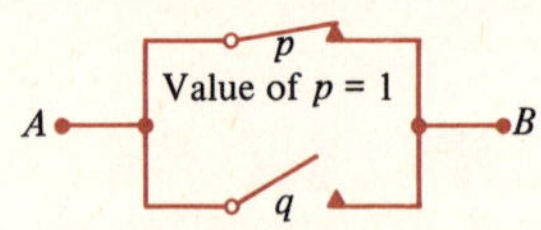

Figure 1.15

These considerations are formalized in the following definition.

▶ Definition 1

For arbitrary switches p and q, the values of $p + q$ and $p \times q$ are defined according to Table 1.1.

Table 1.1. DEFINITION OF VALUES OF PARALLEL AND SERIES CONNECTIONS.

p	q	$p + q$	$p \times q$
1	1	1	1
1	0	1	0
0	1	1	0
0	0	0	0

We are now in a position to evaluate more complicated compound switches by using the definition just given.

Example 1

Evaluate the compound switch $[p \times (p + q)] + q$, shown in Figure 1.16, for all possible arrangements of the values of p and q. The evaluation is made by means of a table similar to that given in Definition 1; see Table 1.2. ▮

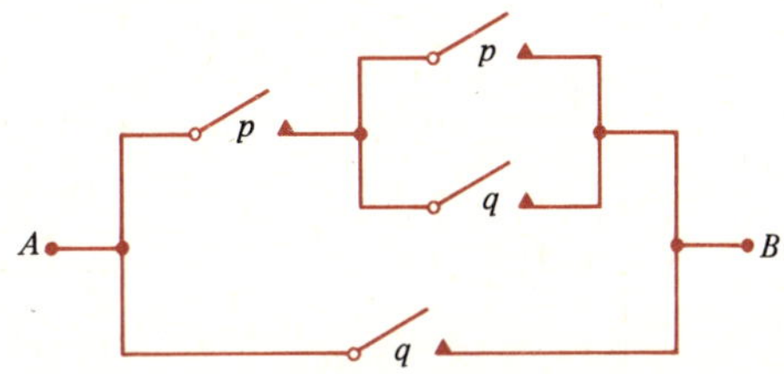

Figure 1.16

Table 1.2. EVALUATION OF $[p \times (p + q)] + q$.

(1)	(2)	(3)	(4)	(5)
		(1) + (2)	(1) × (3)	(4) + (2)
p	q	$p + q$	$p \times (p + q)$	$[p \times (p + q)] + q$
1	1	1	1	1
1	0	1	1	1
0	1	1	0	1
0	0	0	0	0

Computer help is available for the evaluation of large compound switches and is illustrated in the appendix of this chapter.

If a switch t is wired in such a way that it opens when switch p is closed and closes when switch p is opened, then switch t is said to be the **opposite** of p, and switch t is designated with the symbol $\sim p$. The values of $\sim p$ are defined formally in Definition 2. The symbol "$\sim$" can be read "the opposite of."

▶ **Definition 2**

For arbitrary switch p the values of $\sim p$ are defined in Table 1.3.

Table 1.3 DEFINITION OF VALUES OF $\sim p$.

p	$\sim p$
1	0
0	1

When a switch is closed it is often referred to as being in a **closed state;** when open it is often referred to as being in an **open state.** Thus if a switch p is in an open state, the opposite of p is in a closed state, and vice versa.

Example 2

Evaluate $(p + q) \times (\sim r)$. Here, three switches, including an opposite switch, are involved; therefore, the evaluation table, Table 1.4, will require 2^3, or 8, rows in order to achieve all possible arrangements of values of the single components. ▮

Table 1.4. EVALUATION OF $(p + q) \times (\sim r)$.

(1) p	(2) q	(3) r	(4) (1) + (2) $(p + q)$	(5) $\sim$(3) $\sim r$	(6) (4) $\times$ (5) $(p + q) \times (\sim r)$
1	1	1	1	0	0
1	1	0	1	1	1
1	0	1	1	0	0
1	0	0	1	1	1
0	1	1	1	0	0
0	1	0	1	1	1
0	0	1	0	0	0
0	0	0	0	1	0

Although for pedagogical reasons our attention has been confined to the evaluation of switches, the reader is again reminded that our primary purpose is to treat a switch as simply one concrete example of the broader notion of a component of a system; hence the definitions of this section can be reconsidered with this view in mind. For example, the eye of a frog can be thought of as a component of a living system. If that eye functions properly according to a predetermined standard, then the component can be evaluated as 1. Otherwise the eye would be evaluated as 0. Both eyes can be thought of as a parallel connection of like components whose value is 1 if and only if one or both eyes have value 1.

APPLICATIONS

Example 3

In this example we illustrate the construction of a voting device that will cause a light to flash if a majority of people vote "yes" simultaneously on a yes–no option. For simplicity this illustration deals with a device for three voters. The possible votes with an indication of majority outcomes are given in Table 1.5.

Let p, q, and r be the yes–no switches operated by P, Q, and R, respectively. Electrical current will flow from terminal A to terminal B if all three switches shown in Figure 1.17 have value 1; this compound switch corresponds to the first row of Table 1.5. Electrical current will flow from

Table 1.5

Voters			
P	Q	R	Majority of yes votes
yes	yes	yes	yes
yes	yes	no	yes
yes	no	yes	yes
no	yes	yes	yes
yes	no	no	no
no	yes	no	no
no	no	yes	no
no	no	no	no

terminal A to terminal B if all three switches shown in Figure 1.18 have value 1; this situation corresponds to the second row of Table 1.5. Two other cases in which electrical current will flow from A to B are shown in Figure 1.19. In both cases all three switches must have value 1; the latter cases correspond to the third and fourth rows of Table 1.5. There will be a majority of yes votes and current will flow from A to B in all four of the situations that have just been described.

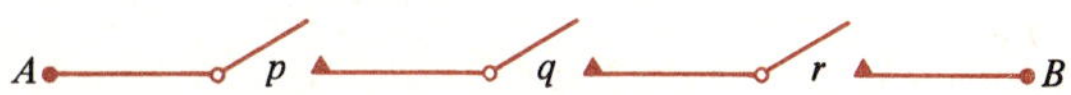

Figure 1.17. *Switch corresponding to first row of Table 1.5.*

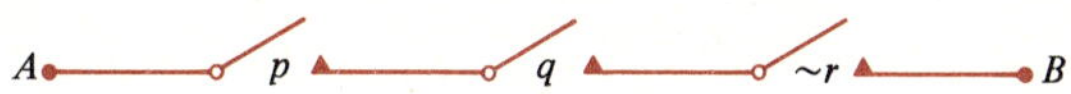

Figure 1.18. *Switch corresponding to the second row of Table 1.5.*

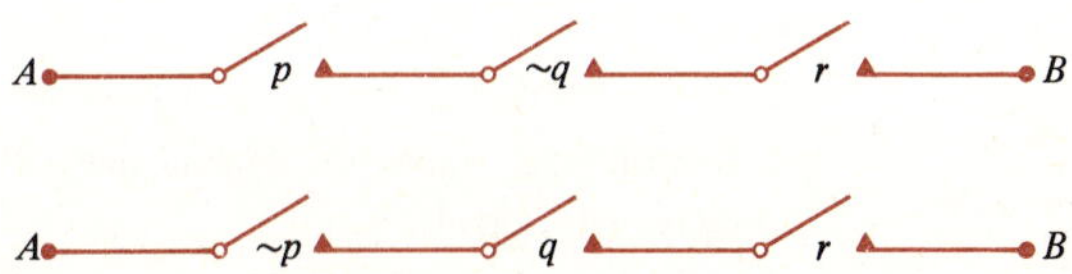

Figure 1.19. *Switches corresponding to the third and fourth rows of Table 1.5.*

If the four compound switches shown in Figures 1.17, 1.18, and 1.19 are connected in parallel, then electrical current will flow from A' to B' as shown in Figure 1.20 if and only if there is a majority of yes votes cast by voters P, Q, and R. The compound switch shown in Figure 1.20 can be represented by s, where we let s be

$$[(p \times q) \times r] + [(p \times q) \times \sim r] + [(p \times \sim q) \times r] + [(\sim p \times q) \times r].$$

In Exercise 22 of Section 1.4 it will be shown that the compound switch s can be simplified; that is, it will be shown that the flow of electricity from A' to B' through s is the same as the flow from A' to B' through the simpler compound switch

$$[p \times (q + r)] + (q \times r)$$

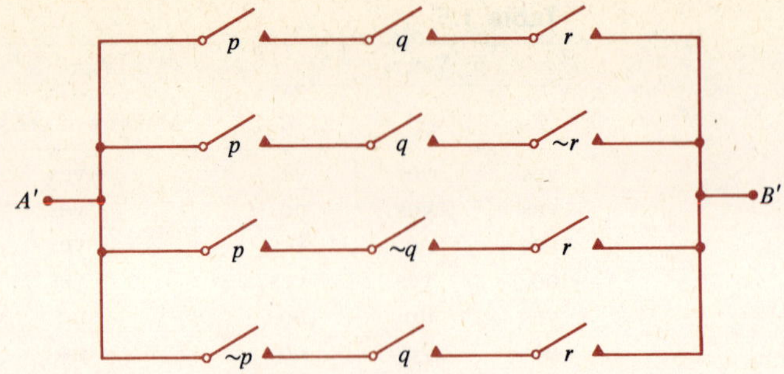

Figure 1.20. *Switch representing a majority or nonmajority of yes votes in Table 1.5.*

shown in Figure 1.21. The evaluation of the former compound switch by means of the computer (using APL) is illustrated in the appendix of this chapter. ∎

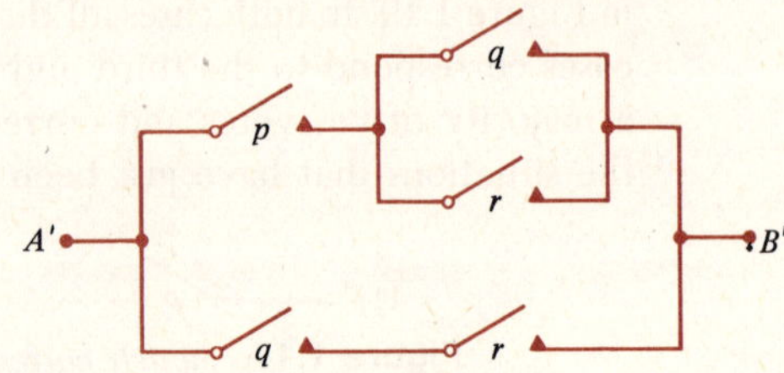

Figure 1.21. *Switch representing a majority or nonmajority of yes votes in Table 1.5.*

Example 4

As civilization has developed, men have found ways to quantify certain events, situations, or arrangements of matter of the real world. Such mathematical representations are often called mathematical models. The compound switch

$$[(p \times q) \times r] + [(p \times q) \times \sim r] + [(p \times \sim q) \times r] + [(\sim p \times q) \times r]$$

that was constructed in Example 3 is another illustration of a simple mathematical model of a small but real voting situation. Mathematical modeling enables a scientist to study the properties of the physical, biological, social, or commercial phenomena by studying the analogous mathematical models. From this study better theoretical understanding of the real situation is gained, and perhaps through this understanding, new theories can be justified. Example 3 further illustrated the concept of modeling in that the reader also was instructed in the construction of a mechanical model of the voting situation.

Mathematical modeling often provides insights into useful application or generalization of a single model in other ways. To show how a model that is constructed to simulate one real situation can be used to originate other applications, we shall generalize Table 1.5 of Example 3 to read "factors" rather than "voters" and "attainment of a goal" rather than "majority of yes votes"; see Table 1.6.

Table 1.6

Factors			
P	Q	R	Attainment of a goal
yes	yes	yes	yes
yes	yes	no	yes
yes	no	yes	yes
no	yes	yes	yes
yes	no	no	no
no	yes	no	no
no	no	yes	no
no	no	no	no

With this generalization one can give a variety of interpretations to the "factors" and the "goals," with each situation having the same basic mathematical model. For example, the "factors" may be environmental factors involved in the "goal" of maintaining the life of a certain organism.

The factors may be indicators (such as the price–earnings ratio) by which stocks are judged; if specific levels of two of the three indicators are satisfied, then a purchase of the stock is indicated. The factors may represent certain desirable conditions for drilling an oil well or building a hotel. If two of the three conditions are met, then a decision to act is in order. Larger versions of the basic model undoubtedly would be needed in realistic applications, such as meaningful environmental problems, stock purchases, or business ventures.

Moreover, the model can be altered to reflect weights of factors such as might be encountered in the election of a presidential candidate by the electoral college.

Further discussion of the construction of mathematical models may be found in Freund [14], pp. 1–6. ∎

Example 5

As mentioned before, the concepts developed in this chapter need not be confined to the flow of electricity. To illustrate, consider an open or closed communication system between two terminals (perhaps nations or people) by way of *other* terminals. Consider a hypothetical situation in which a citizen wishes to communicate a message to the president of a large corporation. Of course, one way is by direct communication. Another alternative is through a friend, who can communicate directly or through one of two influential people who have the ear of the president. A diagram of this communication flow is given in Figure 1.22. The algebraic expression corresponding to Figure 1.22 is

$$p + [q \times (r + \{[u \times v] + [s \times t]\})]. ∎$$

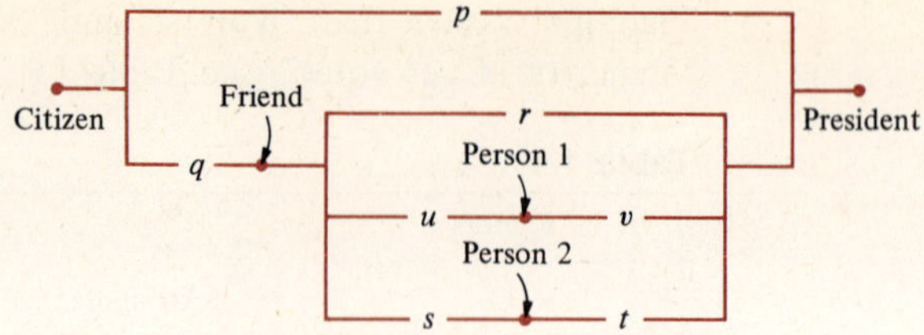

Figure 1.22. *Influence diagram.*

EXERCISES

Suggested minimum assignment: Exercises 1, 3, 5, 8, and 9.

In each of Exercises 1–8 evaluate the given compound switch for all possible arrangements of the values of the component single switches.

1. $p \times (p + q)$.
2. $p + (\sim p \times q)$.
3. $(p + \sim q) + (q + \sim p)$.
4. $(q \times p) \times (q + \sim p)$.
5. $(p + r) \times \sim q$. (Eight rows will be needed in evaluation table.)
6. $(r + q) \times (\sim p + \sim q)$. (Eight rows will be needed in evaluation table.)
7. $(p \times q) + (r \times \sim q)$.
8. $p + [\sim p \times (r + q)]$.
9. Using p, q, s, t, $+$, and $\times$, write an expression that represents the arrangement of switches shown in Figure 1.23. Draw another diagram corresponding to your answer.

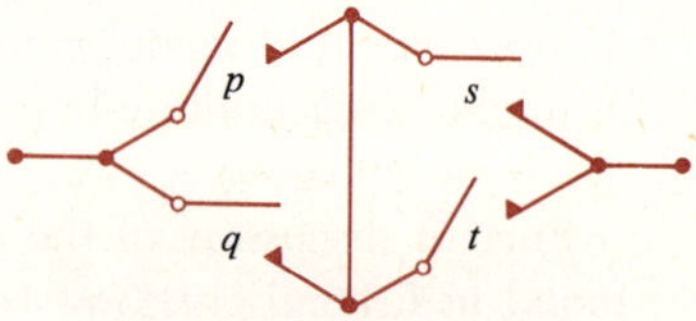

Figure 1.23

10. Using p, q, s, r, t, $+$, and $\times$, write an expression that represents the arrangement of switches shown in Figure 1.24. Draw another diagram corresponding to your answer.

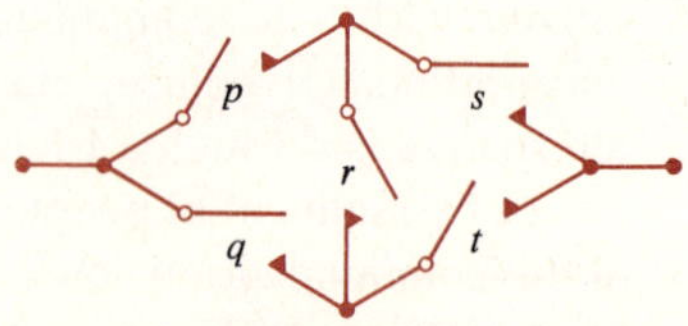

Figure 1.24

1.3 Equivalent Components

Consider the compound switch $p \times (p + q)$ shown in Figure 1.25 and the switch p shown in Figure 1.26. It turns out that when p is closed, $p \times (p + q)$ is closed, and when p is open, $p \times (p + q)$ is open. In other words, as far as the flow of electricity (or information, or money) is concerned, $p \times (p + q)$ and p are the same.

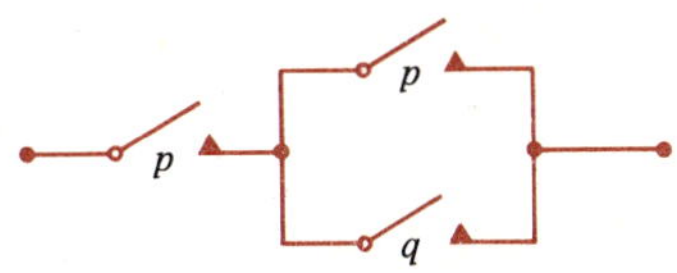

Figure 1.25

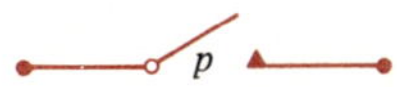

Figure 1.26

▶ **Definition 3**

Two switches are **equivalent** when they have the same value for all possible arrangements of states of the component single switches. Equivalence will be denoted by the symbol "≡."

Example 1

The compound switches $p \times (q + r)$ and $(p \times q) + (p \times r)$, illustrated in Figure 1.27, are equivalent according to Definition 3 because columns (5) and (8) of the evaluation table (Table 1.7) are the same. The labor involved in determining columns (5) and (8) of Table 1.7 can be accomplished by means of the computer; one procedure using APL is illustrated in the appendix to this chapter. ▮

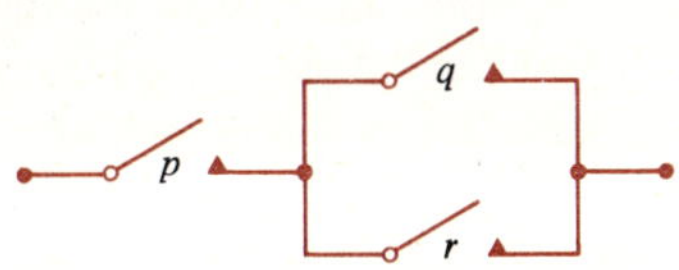

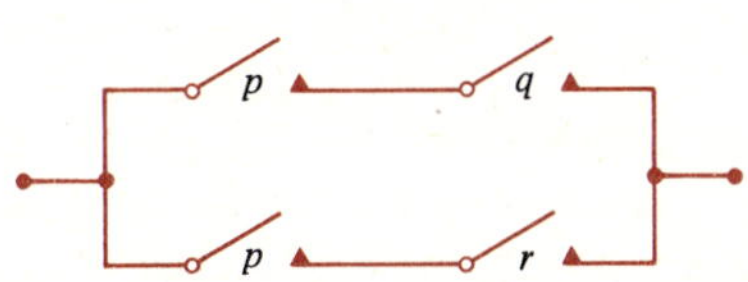

Figure 1.27. *Pair of equivalent switches.*

Example 2

The compound switch $\{p \times (p + q)\} + q$ is equivalent to the compound switch $p + q$. This can be observed from the evaluation table of Example 1 of Section 1.2. ▮

Table 1.7. EVALUATION SHOWING EQUIVALENCE OF $p \times (q + r)$ AND $(p \times q) + (p \times r)$.

(1) p	(2) q	(3) r	(4) (2) + (3) $q + r$	(5) (1) × (4) $p \times (q + r)$	(6) (1) × (2) $p \times q$	(7) (1) × (3) $p \times r$	(8) (6) + (7) $(p \times q) + (p \times r)$
1	1	1	1	1	1	1	1
1	1	0	1	1	1	0	1
1	0	1	1	1	0	1	1
1	0	0	0	0	0	0	0
0	1	1	1	0	0	0	0
0	1	0	1	0	0	0	0
0	0	1	1	0	0	0	0
0	0	0	0	0	0	0	0

Next we consider a set of switches that are always closed regardless of the state of the component single switches. For example, the compound switch $p + \sim p$, diagrammed in Figure 1.28, is always closed; notice that such a switch always has a value 1 and will be called a **unit switch.**

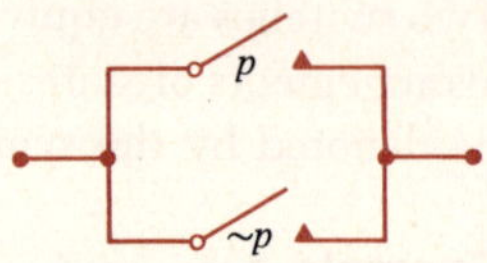

Figure 1.28. *Unit switch* $p + \sim p$.

▶ **Definition 4**

A compound switch that always has a value of 1, regardless of the state of its single switches is called a **unit switch** and is denoted by U.

There are certain compound switches, such as $p \times \sim p$ (Figure 1.29), that are always open regardless of the state of the component single switches; a switch whose value is always 0 will be called a **zero switch.**

Figure 1.29. *Zero switch* $p \times \sim p$.

▶ **Definition 5**

A compound switch that always has a value of 0, regardless of the state of its single switches, is called a **zero switch** and is denoted by Z.

Example 3

For an arbitrary switch p (single or compound) we can show that

$$p + Z \equiv p \text{ and } p \times U \equiv p$$

by using Tables 1.8 and 1.9 and Definition 3. ▮

Table 1.8. EVALUATION SHOWING $p + Z \equiv p$.

p	Z	$p + Z$
1	0	1
0	0	0

Table 1.9. EVALUATION SHOWING $p \times U \equiv p$.

p	U	$p \times U$
1	1	1
0	1	0

Recall that the real number 0 was called the identity element for the addition of real numbers because for an arbitrary real number x, we have $x + 0 = 0 + x = x$. From Example 3 and the fact that also $Z + p \equiv p$, we see that Z is the identity element for the operation $+$ just as the real number 0 is the identity element for the addition of real numbers. U is the identity element for the operation $\times$ just as the real number 1 is the identity element for the multiplication of real numbers.

APPLICATIONS

Example 4

The concepts presented so far are very useful in a formal study of logic. For example, each switch (component) can be thought of as a verbal statement that can be judged as true or false (evaluated as 1 or 0). These statements in turn can be connected by the words "and" and "or," which correspond to the switch operations $\times$ and $+$. Compound statements can then be constructed, analyzed, evaluated as true or false, and perhaps simplified. Such activities can be useful in legal analysis, as illustrated in the next example. ∎

Example 5

The purpose of this application is to illustrate the use of the material of this chapter in legal analysis. Consider Section 6731(*e*) of the Business and Professions Code of California: "Nothing in this chapter shall prohibit the preparation of plans, drawings, specifications, estimates, or instruments of service for single or multiple dwellings not more than two stories and basement in height; garages or other structures appurtenant to such dwellings; farm or ranch buildings; or any other buildings, except steel frame and concrete buildings, not over one story in height, where the span between bearing walls does not exceed twenty-five (25) feet."

On the basis of the statutory provision this question arises: What kind of papers pertaining to buildings are not prohibited? To answer the question and to analyze Section 6731(*e*), the code can be considered as a system of words consisting of component phrases with the connectives "or," "and," and "not." The connective "or" will be considered as a parallel connection of the component phrases; the word "and" will be considered as a series connection of phrases, and the word "not" will call for the opposite state.

On the basis of the following identification of component phrases, Section 6731 (*e*) can be diagrammed as shown in Figure 1.30: "Nothing in this chapter shall prohibit the preparation of $\overbrace{\text{plans}}^{a}$, $\overbrace{\text{drawings}}^{b}$, $\overbrace{\text{specifications}}^{c}$, $\overbrace{\text{estimates}}^{d}$, or $\overbrace{\text{instruments of service}}^{e}$ for $\overbrace{\text{single}}^{f}$ or $\overbrace{\text{multiple}}^{g}$ dwellings not $\overbrace{\text{more than two stories and basement in height}}^{h}$; $\overbrace{\text{garages}}^{i}$ or $\overbrace{\text{other structures}}^{j}$ appurtenant to $\overbrace{\text{such dwellings}}^{(f+g)\times\sim h}$; $\overbrace{\text{farm}}^{k}$ or $\overbrace{\text{ranch}}^{l}$ buildings; or $\overbrace{\text{any other buildings}}^{m}$, except $\overbrace{\text{steel frame}}^{n}$ and $\overbrace{\text{concrete}}^{o}$ buildings, $\overbrace{\text{not over one story in height}}^{p}$, $\overbrace{\text{where the span between the bearing walls does not exceed twenty-five (25) feet}}^{q}$."

Figure 1.30 provides a diagram of the code for purposes of analysis. Such a diagram may be particularly appropriate in many problems involving statutory interpretation, since it makes visible to whom or to what cumbersome statutory provisions apply. In this particular example, any paper pertaining to buildings will not be prohibited, provided a closed path may be found from A to B in Figure 1.30. Section 6731(e) is analyzed in a slightly different way by Allen and Caldwell [1]. This reference, which deals with logic and judicial decision making, will be discussed later in Chapter 2. ∎

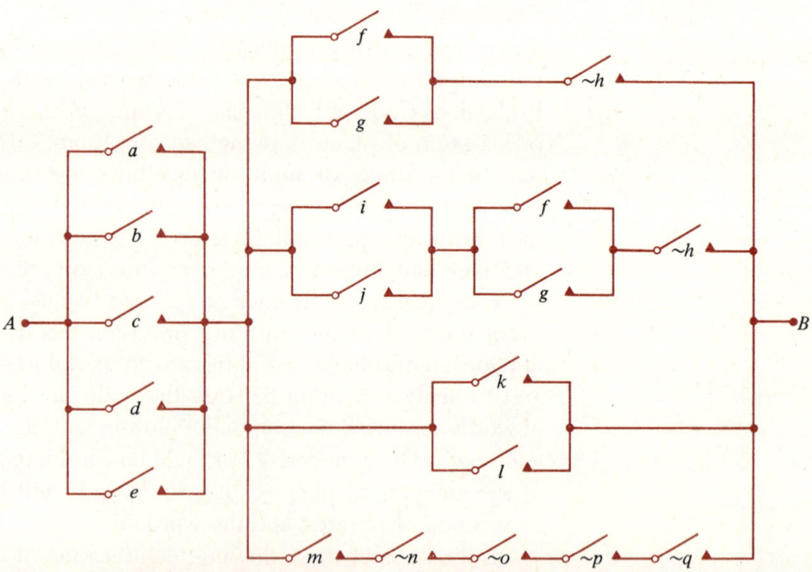

Figure 1.30. *Section of business and professions code.*

Example 6

The concepts presented so far in this chapter play a very important part in the design of electronic computers as well as in certain communications and control systems. An introduction to some ideas in this regard are presented by Hohn [18]; a more elementary but less comprehensive discussion may be found in [21]. The latter reference is a programmed self-instructional manual. A very short elementary introduction to the use of the algebra of switches to computer design may be found in Arnold [3], pp. 128–131.

To give the reader an intuitive idea of the connection between compound switches and computer design we will design a pair of compound switches that can be used to add a certain pair of integers. First we must be able to express integers as binary numerals involving only the digits 0 and 1; this means to write the integers using the base 2 rather than the base 10. For those not familiar with numerals using the base 2, the binary numerals shown in Table 1.10 may be thought of as a "code" of the corresponding integers. (A more comprehensive presentation of numerals using base 2 may be found in Freund [14], pp. 49–54.)

Table 1.10

Integer	Corresponding binary numerals
0	000
1	001
2	010
3	011
4	100
5	101
6	110
7	111

To perform the addition

$$\begin{array}{r} 1 \\ +5 \\ \hline 6 \end{array}$$

we must add the binary equivalents (found in Table 1.10),

$$\begin{array}{r} 001 \\ +101 \\ \hline 110. \end{array}$$

Although it appears that we are adding pairs of digits in each column, in essence we are adding three digits in each column (except the right) because of the "carry" aspect of addition. In other words, we can think of the sum

$$\begin{array}{r} 001 \\ +101 \\ \hline 110 \end{array}$$

as the succession of sums (reading from right to left)

$$\begin{array}{ccc} 0 & 1 & 0 \\ 0 & 0 & 1 \\ +1 & +0 & +1 \\ \hline 1 & 1 & 0 \end{array}$$

where the top digit in each sum is the carried digit from the previous sum. (The binary addition table is given in Table 1.11.)

Table 1.11 BINARY ADDITION.

+	0	1
0	0	1
1	1	10

When adding three binary digits there are eight possibilities, as shown in Table 1.12.

Table 1.12. EIGHT POSSIBILITIES WHEN ADDING THREE BINARY DIGITS.

	Digits to be added				
	c	p	q	Recorded digit	Digit to be carried
(1)	1	1	1	1	1
(2)	1	1	0	0	1
(3)	1	0	1	0	1
(4)	1	0	0	1	0
(5)	0	1	1	0	1
(6)	0	1	0	1	0
(7)	0	0	1	1	0
(8)	0	0	0	0	0

The next step is to construct† a compound switch whose evaluation will be that of the fourth column of Table 1.12; the switch shown in Figure 1.31 is such a switch. Then we construct a compound switch whose evaluation will be that of the fifth column of Table 1.12; Figure 1.32 illustrates such a switch.

The subscript i in c_i, p_i, and q_i represents the column number starting from the right in a sum such as

$$\begin{array}{ccc} 0 & 0 & 1 \\ 1 & 0 & 1 \\ \hline \end{array}.$$

† One way of constructing such a switch is to consider only those rows of Table 1.12 where the desired evaluation in the fourth column is 1 (that is, rows 1, 4, 6, and 7). Within the first column, replace each 1 by c and each 0 by $\sim c$. Within the second column, replace each 1 by p and each 0 by $\sim p$. Within the third column, replace each 1 by q and each 0 by $\sim q$. Then treat each row as a series connection of the three component switches and connect the four rows in parallel.

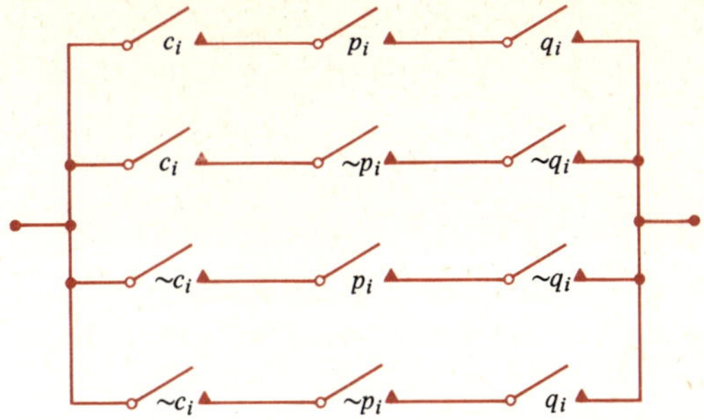

Figure 1.31. *Switch representing recorded digit of Table 1.12.*

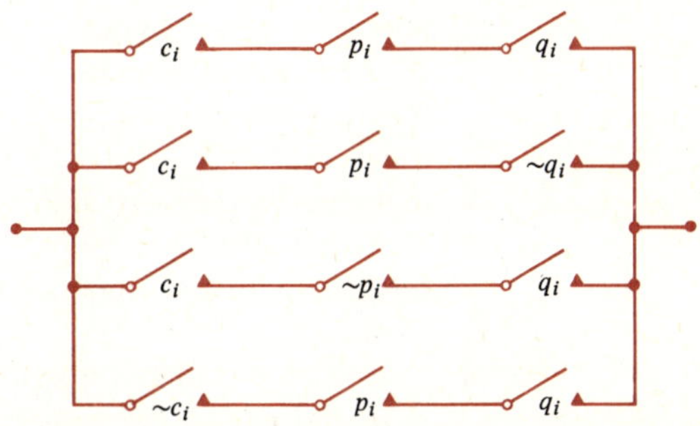

Figure 1.32. *Switch representing carried digit of Table 1.12.*

In the right column of this particular sum both p_1 and q_1 are 1, while c_1 (the carried digit) is 0 at the outset; hence the compound switch of Figure 1.31 has value 0, and the compound switch of Figure 1.32 has value 1. Therefore, the recorded digit is 0, and the carried digit is 1. We now have

$$\begin{array}{r} 1 \\ 0\ \ 0\ \ 1 \\ +1\ \ 0\ \ 1 \\ \hline 0 \end{array},$$

where now in the second column c_2 is 1, and p_2 and q_2 are 0. The process is repeated until terminated. The sum is then decoded (changed back to base 10). ▮

EXERCISES

Suggested minimum assignment: Exercises 1, 4, 5, 7, 8, 9, and 11.

In each of Exercises 1–4 determine which of the following compound switches are unit switches and which are zero switches and which are neither:

1. $\{q + (p \times \sim p) + \sim q\}$.
2. $(\sim p \times \sim q) + (p \times q)$.

3. $\{(p \times \sim p) + q\} \times \sim q$.
4. $\{(\sim p \times q) + (p \times q)\} \times \sim q$.

In each of Exercises 5–8 determine whether the given pair of compound switches are equivalent:

5. $p + (p \times q)$, p.
6. $\sim(p + q)$, $(\sim p \times \sim q)$.
7. $(p \times q) + (p \times \sim q)$, $(p + q)$.
8. $(p + \sim q) \times (\sim p + q)$, $(p \times \sim q)$.

9. An element $\in$ in a set of elements A such that $a \circ \in = \in \circ a = a$ for every a in A is called an identity element for the operation "$\circ$."
   (*a*) Show that a unit switch U is an identity element for the operation $\times$ over a set of compound switches composed of n single switches together with all compound switches that can be constructed from them.
   (*b*) Show that a zero switch Z is an identity element for the operation $+$ over a set of compound switches composed of n single switches together with all compound switches that can be constructed from them.
10. (*a*) Construct a compound switch that will be closed when switches p, q, and r have the values shown in the *first* row of Table 1.13 and which is open otherwise. *Hint:* Consider a series connection of some three single switches.

Table 1.13

p	q	r
0	0	1
1	0	1

   (*b*) Construct a compound switch that will be closed when switches p, q, and r have the values shown in the *second* row of Table 1.13 and which is open otherwise.
   (*c*) Construct and sketch a diagram of a compound switch that will be closed when switches p, q, and r have the values shown in *either* row of Table 1.13 and which is open otherwise. *Hint:* Consider the parallel connection of your answers to parts (*a*) and (*b*).
   (*d*) Determine whether your answer to part (*c*) is equivalent to the switch shown in Figure 1.33.

Figure 1.33

11. (*a*) Construct a compound switch that will be closed when switches p and q have the values shown in the *first* row of Table 1.14 and which is open otherwise. *Hint:* Consider a series connection of some two single switches.

Table 1.14

p	q
1	1
0	1

(*b*) Construct a compound switch that will be closed when switches p and q have the values shown in the second row of Table 1.14 and which is open otherwise.

(*c*) Construct and sketch a diagram of a compound switch that will be closed when switches p and q have the values shown in either row of Table 1.14 and which is open otherwise. *Hint:* Consider a parallel connection of your answer to parts (*a*) and (*b*).

(*d*) Determine whether your answer to part (*c*) is equivalent to the switch q.

1.4 Properties of the Algebra of Components

So far we have laid the foundation for a new mathematical structure which we have called the Algebra of Switches (components). The elements of this structure are single or compound switches (components); we have a relation between switches called equivalence, and we have the operations of $+$, $\times$, and $\sim$ defined over the elements. Once such a foundation has been laid, it is natural to formulate conjectures involving the elements, operations, and relations, and then to try to prove or disprove these conjectures. As the reader is probably aware, those conjectures that we can prove are called theorems and in a sense constitute part of the superstructure of our mathematical structure. These theorems can be used to prove later theorems or can be applied directly to solve practical problems. Some exercises will point out the usefulness of the theorems in simplifying compound switches and we can thereby achieve a greater efficiency in approaching those situations that employ compound switches.

► **Theorem 1**

(Commutative Properties) If p and q are arbitrary switches, then

$$(i)\quad p + q \equiv q + p$$

and

$$(ii)\quad p \times q \equiv q \times p.$$

The proof of Theorem 1 is easily accomplished by use of evaluation tables and is left as Exercises 18 and 19. ◄

► **Theorem 2**

(Distributive Properties) If p, q, and r are arbitrary switches, then

$$(i)\quad p \times (q + r) \equiv (p \times q) + (p \times r)$$

and

$$(ii)\quad p + (q \times r) \equiv (p + q) \times (p + r).$$

The proof of Theorem 2(*i*) is given in Example 1 of Section 1.3. The proof of Theorem 2(*ii*) is left as Exercise 20. ◀

▶ **Theorem 3**

(Identity Properties) If Z is a zero switch, U is a unit switch, and p is an arbitrary switch, then for every p,

$$(i) \quad p + Z \equiv p$$

and

$$(ii) \quad p \times U \equiv p.$$

The proof of Theorem 3 is given in Example 3 of Section 1.3. ◀

▶ **Theorem 4**

(Complement Properties) If Z is a zero switch, U is a unit switch, and p is an arbitrary switch, then for every p

$$(i) \quad p \times \sim p \equiv Z$$

and

$$(ii) \quad p + \sim p \equiv U.$$

Proof: From Definition 2, there is a switch $\sim p$ such that $p \times \sim p$ and $p + \sim p$ can be evaluated according to Table 1.15. Then by Definition 5, we have (*i*), and by Definition 4, we have (*ii*). ◀

Table 1.15. EVALUATION OF $p \times \sim p$ AND $p + \sim p$.

p	$\sim p$	$p \times \sim p$	$p + \sim p$
1	0	0	1
0	1	0	1

The four theorems presented so far can be proved using only the basic definitions. These four theorems were carefully chosen, however, so that the later theory can be based entirely on them. In other words, later theorems may be proved by using only preceding theorems, and evaluation tables are no longer required. Of course, one may choose to use an evaluation table for a proof if convenient. To illustrate the alternative methods of proof that are available from here on, we provide two proofs of the following theorem.

▶ **Theorem 5**

If Z is a zero switch, U is a unit switch, and p is an arbitrary switch, then for every p

$$(i) \quad p + U \equiv U$$

and

$$(ii) \quad p \times Z \equiv Z.$$

Proof of (*i*)

	Statement	Reason
1.	$p + U \equiv (p + U) \times U$	1. Theorem 3.
2.	$\equiv U \times (p + U)$	2. Theorem 1.
3.	$\equiv (p + \sim p) \times (p + U)$	3. Theorem 4.
4.	$\equiv p + (\sim p \times U)$	4. Theorem 2.
5.	$\equiv p + \sim p$	5. Theorem 3.
6.	$\equiv U$	6. Theorem 4.

Alternative proof of (*i*): By Definitions 1 and 4 we have the evaluation table shown in Table 1.16. By Definition 3, applied to columns (2) and (3), the conclusion follows.

A proof of (*ii*) is left as Exercise 5. ◀

Table 1.16. EVALUATION OF $p + U$.

(1) p	(2) U	(3) $p + U$
1	1	1
0	1	1

For Theorem 5 the proof by evaluation table is probably easier, but this is not always the case, and the reader should study the other method as well.

The theorems can be particularly useful in simplifying compound switches, as illustrated in the following example.

Example 1

The compound switch $(\sim q \times p) + \{p \times (\sim p + q)\}$ (shown in Figure 1.34) is equivalent to switch p.

$(\sim q \times p) + \{p \times (\sim p + q)\}$

$\equiv (\sim q \times p) + \{(p \times \sim p) + (p \times q)\}$	by Theorem 2,
$\equiv (\sim q \times p) + \{Z + (p \times q)\}$	by Theorem 4,
$\equiv (\sim q \times p) + (p \times q)$	by Theorems 1, 3,
$\equiv (p \times \sim q) + (p \times q)$	by Theorem 1,
$\equiv p \times (\sim q + q)$	by Theorem 2,
$\equiv p \times U$	by Theorems 1, 4,
$\equiv p$	by Theorem 3. ▮

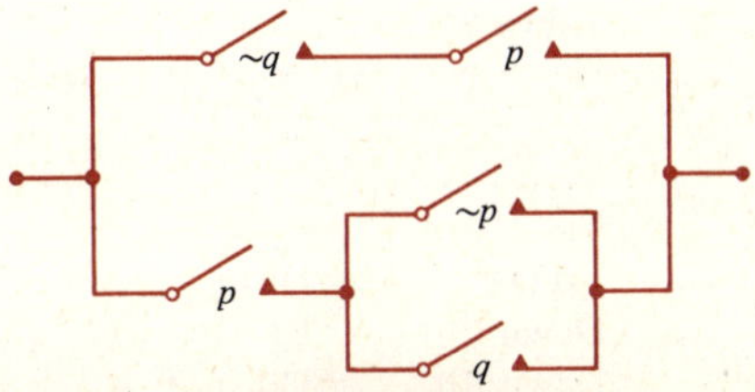

Figure 1.34

A few of the many other theorems that may be useful in applied problems follow.

Theorems 6–17

If p, q, and r are arbitrary switches, and if Z is a zero switch and U is a unit switch, then

6. For every p, $p + p \equiv p$.
7. For every p, $p \times p \equiv p$.
8. If x is a switch such that $p + x \equiv p$ for every p, then $x \equiv Z$.
9. If x is a switch such that $p \times x \equiv p$ for every p, then $x \equiv U$.
10. $p + (p \times q) \equiv p$ (Absorption Property).
11. $p \times (p + q) \equiv p$ (Absorption Property).
12. $p \times q \equiv p$ if and only if $p + q \equiv q$.
13. If $p + q \equiv p + r$ and if $p \times q \equiv p \times r$, then $q \equiv r$.
14. $p + (q + r) \equiv (p + q) + r$ (Associative Property).
15. $p \times (q \times r) \equiv (p \times q) \times r$ (Associative Property).
16. $\sim(p + q) \equiv \sim p \times \sim q$ (DeMorgan's Property).
17. $\sim(p \times q) \equiv \sim p + \sim q$ (DeMorgan's Property).

Proofs of each of Theorems 6–17 are not difficult and are left as Exercises 6–17. ◀

APPLICATIONS

Example 2

In this example we illustrate the construction of a system whose value will be changed if the value of any one simple component is changed. (In fact, the value of the system will be changed if the value of an odd number of the simple components is changed.) One common use of this mathematical model occurs in the electrical wiring of a modern home. Although the lighting in many rooms is controlled by a single on–off switch, there are also many rooms with lighting that is controlled by either of two individual switches. We want to construct a compound switch whose evaluation is shown in Table 1.17. Considering those rows where the desired evaluation is 1, we find that such an evaluation corresponding to Table 1.17 can be obtained by the compound switch

$$(p \times q) + (\sim p \times \sim q).$$

Table 1.17

p	q	Compound switch
Up (1)	Up (1)	1
Up (1)	Down (0)	0
Down (0)	Up (1)	0
Down (0)	Down (0)	1

A diagram of this compound switch is given in Figure 1.35. This switch can be changed to coincide with the practical connection of wires by an electrician as shown in Figure 1.36.

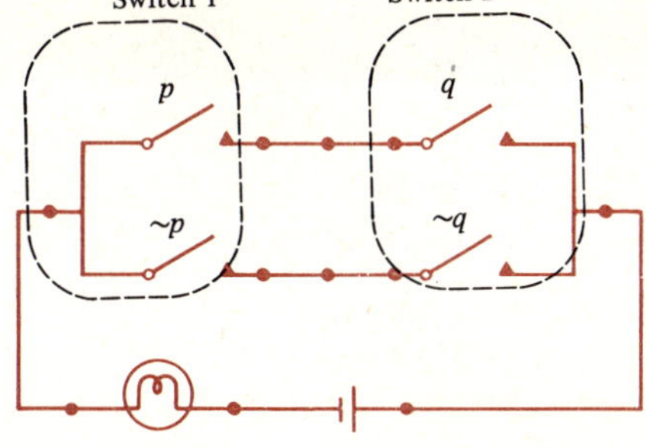

Figure 1.35. *Theoretical diagram of two switches controlling a single light.*

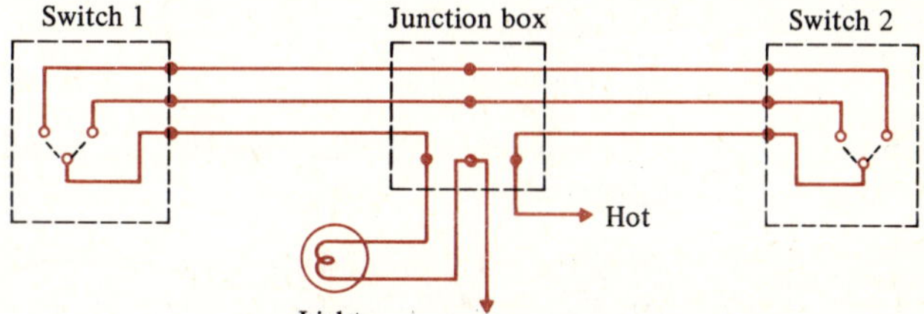

Figure 1.36. *Practical diagram of two switches controlling a single light.*

The theorems of this section can be used to show that the latter compound switch is equivalent to

$$(p + \sim q) \times (q + \sim p).$$

Next, suppose that three individual switches are used to control a single light in such a way that a change in state of any single switch will alter the state of the compound switch. The following compound switch will do the job:

$$[(p \times q) \times r] + [(p \times \sim q) \times \sim r] + [(\sim p \times \sim q) \times r] + [(\sim p \times q) \times \sim r].$$

The reader may illustrate, analyze, and try to simplify the last compound switch, as well as consider a comparable compound switch with four single switches. ▮

EXERCISES

Suggested minimum assignment: Exercises 1, 3, 4, 5, 7, and 18.

In Exercises 1–4 simplify each of the following compound switches using only Theorems 1–5.

1. $\sim p \times (q + p)$.
2. $\sim p + (q \times p)$.
3. $(p \times \sim p) + \{(q + \sim q) \times p\}$.
4. $p + \{q + (p \times \sim p)\}$.

5. Prove Theorem 5(*ii*).

In each of Exercises 6–17 prove the indicated theorem by using evaluation tables or by using any theorem that precedes the one in question. Assume that p, q, and r are arbitrary switches.

6. Theorem 6.
7. Theorem 7.
8. Theorem 8.
9. Theorem 9.
10. Theorem 10.
11. Theorem 11.
12. Theorem 12.
13. Theorem 13.
14. Theorem 14. *Hint:* Use evaluation tables—very difficult otherwise.
15. Theorem 15. *Hint:* Use evaluation tables—very difficult otherwise.
16. Theorem 16.
17. Theorem 17.

In Exercises 18–20 prove the indicated theorem by means of evaluation tables.

18. Prove Theorem 1(*i*).
19. Prove Theorem 1(*ii*).
20. Prove Theorem 2(*ii*).
21. Simplify the following compound switch using Theorems 1–5 and then draw the diagram for both the original and the simplified compound switch:

$$(\sim q \times p) + \{(p \times q) + (p \times \sim p)\}.$$

22. Use the theorems of this section to show that the compound switch

$$s = [(p \times q) \times r] + [(p \times q) \times \sim r] + [(p \times \sim q) \times r] + [(\sim p \times q) \times r]$$

that was developed in the application Example 3 of Section 1.2 is equivalent to

$$[p \times (q + r)] + (q \times r).$$

Appendix: APL
A Programming Language

Introduction

The computer language known as APL was developed primarily by Kenneth Iverson of Harvard and first released in 1962 in his book, *A Programming Language* [22]. APL is a *terminal* language; that is, instructions are sent to a computer through a typewriter console that is linked to the computer by telephone. Replies from the computer are typed automatically by the console in a manner similar to a teletype machine. For this reason APL is quite easy to use, and responses from the computer are almost instantaneous. The APL keyboard is shown in Figure 1.37.

Figure 1.37. *APL keyboard.*

APL is, of course, just one of many computer languages, and all the applications that appear in this text could have been presented in other languages. However, we have chosen to use APL primarily because it is relatively easy to learn, it is particularly adaptable to matrix operations which are to be studied in Chapter 5, and it is a terminal language with the built-in advantage of instantaneous two-way communication with the computer.

Communication with a computer using APL can be divided into three basic steps:

1. Signing on.
2. Computations or other activities.
3. Signing off.

Signing on involves acquiring a computer terminal, dialing into the main computer (by telephone), and typing in a special code to gain access. Detailed instructions for this and other activities described in this appendix may be found in APL/360 User's Manual (IBM, 1970).

The computations or other activities involve two-way communication with the computer by means of the APL programming language. Examples of procedures for performing some of these operations will be described later in this section.

Signing off involves typing some special characters in order to sever communication with the computer.

Some Basics of APL

The most basic of all APL operators is the "assign." The symbol, $\leftarrow$, always assigns what follows the symbol to that which precedes it. For example, $A\leftarrow 3$ means that A is assigned the value 3.

Other APL operators that we shall use are $\vee$, $\wedge$, and $\sim$. In APL these symbols are used to denote connectives of the *evaluations of components* rather than as connectives of the components themselves. For example, the APL notation

$$P\vee Q$$

means the *evaluation* of the compound component $p + q$, and hence $P\vee Q$ will produce either 0 or 1 as an answer.

The RETURN key on the terminal has two functions:

1. Pressing the RETURN key will return the carriage and space one line, exactly as a RETURN key would on an ordinary electric typewriter.
2. Pressing the RETURN key will ask the computer to print out an answer whenever any "nonequation" has been typed in.

Examples of RETURN

1. Type: $P\leftarrow 1\wedge 1$
   (press RETURN)

The computer returns carriage and spaces down one line but *does not* type anything. We did not ask for anything! $P\leftarrow 1\wedge 1$ is classified as an equation. We simply assigned to P the value of $1\wedge 1$.

2. Now type: P
   (press RETURN)

The computer will return the carriage, space down one line, and type: 1 (which is of course the current value of P).

3. Type: $1\vee 0$
   (press RETURN)

The computer returns carriage, spaces up one line, and types: 1.

The Use of APL in Examples of This Chapter

Example 1 of Section 1.2

We were asked to find the evaluations of $\{[p \times (p + q)] + q\}$, and in Section 1.2, this had to be done by constructing evaluation tables. Here we shall illustrate the evaluations by use of APL.

Type: $P\leftarrow 0$ (push RETURN)
Type: $Q\leftarrow 0$ (push RETURN)

Type: $S \leftarrow (P \wedge (P \vee Q)) \vee Q$ (push RETURN)
Type: S (push RETURN)
Computer types: 0
Type: $Q \leftarrow 1$ (push RETURN) (Note that the value assigned to P remains unchanged from the previous step)
Type: $S \leftarrow (P \wedge (P \vee Q)) \vee Q$ (push RETURN)
Type: S (push RETURN)
Computer types: 1
Type: $P \leftarrow 1$ (push RETURN)
Type: $Q \leftarrow 0$ (push RETURN)
Type: $S \leftarrow (P \wedge (P \vee Q)) \vee Q$ (push RETURN)
Type: S (push RETURN)
Computer types: 1
Type: $Q \leftarrow 1$
Type: $S \leftarrow (P \wedge (P \vee Q)) \vee Q$ (push RETURN)
Type: S
Computer types: 1

In problems involving a large number of simple components, special programs can be written to streamline the procedure outlined above.

Example 3 of Section 1.2

Let us assume that a 0 means a no vote and that a 1 means a yes vote. As is explained in the example, the compound switch shown in Figure 1.20 can be represented by s, where s is

$$[(p \times q) \times r] + [(p \times q) \times \sim r] + [(p \times \sim q) \times r] + [(\sim p \times q) \times r].$$

To use APL to evaluate the outcomes of the various votes type the following:

$P \leftarrow 1$ (push RETURN)
$Q \leftarrow 1$ (push RETURN)
$R \leftarrow 0$ (push RETURN)
$S \leftarrow ((P \wedge Q) \wedge R) \vee ((P \wedge Q) \wedge \sim R) \vee$
$((P \wedge \sim Q) \wedge R) \vee ((\sim P \wedge Q) \wedge R)$ (push RETURN)
S (push RETURN)
The computer types: 1

Similarly, all the possible values could be keyed in for P, Q, and R to obtain all the possible vote outcomes shown in Table 1.5.

Example 1 of Section 1.3

In this example the object is to show that two compound switches are equivalent. To do this, one can show that the evaluations are the same for all possible arrangements of the simple components. In Section 1.3 this was accomplished by the construction of an evaluation table. For compound switches with many simple components, this construction is a tedious task. An alternative using APL follows. Type:

$P \leftarrow 0$
$Q \leftarrow 0$

$R \leftarrow 0$
$S \leftarrow P \wedge (Q \vee R)$
$W \leftarrow (P \wedge Q) \vee (P \wedge R)$
S
0 (typed by computer)
W
0 (typed by computer)
$R \leftarrow 1$ (Remember that the values assigned to P and Q remain unchanged)
$S \leftarrow P \wedge (Q \vee R)$
$W \leftarrow (P \wedge Q) \vee (P \wedge R)$
S
0 (typed by computer)
W
0 (typed by computer)
$Q \leftarrow 1$
$S \leftarrow P \wedge (Q \vee R)$
$W \leftarrow (P \wedge Q) \vee (P \wedge R)$
S
0 (typed by computer)
W
0 (typed by computer)
$P \leftarrow 1$
$S \leftarrow P \wedge (Q \vee R)$
$W \leftarrow (P \wedge Q) \vee (P \wedge R)$
S
1 (typed by computer)
W
1 (typed by computer)

Similarly, the other four possibilities can be evaluated to show that S and W are the same in all cases and thereby show that the two compound switches are equivalent. Here again, for large problems special programs can be developed to streamline the above procedure.

Obviously, for complicated compound switches the computer becomes more attractive as a "work-saving" device.

NEW VOCABULARY

system 1.1
components 1.1
switch 1.1
compound switch 1.1
parallel connection 1.1
series connection 1.1
algebra of switches 1.1
value of a switch 1.2
opposite of a switch 1.2
closed state 1.2
open state 1.2
equivalent switches 1.3
unit switch 1.3
zero switch 1.3
commutative properties 1.4
distributive properties 1.4

2 Logic and the Algebra of Statement Forms

The primary purpose of this chapter is best described in the words of Charles Dodgson, a 19th-century mathematician who was better known as Lewis Carroll, the author of Alice in Wonderland. *In his book* Symbolic Logic, *Dodgson made the following ageless observation: "It [symbolic logic] will give you clearness of thought—the ability to see your way through a puzzle—the habit of arranging your ideas in an orderly and get-at-able form—and, more valuable than all, the power to detect fallacies, and to tear to pieces the flimsy illogical arguments which you will so continually encounter in books, in newspapers, in speeches . . . and which so easily delude those who have never taken the trouble to master this fascinating art." In the process of accomplishing this primary purpose, we shall (1) introduce another mathematical structure in which elements other than numbers are combined by various operations, and (2) emphasize and use the analogies between the mathematical structures called the Algebra of Switches and the Algebra of Statement Forms.*

Prerequisites: 1.1, 1.2
Suggested Sections for Moderate Emphasis: 2.1–2.3
Suggested Sections for Minimum Emphasis: 2.1, 2.2

2.1 Statement Forms and Connectives

In the ordinary usage of language there are certain declarative expressions for which it is meaningful to say that they are either true or false (but not both); in the study of mathematical logic such expressions often are called **statements.**†

▶ **Definition 1**

A **statement** is a declarative sentence which can be assigned exactly one of the values, True or False.

† The reader should be aware that the terms applied to some of the concepts of mathematical logic vary from one textbook to another.

Example 1

The following expressions are specific examples of statements.

1. The Consumer Price Index rose last month.
2. Both Senators from Florida are now Republicans.
3. There is a man by the name of William Howard Palm who is a professional engineer.
4. There is a swinging footbridge in Hardwick, Vermont.
5. $x^2 - 4 = (x - 2)(x + 2)$ for all real numbers x.
6. $1 + 3 = 9$. ▮

Example 2

The following expressions are not statements because they are *not* declarative assertions for which it is meaningful to say that they are either true or false.

1. Run home.
2. $x^2 + 1$.
3. How high is the sky? ▮

Hereafter we will be concerned primarily with symbolic representations of statements rather than with specific statements. We immediately encounter a problem in terminology, however, because the unspecified letters p, q, r, and so on, that we will use do not in themselves constitute a statement according to Definition 1; that definition requires that exactly *one* value, true or false, be assigned to a statement, and, without knowing what p, q, r, and so on, represent, this cannot be done. To escape this dilemma we introduce a new term; the unspecified letters p, q, r, and so on, used in symbolic representations of statements will be called **statement forms.** These letters are not statements but are simply placeholders for statements.

▶ Definition 2

A symbolic representation p that becomes a statement when the placeholder p is specified is called a **statement form.**

When p is specified, the resulting statement is called an **instance** of the statement form. The distinction between a statement and a statement form can be illustrated by considering an analogy with switches. A specific open switch corresponds to a false statement, a specific closed switch corresponds to a true statement, and an unspecified switch in an unspecified state corresponds to a statement form.

The advantage of studying statement forms rather than statements is that we can develop properties for the statement forms that will be valid regardless of specific interpretations that might be given. These properties furnish us with rules of logic that are very useful in such areas as deductive reasoning, legal analysis (see Kort [29]), and computer science.

Recall that we constructed compound switches by joining simple switches in certain ways. Similarly, compound statement forms will be formed by using the words "or," "and," "if . . . then," "if and only if," and "not." Symbols that will be used to represent these words are given in Table 2.1.

Table 2.1. SYMBOLS OF SPECIFIED CONNECTIVES.

or	$\vee$
and	$\wedge$
if . . . then	$\rightarrow$
if and only if	$\leftrightarrow$
not	$\sim$

Thus, if we are given statement forms p, q, the compound form using the connective "**and**" will be designated by $p \wedge q$. If we let p represent the statement

"John Jones is mayor"

and if we let q represent the statement

"Taxes have increased,"

then the statement

"John Jones is mayor *and* taxes have increased"

is one instance of the statement form $p \wedge q$.

In using the connective "**or**," denoted by the symbol $\vee$, we must be careful because in ordinary usage two interpretations are common. For example, when we say

"John Jones is mayor or taxes have increased,"

do we mean to include or exclude the possibility of both? In mathematical logic it is customary to adopt the inclusive interpretation unless specified otherwise. In other words, $p \vee q$ means either p or q *or both*. The exclusive "or" will be discussed in Exercise 11.

The statement

"John Jones is mayor *or* taxes have increased"

is one instance of the statement form $p \vee q$.

Statement forms that do not include any of the connectives of Table 2.1 are known as **simple statement forms.** When two or more simple statement forms are connected by connectives or when the symbol $\sim$ is used, the resulting statement form is called a **compound statement form.** Hereafter the phrase *statement form* will mean either a simple or a compound statement form. When the forms p, q, r, and so on, are specified, then simple and compound statement forms become *simple* and *compound statements*.

Any simple statement form or connection of two or more simple statement forms that make up a part or all of a compound statement form is a **component** of the compound statement form. Large and complicated compound statements are used frequently in normal communications such as legal documents. The analysis, and perhaps simplification, of compound statements through symbolic representation and the rules of logic can be very helpful in some cases.

Other connectives, "if . . . then," "if and only if," and "not," as well as the two already discussed, are presented and illustrated in Table 2.2; this table should be studied carefully.

Table 2.2. SUMMARY OF FIVE CONNECTIVES.

Connective	Symbol	Statement form	Name	Instance
and	$\wedge$	$p \wedge q$	conjunction	John Jones is mayor **and** taxes have increased.
or	$\vee$	$p \vee q$	disjunction	John Jones is mayor **or** taxes have increased.
if . . . then	$\rightarrow$	$p \rightarrow q$	conditional	**If** John Jones is mayor, **then** taxes have increased.
if and only if	$\leftrightarrow$	$p \leftrightarrow q$	biconditional	John Jones is mayor **if and only if** taxes have increased.
not	$\sim$	$\sim p$	negation	It is **not** the case that John Jones is mayor.

Example 3

Consider the compound statement: "If income exceeds \$40,000 and expenses do not exceed 50 per cent of income, then an extra salesman can be hired." Let "income exceeds \$40,000" be an instance of p; let "expenses exceed 50 per cent of income" be an instance of q; let "an extra salesman can be hired" be an instance of r. The original compound statement then is an instance of the compound statement form

$$(p \wedge \sim q) \rightarrow r;$$

the statement forms p, q, r, $\sim q$, $(p \wedge \sim q)$, $\{(p \wedge \sim q) \rightarrow r\}$ are all components of this compound statement form. The first three components are simple and the last three are compound. ▮

In summary, we have begun the construction of a second structure called the Algebra of Statement Forms. The elements or building blocks of this structure are the statement forms p, q, r, and so on. Over this set of statement forms we have a set of operations that we have identified as the connectives $\wedge$, $\vee$, $\rightarrow$, $\leftrightarrow$, and $\sim$. We continue our construction in the remaining sections of this chapter.

APPLICATIONS

Example 4

In recent years, scholars, both in and out of the legal profession, have begun to explore the relationship between modern mathematics and law. The use of mathematical logic in this study has proved particularly promising. Allen and Caldwell [1] provide a discussion of the use of logic in legal analysis as well as many references on the subject. Part II, beginning on page 234, is particulary relevant to the material of this chapter, although the interested reader is warned that the notation and diagrams differ somewhat from this text. As Allen and Caldwell point out, many assertions in legal writings—whether by design or accident—are syntactically ambiguous. They then illustrate this fact with a portion of the 1938 Federal Food and Drug Act:

"Sec. 343. Misbranded Food

"A food shall be deemed to be misbranded . . .

"(*c*) If it is an imitation of another food, unless its label bears, in type of uniform size and prominence, the word 'imitation' and, immediately thereafter, the name of the food imitated."

The question is whether such an assertion is appropriately interpreted as

(1) If a food is an imitation of another food, then it shall be deemed to be misbranded *if* its label *does not* bear, in type of uniform size and prominence, the word "imitation" and, immediately thereafter, the name of the food imitated;

or whether it is more appropriately interpreted as

(2) If a food is an imitation of another food, then it shall be deemed to be misbranded *if and only if* its label *does not* bear, in type of uniform size and prominence, the word "imitation" and, immediately thereafter, the name of the food imitated.

Allen and Caldwell then use some of the ideas of logic to describe and illustrate the method for representing and detecting alternative interpretations of syntactically ambiguous legal assertions. ∎

EXERCISES

Suggested minimum assignment: Exercises 1, 4, 5, 7, 8, 9, and 11.

In each of Exercises 1–4 state which sentences are statements and which are not.

1. (*a*) Fix the car.
   (*b*) The moon is visible now.
   (*c*) The Republicans hold exactly one of the U.S. Senate seats from Florida.
2. (*a*) Where is John?
   (*b*) Eufaula is a town in Alabama.
   (*c*) Ford Motor Company is one of the three largest manufacturers of automobiles in the United States.

3. (*a*) George Washington lived in California for 10 years.
   (*b*) $6 > 2$.
   (*c*) The painting is good.
4. (*a*) $3 + 4 = 13$.
   (*b*) Solve the equation $7x + 6 = 8$.
   (*c*) For some rational values of x, $x^2 + 2 = 6$.

In each of Exercises 5–8 write the statement form for which the given statement is an instance. Also state the primary connective and the name of the compound statement form.

5. (*a*) If the object glitters, then it is gold.
   (*b*) If the vote is favorable, then the labor dispute will be settled and the workers will return to work.
   (*c*) The following is not the case: "Inflation will continue, and unemployment will increase."
6. (*a*) The productivity of the people is lower, or there has not been a greater efficiency.
   (*b*) It is not the case that if we institute an austerity program, then profits will increase.
   (*c*) If the European common market includes England, then U.S. exports will be affected and the balance of payments may change.
7. (*a*) The cash reserves of the First State Bank of Eustis, Florida, have increased, and it is not true that their dividend will decrease this year.
   (*b*) If wishes were horses, then we would all take a ride.
   (*c*) The real estate prices in Blacksburg, Virginia, will increase next year if and only if the population increases and the University expands.
8. (*a*) Federal taxes increased, and money became tighter.
   (*b*) The school bond issue will be approved by the voters, or there will be no new schools next year.
   (*c*) Greater resources can be allocated to education if and only if the voters support the educational institutions in the referendum and the money is available.

In each of Exercises 9–10 give instances of p, q and write statements that are instances of the given statement forms. State the connective and the name of the compound statement form.

9. (*a*) $p \vee q$; (*b*) $p \rightarrow q$; (*c*) $\sim p$.
10. (*a*) $p \wedge q$; (*b*) $p \leftrightarrow q$; (*c*) $\sim p \rightarrow q$.
11. As mentioned in the text, it is customary in mathematical logic to use the connective "or" in the inclusive sense; legal documents often use the symbol and/or to denote this inclusive intent. The exclusive "or" will be denoted by the symbol $\underline{\vee}$, and the compound statement form $p \underline{\vee} q$ means p or q but not both; the arrangement "either . . . or . . ."

is often used to denote this exclusive intent. Let "The computer is available" be an instance of p and "I will get my project finished" be an instance of q; write an instance of $p \veebar \sim q$.

2.2 Evaluation of Statement Forms

In Chapter 1 we assigned values to switches according to whether they were closed or open. Similarly, we will assign values to statements according to the truth or falsity of the statement. We assign the value T to a true statement and a value F to a false statement. We could have assigned values of 1 and 0 rather than T and F; some authors do, and it is permissible for the reader to read a T as a 1 and an F as a 0 if he prefers. The value of a simple statement form may be T or F but not both simultaneously. The values for compound statement forms are defined next.

▶ **Definition 3**

For statement forms p, q, the values of the compound statement forms $p \wedge q$, $p \vee q$, $p \rightarrow q$, $p \leftrightarrow q$, and $\sim p$ are defined according to Tables 2.3–2.7.

Table 2.3. DEFINITION OF VALUES OF CONJUNCTION.

p	q	$p \wedge q$
T	T	T
T	F	F
F	T	F
F	F	F

Table 2.4. DEFINITION OF VALUES OF DISJUNCTION.

p	q	$p \vee q$
T	T	T
T	F	T
F	T	T
F	F	F

Table 2.5. DEFINITION OF VALUES OF CONDITIONAL.

p	q	$p \rightarrow q$
T	T	T
T	F	F
F	T	T
F	F	T

Table 2.6. DEFINITION OF VALUES OF BICONDITIONAL.

p	q	$p \leftrightarrow q$
T	T	T
T	F	F
F	T	F
F	F	T

Table 2.7 DEFINITION OF VALUES OF NEGATION.

p	$\sim p$
T	F
F	T

The definitions given in Tables 2.3–2.7 are compatible with common usage of the English language; it is important, however, that the reader convince himself of this fact. The only definition likely to cause much difficulty is the conditional (Table 2.5), which we discuss in the next example.

Example 1

The purpose of this example is to attempt the sometimes difficult task of convincing the reader that the values assigned to the conditional $p \rightarrow q$ are plausible. Consider the following specific statement made by a candidate for public office. "If I am elected, then the expressway will be completed." Under what circumstances will the candidate's statement be a lie and under what circumstances will the candidate not be lying (that is, telling the truth in a two-value logic)? The alternatives are listed in Table 2.8; notice that Table 2.8 is consistent with Table 2.5. The difficulty that students have in accepting the plausibility of the definition of the values of $p \rightarrow q$ usually centers on the third and fourth lines of Tables 2.5 and 2.8. "How can $p \rightarrow q$ be true if p is false?" Notice in the illustration that if the candidate is not elected, that is, if p is false, then there is no way that his statement can be untrue regardless of what q represents or whether q is true or false. If his statement is not untrue, then it must be true in a two-value logic. ▮

Table 2.8

Candidate elected	Expressway completed	Candidate told truth (that is, did not lie)
yes	yes	yes
yes	no	no
no	yes	yes
no	no	yes

We now turn our attention to a method of evaluation of more complicated compound statement forms. Such an evaluation can be made by use of evaluation tables and repeated use of Definition 3 in a way similar to that employed in the last chapter.

Example 2

Evaluate the compound statement form

$$[(\sim p) \vee q] \wedge (q \rightarrow p).$$

Table 2.9 illustrates an evaluation procedure. We conclude from Table 2.9 that the given compound statement form is true when both p and q are true or when both p and q are false. ▮

Table 2.9. EVALUATION OF $[(\sim p) \vee q] \wedge (q \to p)$.

(1) p	(2) q	(3) $\sim p$	(4) $\sim p \vee q$	(5) $q \to p$	(4) $\wedge$ (5) $[(\sim p) \vee q] \wedge (q \to p)$
T	T	F	T	T	T
T	F	F	F	T	F
F	T	T	T	F	F
F	F	T	T	T	T

Example 3

We now illustrate a compound statement form that is always true, namely $(p \to q) \leftrightarrow [(\sim p) \vee q]$. (See Table 2.10.)

Table 2.10. EVALUATION OF $(p \to q) \leftrightarrow [(\sim p) \vee q]$.

(1) p	(2) q	(3) $\sim p$	(4) $p \to q$	(5) $(\sim p) \vee q$	(4) $\leftrightarrow$ (5) $(p \to q) \leftrightarrow [(\sim p) \vee q]$
T	T	F	T	T	T
T	F	F	F	F	T
F	T	T	T	T	T
F	F	T	T	T	T

Statement forms that are always true are very important and will be discussed further in the next section. ∎

We conclude by observing the analogy between the material of this section and that of Section 1.2. If the value T is read as a 1, and if the value F is read as a 0, then the analogy between the operation $\vee$ of statement forms and the operation $+$ of switches becomes apparent. (See Table 2.11.) There is a similar analogy between $\wedge$ and $\sim$ of statement forms and $\times$ and $\sim$ of switches. (See Table 2.11.)

Table 2.11. SUMMARY OF EVALUATIONS OF SWITCH AND STATEMENT FORM OPERATIONS.

Statement forms					Switches				
p	q	$p \vee q$	$p \wedge q$	$\sim p$	p	q	$p + q$	$p \times q$	$\sim p$
1 (T)	1 (T)	1	1	0	1	1	1	1	0
1 (T)	0 (F)	1	0	0	1	0	1	0	0
0 (F)	1 (T)	1	0	1	0	1	1	0	1
0 (F)	0 (F)	0	0	1	0	0	0	0	1

These analogies furnish the means for diagramming compound statement forms involving the connectives $\vee$ and $\wedge$. For example, the compound statement form $[p \wedge (p \vee q)]$ has $[p \times (p + q)]$ as its switch analog, and the latter can be diagrammed as shown in Figure 2.1.

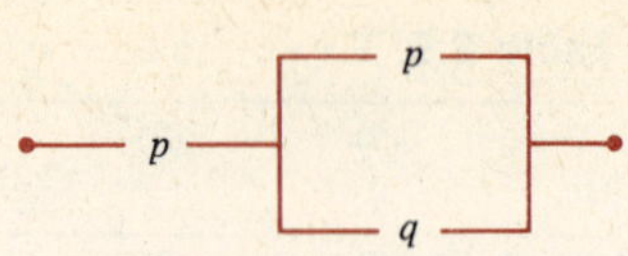

Figure 2.1. *Diagram corresponding to* $p \wedge (p \vee q)$.

APPLICATIONS

Example 4

An analysis of the First Amendment of the U.S. Constitution can be made as follows: "Congress shall make *no* law {[($\overbrace{\text{respecting an establishment of religion}}^{p}$), *or* ($\overbrace{\text{prohibiting the free exercise thereof}}^{q}$)]; *or* [($\overbrace{\text{abridging the freedom of speech}}^{r}$) *or* ($\overbrace{\text{of the press}}^{s}$)]; *or* [($\overbrace{\text{of the right of the people to peaceably assemble}}^{t}$) *and* ($\overbrace{\text{to petition the Government for a redress of grievances}}^{u}$)]}." As a statement form, the First Amendment can be expressed as

$$\sim\{[p \vee q] \vee [r \vee s] \vee [t \wedge u]\}.$$

A violation of the First Amendment occurs when the statement form is false. An evaluation table is one way of determining when this happens.

Later in this chapter we shall find that the statement form above has the same truth values as

$$(\sim p) \wedge (\sim q) \wedge (\sim r) \wedge (\sim s) \wedge (\sim t \vee \sim u).$$

The comparable compound switch is

$$(\sim p) \times (\sim q) \times (\sim r) \times (\sim s) \times (\sim t + \sim u),$$

whose diagram is given in Figure 2.2. The diagram illustrates that a violation of the First Amendment will occur if any one of the components in series is open (for example, a violation occurs if $\sim s$ has value 0, which means that s has value 1, which means that Congress made a law that abridged the freedom of the press).

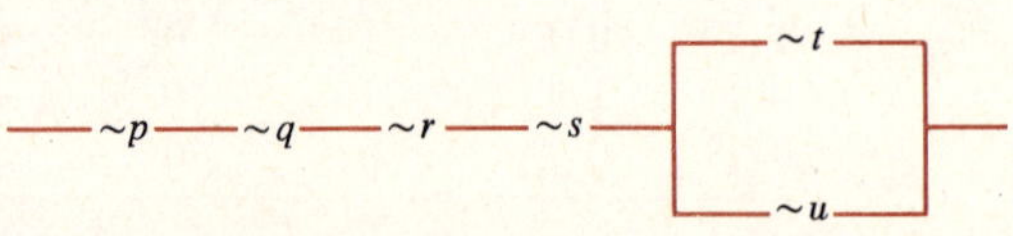

Figure 2.2. *Diagram of first amendment.*

An interesting observation can be made by noticing the effect of the connective "and" in the statement of the amendment. As seen in the diagram, a violation of $(\sim t \vee \sim u)$ occurs only if both t and u have value 1. This analysis shows an interpretation that may *not* have been intended by its authors and that has *not* been consistent with interpretations made by the courts. Thus we have here an excellent example showing how mathematical analysis can expose flaws in legal draftsmanship or in legal interpretation. ∎

EXERCISES

Suggested minimum assignment: Exercises 1, 3, 6, 9, 11, and 13.

In each of Exercises 1–8 evaluate the given compound statement forms for all possible arrangements of values of the placeholders (or variables) p, q, r:

1. $\{p \to (\sim q)\} \wedge p$.
2. $\{p \to q\} \wedge \sim q$.
3. $[(\sim p) \vee (\sim q)] \leftrightarrow \sim (p \wedge q)$.
4. $[(\sim p) \wedge \sim q] \leftrightarrow \sim (p \vee q)$.
5. $[(\sim r) \vee q] \vee (p \wedge r)$.
6. $(p \wedge \sim q) \vee (r \wedge q)$.
7. $\{(p \vee r) \vee q\} \to (p \wedge q)$.
8. $\{(p \wedge r) \wedge q\} \to (p \vee r)$.

In each of Exercises 9 and 10 state a compound statement form that has the given evaluation table:

9. **Table 2.12**

p	q	?
T	T	T
T	F	T
F	T	F
F	F	T

10. **Table 2.13**

p	q	?
T	T	F
T	F	F
F	T	T
F	F	F

11. If we have a compound statement form involving four placeholders (or variables) p, q, r, and s, how many lines will be required in an evaluation table?
12. Refer to Exercise 11 of Section 2.1 and write an evaluation table defining the truth values of the exclusive "or" that are consistent with common usage.
13. Using switch operations, write an expression that corresponds to the compound statement form

$$[p \vee (q \wedge \sim p)] \vee \sim q,$$

and diagram the resulting expression.
14. Using switch operations, write an expression that corresponds to the compound statement form

$$[(p \wedge \sim q) \vee (r \vee q)] \wedge (\sim p),$$

and diagram the resulting expression.

2.3 Equivalent Statement Forms

Consider the compound statement form $p \wedge (p \vee q)$ and the simple statement form p. From Table 2.14 it can be observed that the truth values are identical. In other words, as far as the truth values are concerned the form $p \wedge (p \vee q)$ and the form p are the same.

Table 2.14. $p \wedge (p \vee q)$ AND p HAVE THE SAME EVALUATION.

p	q	$p \vee q$	$p \wedge (p \vee q)$
T	T	T	T
T	F	T	T
F	T	T	F
F	F	F	F

▶ **Definition 4**

Two statement forms are **equivalent** when they have the same value for all possible arrangements of values of the simple component statement forms. Equivalence will be denoted by the symbol "$\equiv$".

Example 1

Table 2.15 shows that $(p \rightarrow q) \equiv [(\sim q) \rightarrow (\sim p)]$ because the values in column (3) are *identical* with those of column (6). This particular equiva-

Table 2.15. $p \rightarrow q$ AND $(\sim q) \rightarrow (\sim p)$ HAVE THE SAME EVALUATION.

(1) p	(2) q	(3) $p \rightarrow q$	(4) $\sim q$	(5) $\sim p$	(6) $\sim q \rightarrow \sim p$
T	T	T	F	F	T
T	F	F	T	F	F
F	T	T	F	T	T
F	F	T	T	T	T

lence is used often enough to warrant a special name. The form $(\sim q) \rightarrow (\sim p)$ is called the **contrapositive** of $p \rightarrow q$. The statement "If penicillin is administered, then the patient will recover" is an instance of $(p \rightarrow q)$ and hence by Table 2.15 is equivalent to the statement "If the patient did not recover, then penicillin was not administered." ∎

There are certain compound statement forms that are always true, regardless of the values of the simple components; such statement forms are also important enough to be given a special name.

▶ **Definition 5**

A compound statement form is a **tautology** if the value of the compound statement form is always T, regardless of the values of its simple components. A tautology will be designated by the symbol U.

Example 2

The compound statement form $p \vee \sim p$ is a tautology because column (3) of Table 2.16 consists of all T's. ▮

Table 2.16. EXAMPLE OF A TAUTOLOGY.

(1) p	(2) $\sim p$	(3) $p \vee \sim p$
T	F	T
F	T	T

Example 3 of Section 2.2 illustrates another tautology.

▶ **Definition 6**

A conditional $p \rightarrow q$ that is a tautology is called an **implication**† and is denoted $p \Rightarrow q$.

Example 3

The conditional $\{(p \rightarrow q) \wedge p\} \rightarrow q$ is a tautology because column (5) of Table 2.17 consists of all T's. Hence the given conditional is an implication and can be designated

$$\{(p \rightarrow q) \wedge p\} \Rightarrow q. \; ▮$$

Table 2.17. EXAMPLE OF A TAUTOLOGY.

(1) p	(2) q	(3) $p \rightarrow q$	(4) (3) $\wedge$ (1) $(p \rightarrow q) \wedge p$	(5) (4) $\rightarrow$ (2) $\{(p \rightarrow q) \wedge p\} \rightarrow q$
T	T	T	T	T
T	F	F	F	T
F	T	T	F	T
F	F	T	F	T

As might be expected, there are also compound statement forms that have false values, regardless of the values of its simple components.

▶ **Definition 7**

A compound statement form is a **contradiction** if the value of the compound statement form is always F, regardless of the values of its simple components. A contradiction will be designated by the symbol C.

Example 4

The compound statement form $p \wedge \sim p$ is a contradiction, as verified in Table 2.18. [Column (3) has all F's.] ▮

† The reader is reminded of the previous warning that the terminology used in the study of logic is not uniform. The term "implication" is an example.

Table 2.18. EXAMPLE OF A CONTRADICTION.

(1)	(2)	(3)
p	$\sim p$	$p \wedge \sim p$
T	F	F
F	T	F

We have seen that when a conditional $p \to q$ is a tautology, then the conditional is called an implication and is denoted by $p \Rightarrow q$. When a biconditional $p \leftrightarrow q$ is a tautology, then p and q must have the same values, and hence p and q are *equivalent*. Thus a biconditional $p \leftrightarrow q$ that is a tautology is an equivalence and often is designated $p \Leftrightarrow q$ because $p \Rightarrow q$ and $q \Rightarrow p$. Thus we have two symbols, $\equiv$ and $\Leftrightarrow$, that are commonly used to represent the very important concept of equivalence.

Example 5

In the evaluation table of Example 3 in Section 2.2 we showed that $(p \to q) \leftrightarrow ((\sim p) \vee q)$ always had a value T. Hence we have a biconditional that is a tautology, and we write

$$(p \to q) \Leftrightarrow ((\sim p) \vee q)$$

or

$$(p \to q) \equiv ((\sim p) \vee q)),$$

both of which are read "$(p \to q)$ is equivalent to $((\sim p) \vee q)$." ▮

The significance of the result of Example 5 should be emphasized. Because $(p \to q)$ is equivalent to $((\sim p) \vee q)$, the form $(p \to q)$ can be replaced by a statement form involving the connectives "or" and "not." We can also show that the connective $\leftrightarrow$ can be expressed in terms of the connectives $\vee$, $\wedge$, and $\sim$. Thus the five connectives introduced in Section 2.1 could be reduced to three connectives; it is more convenient, however, to keep all five.

In summary, a pair of statement forms can be shown to be logically equivalent by showing that their truth values are the same for all possible arrangements of values of the simple components; this is the same as showing that the biconditional of the two expressions is a tautology. Also, one statement form can be shown to imply another by showing that the associated conditional is a tautology.

APPLICATIONS

Example 6

An illustration using the algebra of statement forms in a problem of production management may be found in an article by Kattsoff and Simone [24]. The problem is concerned with the production of an item with m components that have been produced from n different materials. ▮

Example 7

Applications of this material for the solution of logical-puzzle-type problems may be found in Arnold [3], pp. 133–137. A simple illustration of such a problem follows. Suppose that a window has been broken and four boys are questioned. They make the following statements: Hippler: "Myers did it." Myers: "Hippler is lying." Palm: "I didn't do it." Stebbins: "Hippler did it." If exactly three of the suspects are lying, then who did it? If exactly one of the suspects is lying, then who did it? Let h, m, p, and s represent the following statements:

h: Hippler did it.
m: Myers did it.
p: Palm did it.
s: Stebbins did it.

The four boys made the following assertions: m, $\sim m$, $\sim p$, h. If three of them are lying, then one of the following statements must be true:

$$m \wedge (\sim\sim m) \wedge (\sim\sim p) \wedge \sim h;$$
$$(\sim m) \wedge (\sim m) \wedge (\sim\sim p) \wedge \sim h;$$
$$(\sim m) \wedge (\sim\sim m) \wedge (\sim p) \wedge \sim h;$$
$$(\sim m) \wedge (\sim\sim m) \wedge (\sim\sim p) \wedge h.$$

By evaluation tables or by the theorems of the next section, the latter four statements can be shown to be equivalent to

$$m \wedge p \wedge \sim h;$$
$$\sim m \wedge p \wedge \sim h;$$
$$C;$$
$$C.$$

Since exactly one of these must have value T, we can say that

$$(m \wedge p \wedge \sim h) \vee (\sim m \wedge p \wedge \sim h) \vee (C) \vee (C)$$

must have value T. The latter statement is equivalent to $(p \wedge \sim h)$, and for this statement to be true, p must be true; hence Palm did it. The solution of the second part (when only one boy is lying) is left as Exercise 28 of Section 2.4. ▌

EXERCISES

Suggested minimum assignment: Exercises 1, 3, 6, 7, 9, 11, and 13.

In each of Exercises 1–8 determine which of the statement forms are tautologies, which are contradictions, and which are neither. Also identify any implications and equivalences.

1. $(p \wedge q) \rightarrow \sim q$.
2. $\sim(p \vee q) \rightarrow p$.

3. $\{p \vee (p \wedge q)\} \leftrightarrow p$.
4. $\{p \wedge (p \vee q)\} \leftrightarrow p$.
5. $(\sim p \vee q) \leftrightarrow (q \wedge p)$.
6. $(q \wedge \sim p) \leftrightarrow (\sim q \vee p)$.
7. $\{(p \vee q) \vee \sim r\} \rightarrow p$.
8. $p \rightarrow \{(p \wedge q) \wedge \sim r\}$.

In each of Exercises 9–12 prove that the given statement forms are equivalent.

9. $\{\sim (p \vee q)\}, \{\sim p \wedge \sim q\}$.
10. $\{\sim (p \wedge q)\}, \{(\sim p) \vee \sim q\}$.
11. $\{(p \rightarrow q) \wedge (q \rightarrow p)\}, \{p \leftrightarrow q\}$.
12. $\{(p \vee q) \vee r\}, \{p \vee (q \vee r)\}$.
13. The exclusive "or" has values defined by Table 2.19. Does $p \veebar (p \wedge q)$ imply $\sim q$? Justify your answer.

Table 2.19

p	q	$p \veebar q$
T	T	F
T	F	T
F	T	T
F	F	F

14. Does $(q \wedge \sim p) \vee q$ imply $p \vee q$? Justify your answer.
15. For the Algebra of Switches, the following terms were introduced: "series connection," "parallel connection," "opposite," "unit switch," "zero switch," and "equivalence." List the corresponding terms for the Algebra of Statement Forms.

2.4 Properties of the Algebra of Statement Forms

So far we have laid the foundation for a new mathematical structure which we have called the Algebra of Statement Forms. The elements of this structure are statement forms p, q, r, and so on (simple or compound). We have two relations between statement forms called equivalence and implication, and we have the operations of $\wedge$, $\vee$, $\rightarrow$, $\leftrightarrow$, and $\sim$ that are defined over the elements. As with switches, once the foundation is laid, it is natural to begin the superstructure; that is, formulate conjectures involving the elements, operations, and relations, and then try to prove or disprove the conjectures. The proved conjectures are called theorems and become very useful laws of logic. In this section we list some of the more important and basic theorems and observe the similarities between the two structures of Chapters 1 and 2; the advantages of the latter observation will be stressed.

▶ **Theorem 1**

(Commutative Properties) If p and q are arbitrary statement forms, then

$$(i) \quad (p \vee q) \equiv (q \vee p)$$

and

$$(ii) \quad (p \wedge q) \equiv (q \wedge p).$$

The proof of Theorem 1 is easily accomplished by the use of evaluation tables and is left as Exercises 18 and 19. ◀

Example 1

The statement "The assets exceed the liabilities or the firm will show a profit this year" is an instance of the form $p \vee q$. Also "The firm will show a profit this year or the assets exceed the liabilities" is an instance of the form $q \vee p$. According to Theorem 1 these two statements are equivalent; we say that the connective $\vee$ is commutative. The connective $\wedge$ is also commutative. ▮

▶ **Theorem 2**

(Distributive Properties) If p, q, and r are arbitrary statement forms, then

$$(i) \quad \{p \wedge (q \vee r)\} \equiv \{(p \wedge q) \vee (p \wedge r)\}$$

and

$$(ii) \quad \{p \vee (q \wedge r)\} \equiv \{(p \vee q) \wedge (p \vee r)\}.$$

The proof of Theorem 2 is left as Exercises 20 and 21. ◀

Example 2

The statement "Sales will increase, and the sales manager or the vice president will be promoted" is an instance of $\{p \wedge (q \vee r)\}$; by Theorem 2(i) this statement is equivalent to the statement "Sales will increase and the sales manager will be promoted, or sales will increase and the vice president will be promoted." ▮

▶ **Theorem 3**

(Identity Properties) If C, U, and p are statement forms, where C is a contradiction and U is a tautology, then for every p

$$(i) \quad (p \vee C) \equiv p$$

and

$$(ii) \quad (p \wedge U) \equiv p.$$

The proof of Theorem 3 is left as Exercises 22 and 23. ◀

Example 3

The statement "If A is elected, then taxes will increase; A is elected; therefore taxes will increase" is an instance of the tautology

$$\{(p \rightarrow q) \wedge p\} \rightarrow q.$$

According to Theorem 3(ii), any statement connected by $\wedge$ with this or any other tautology is equivalent to the statement itself. ∎

▶ **Theorem 4**

(Complement Properties) If C, U, and p are statement forms, where C is a contradiction and U is a tautology, then for every p

$$(i) \quad (p \vee \sim p) \equiv U$$

and

$$(ii) \quad (p \wedge \sim p) \equiv C.$$

The proof of Theorem 4 is left as Exercises 24 and 25. ◀

Example 4

The statement "Candidate A will win or Candidate A will lose" is an instance of the form $(p \vee \sim p)$ and according to Theorem 4(i) is a tautology; that is, the given statement is always true. Theorem 4(i) is often referred to as the law of the excluded middle because $p \vee \sim p$ must be true. ∎

At this point the reader should compare Theorems 1–4 of this chapter with Theorems 1–4 of Chapter 1. As in Chapter 1, Theorem 5, which follows, can be proved by using the Theorems 1–4 or by using evaluation tables.

▶ **Theorem 5**

If C, U, and p are statement forms, where C is a contradiction and U is a tautology, then for every p

$$(i) \quad (p \vee U) \equiv U$$

and

$$(ii) \quad (p \wedge C) \equiv C.$$

The proof of Theorem 5 is left as Exercises 26 and 27. ◀

On page 26 of Section 1.4 twelve theorems involving switches were stated. We now state analogous theorems for statement forms. Although Theorems 1–4 were proved using evaluation tables, the following theorems may be proved using only those theorems that precede the one in question; therefore, we may think of Theorems 1–4 as a basis or a foundation upon which Theorems 5–17 can be built. Of course, Theorems 5–17 also may be proved by using evaluation tables, if it is more convenient to do so.

▶ **Theorems 6–17**

If C, U, p, q, and r are statement forms, where C is a contradiction and U is a tautology, then

6. For every p, $(p \vee p) \equiv p$.
7. For every p, $(p \wedge p) \equiv p$.
8. If x is a statement form such that $(p \vee x) \equiv p$ for every p, then $x \equiv C$.

9. If x is a statement form such that $(p \wedge x) \equiv p$ for every p, then $x \equiv U$.
10. $\{p \vee (p \wedge q)\} \equiv p$ (Absorption Property).
11. $\{p \wedge (p \vee q)\} \equiv p$ (Absorption Property).
12. $\{(p \wedge q) \equiv p\}$ if and only if $\{(p \vee q) \equiv q\}$.
13. If $(p \vee q) \equiv (p \vee r)$ and if $(p \wedge q) \equiv (p \wedge r)$, then $q \equiv r$.
14. $\{p \vee (q \vee r)\} \equiv \{(p \vee q) \vee r\}$ (Associative Property).
15. $\{p \wedge (q \wedge r)\} \equiv \{(p \wedge q) \wedge r\}$ (Associative Property).
16. $\{\sim(p \vee q)\} \equiv \{(\sim p) \wedge (\sim q)\}$ (DeMorgan's Property).
17. $\{\sim(p \wedge q)\} \equiv \{(\sim p) \vee (\sim q)\}$ (DeMorgan's Property). ◀

APPLICATIONS

Example 5

In the algebra of switches we used the properties of the structure to simplify lengthy compound switches. Complicated compound statements also can be simplified by using the properties of the algebra of statement forms developed in this section.

For example, consider the following quotation.

"In case of death:
The interest paid to the party of the first part must exceed $6000;
AND
the party of the second part will sell its interest in the company, or the interest paid to the party of the first part must not exceed $6000."

The quotation can be expressed symbolically as

$$p \wedge (q \vee \sim p),$$

which by Theorems 2, 3, and 4 is logically equivalent to

$$(p \wedge q).$$

Therefore, the quotation written above has the same meaning as

"In case of death:
The interest paid to the party of the first part must exceed $6000;
AND
the party of the second part will sell its interest in the company." ▮

Example 6

To illustrate the analysis of conflict by means of material presented so far, we consider the following hypothetical situation. (Other situations involving conflict, such as labor–management negotiations, could have been chosen.)

Suppose there is general agreement among certain powers at a peace conference that there is a need for a peacekeeping force consisting of forces from more than two nations. The conferees have agreed that such a force must come from among five nations, but they have imposed certain conditions upon the makeup of the peacekeeping force. These conditions are

1. Nations Q and R cannot serve together.

2. If nation X is excluded, then nation P cannot serve.
3. Nations Q and S cannot serve together.
4. Nations X and R cannot serve together.
5. Nation P or nation Q must serve.
6. Nation R or nation S must serve.

Is the peace conference hopelessly deadlocked or is there a solution that will satisfy all six conditions?

Let p represent the statement "Nation P serves"; let q represent the statement "Nation Q serves"; and so on. Conditions 1 through 6 can be restated as:

1. $(q \wedge r) \equiv C$.
2. $\{(\sim x) \wedge p\} \equiv C$ (derived from $\sim x \Rightarrow \sim p$).
3. $(q \wedge s) \equiv C$.
4. $(x \wedge r) \equiv C$.
5. $(p \vee q) \equiv U$.
6. $(r \vee s) \equiv U$.

Using the definitions and theorems stated so far, we can make the following deductions: From 1 and 3 we have

$$\{(q \wedge r) \vee (q \wedge s)\} \equiv C,$$

or

$$\{q \wedge (r \vee s)\} \equiv C.$$

From condition 6 the latter expression becomes

$$(q \wedge U) \equiv C.$$

Therefore, $q \equiv C$. From 5 we can now say that $p \equiv U$. Now, from 2, $(\sim x) \equiv C$ and hence $x \equiv U$. From 4, $r \equiv C$ and, from 6, $s \equiv U$. Therefore, nations P, S, and X can serve under the imposed conditions. Q and R cannot serve under the imposed conditions. ∎

EXERCISES

Suggested minimum assignment: Exercises 1, 3, 11, 14, and 17.

In each of Exercises 1–5 simplify the given compound statement form by using the Theorems 1–5 of this section.

1. $(\sim p) \wedge (q \vee p)$.
2. $(\sim p) \vee (q \wedge p)$.
3. $(p \wedge \sim p) \vee \{(q \vee \sim q) \wedge p\}$.
4. $p \vee \{q \vee (p \wedge \sim p)\}$.
5. $\{(\sim p) \wedge (p \vee q)\} \vee \{(q \vee p) \wedge ((\sim q) \vee p)\}$.

In each of Exercises 6–17 prove the indicated theorem using evaluation tables or any preceding theorems.

6. Theorem 6.
7. Theorem 7.
8. Theorem 8.
9. Theorem 9.

10. Theorem 10.
11. Theorem 11.
12. Theorem 12.
13. Theorem 13.
14. Theorem 14. *Hint:* Use evaluation tables—very difficult otherwise.
15. Theorem 15. *Hint:* Use evaluation tables—very difficult otherwise.
16. Theorem 16.
17. Theorem 17.

In each of Exercises 18–25 use evaluation tables to prove the theorem.

18. Prove Theorem 1(*i*).
19. Prove Theorem 1(*ii*).
20. Prove Theorem 2(*i*).
21. Prove Theorem 2(*ii*).
22. Prove Theorem 3(*i*).
23. Prove Theorem 3(*ii*).
24. Prove Theorem 4(*i*).
25. Prove Theorem 4(*ii*).

26. Prove Theorem 5(*i*) by using Theorems 1–4.
27. Prove Theorem 5(*ii*) by using Theorems 1–4.
28. Solve the second part of Example 7 of Section 2.3.

2.5 Arguments (Optional)

Most of us in our normal lives are subjected to a steady stream of arguments from various sources, such as advertisements, candidates for office, public officials, parents, children, teachers, students, peers, revolutionaries, and various other dispensers of rhetoric.

The purpose of this section is to apply the rules of logic (properties of our structure) to analyze the forms of arguments, to classify them as valid or invalid, and to distinguish arguments from nonarguments.

We shall begin with the following definition of an **argument form,** which is simply a special type of statement form.

▶ **Definition 8**

An **argument form** is a conditional of the form

$$(p_1 \wedge p_2 \wedge \cdots \wedge p_n) \rightarrow q.$$

The statement forms $p_1, \ldots, p_n$ are called **premises** and q is called the **conclusion.** An argument form is said to be **valid** if it is a tautology; otherwise the argument form is a **fallacy.**

In other words, if the argument form (or conditional) is an implication, then the argument form is said to be valid; if it is not an implication, it is said to be a fallacy.

Example 1

The conditional

$$[(p \rightarrow q) \wedge p] \rightarrow q$$

has premise $(p \to q)$ and premise p. The conclusion is q. The conditional is a valid argument form because the conditional is a tautology as verified in Example 3 of Section 2.3 and hence an implication. On the other hand, the conditional

$$[(p \to q) \wedge q] \to p,$$

with premise $(p \to q)$, premise q, and conclusion p, is a fallacy because the conditional is not a tautology, as can be demonstrated in an evaluation table (Table 2.20). ▮

Table 2.20. EXAMPLE OF A FALLACY.

(1) p	(2) q	(3) $p \to q$	(4) $(p \to q) \wedge q$	$(4) \to (1)$
T	T	T	T	T
T	F	F	F	T
F	T	T	T	$\boxed{F}$
F	F	T	F	T

For reference purposes we give a list of basic theorems that are particularly useful in establishing the validity or fallacy of complicated argument forms.

▶ Theorems 18–24

Let p, q, and r be statement forms.

18. (Double Negation Property) $[\sim(\sim p)] \Leftrightarrow p$.
19. (Conjunction Property) $(p \wedge q) \Rightarrow p$.
20. (Disjunction Property) $p \Rightarrow (p \vee q)$.
21. (Detachment Property) $[(p \to q) \wedge p] \Rightarrow q$.
22. (Contrapositive Property) $[(p \to q) \wedge \sim q] \Rightarrow \sim p$.
23. (Disjunction Detachment Property) $[(p \vee q) \wedge \sim p] \Rightarrow q$.
24. (Chain Rule Property) $[(p \to q) \wedge (q \to r)] \Rightarrow (p \to r)$.

The proofs of Theorems 18–24 are left as Exercises 15–21. These proofs can be accomplished by means of evaluation tables or by means of the theorems of the last section. If the latter approach is used, one must recognize that each proof will involve showing that a conditional is a tautology and hence an implication; one can show this by remembering that a conditional $r \to s$ is equivalent to $[(\sim r) \vee s]$ and then showing, if possible, that the latter expression is equivalent to U. ◀

In Example 1, we showed how argument forms could be shown to be valid or fallacious by use of evaluation tables. We next illustrate how argument forms may be determined to be valid by use of the theorems that have been stated so far.

Example 2

Show whether or not the argument form

$$\{[(p \vee q) \wedge \sim p] \wedge [(\sim p) \rightarrow \sim q]\} \rightarrow p$$

is valid.

$$\begin{aligned}
&\{[(p \vee q) \wedge \sim p] \wedge [(\sim p) \rightarrow \sim q]\} \\
&\qquad \Rightarrow \{[q] \wedge [(\sim p) \rightarrow \sim q]\} && \text{by Theorem 23,} \\
&\qquad \Rightarrow \{[\sim(\sim q)] \wedge [(\sim p) \rightarrow \sim q]\} && \text{by Theorem 18,} \\
&\qquad \Rightarrow \sim(\sim p) && \text{by Theorem 1} \\
&&& \text{and Theorem 22,} \\
&\qquad \Rightarrow p && \text{by Theorem 18.}
\end{aligned}$$ ∎

Example 3

The argument forms

1. $(q \rightarrow p) \rightarrow (p \rightarrow q)$,
2. $[(q \rightarrow p) \wedge \sim q] \rightarrow \sim p$,
3. $[(p \rightarrow q) \wedge (q \rightarrow r)] \rightarrow (r \rightarrow p)$,

are common argument forms that are not valid. Unfortunately, they often are treated as valid, and grandiose false arguments are thereby perpetrated on the unwary. The first of the three forms deserves special attention: $(p \rightarrow q)$ is called the **converse** of $(q \rightarrow p)$, but the truth of $(q \rightarrow p)$ does not ensure the truth of $(p \rightarrow q)$. Argument forms such as these can be shown to be fallacious by means of evaluation tables. Illustrations of the fallacious forms presented above are given in Example 6 of the Applications. ∎

We now turn our attention to specific arguments. Because an argument form is a special type of statement form, we might expect an argument to be a special type of statement.

► **Definition 9**

An argument form in which each placeholder p, q, and so on, is a specified statement is known as an **argument**.

Example 4

Arguments are often written in the following manner:

1. If candidate A wins, then taxes will increase.
2. Candidate A wins.

C. Therefore, taxes will increase.

The argument above is an instance of the following form (argument form):

$$\left[\begin{array}{l} p \rightarrow q \\ p \\ \hline \text{Therefore, } q. \end{array}\right] \quad \text{or} \quad [\{(p \rightarrow q) \wedge p\} \rightarrow q],$$

which is valid according to Theorem 21. ∎

In Example 4, the fact that the form of the argument is valid does not necessarily mean that the conclusion is true. A valid form simply means that the rules of reasoning are correct and says nothing about the truth or falsity of the premises and the conclusion. For example, if either one of the two premises in Example 4 is false, then the conclusion may or may not be true. If, however, the premises of an argument are true and the corresponding argument form is valid, then the conclusion must be true.

The reader is reminded that statements that are not conditionals are not arguments.

APPLICATIONS

Example 5

An argument having a valid form and true premises is called a **sound argument.** An argument having a fallacious form or at least one false premise is called an **unsound argument.** Consider the following argument:

1. If the Florida Power Corporation issued rights in 1971, then stockholders with at least 10 shares had an option to buy more stock at $45.25 per share.
2. The Florida Power Corporation issued rights in 1971.

C. Therefore, stockholders with at least 10 shares had an option to buy more stock at $45.25 per share.

The argument has the valid form $[(p \to q) \wedge p] \Rightarrow q$ (Theorem 21), and the premises are true; hence the argument is sound.

Consider another argument:

1. If the population of Mexico increased last year, then the moon had trees.
2. If the moon had trees, wood could be brought to earth from the moon.

C. Therefore, if the population of Mexico increased last year, then wood could be brought to the earth from the moon.

The argument has the valid form $[(p \to q) \wedge (q \to r)] \Rightarrow (p \to r)$, but the first premise is not true; hence the argument is unsound.

Consider a third argument:

1. If candidate A gets a majority of the votes in the general election, then he is elected to office.
2. Candidate A does not get a majority of the votes.

C. Therefore, candidate A is not elected to office.

The form, however, is the fallacious form $[(p \to q) \wedge (\sim p)] \to \sim q$, and hence the argument is unsound, regardless of the truth or falsity of the premises. In other words, the conclusion does not follow logically from the premises even though these premises may be true. Evaluation tables can be used to prove that a form is fallacious.

Finally, we point out that there are some arguments that cannot be labeled as sound or unsound because the truth or falsity of the premises is impossible to ascertain. Consider the assertion:

1. If there are bacteria on Saturn, then there is life on Saturn.
2. There are bacteria on Saturn.

C. Therefore, there is life on Saturn.

Although this argument has a valid form, we cannot say whether the second sentence is true or false, although we know that it is one or the other. Thus we cannot say that the argument is sound or unsound. ▮

Example 6

In everyday life the ordinary citizen is subject to a never-ending barrage of arguments from a variety of sources. Some of these arguments are *not* sound because their premises are not true. Others, however, are not sound simply because their forms are not valid. Instances of the fallacious argument forms of Example 3 follow. We emphasize that in these instances, the truth or falsity of the premises is not considered; rather it is the form that is faulty.

1. If candidate A is elected, there will be peace for 20 years. To have peace for 20 years it is necessary that candidate A is elected.

2. If a system provides the means for peaceful change, then no cause justifies violence in the name of change. No means for peaceful change are provided; therefore, violence is justifiable.

3. If brand X is used, event A will take place, and if event A takes place, happiness will prevail. Happiness prevails; therefore, brand X is used. ▮

EXERCISES

Suggested minimum assignment: Exercises 1, 3, 5, 8, 9, 11, 13, and 17.

In each of Exercises 1–8 determine whether the given statement forms are valid argument forms, fallacious argument forms, or nonargument forms. Give reasons for your answer. Identify the conclusion and each premise.

1. $\{(p \vee q) \wedge \sim p\} \rightarrow q$.
2. $\{\sim q \wedge (p \rightarrow q)\} \rightarrow p$.
3. $\{(p \vee q) \wedge q\} \rightarrow p$.
4. $\{q \wedge (q \rightarrow p)\} \rightarrow \sim p$.
5. $\{q \wedge (p \vee q)\}$.
6. $\{q \vee (p \wedge \sim q)\}$.
7. $[(p \vee \sim q) \wedge \{(p \vee q) \wedge q\}] \rightarrow (p \wedge q)$.
8. $[(p \wedge \sim q) \wedge \{(p \rightarrow q) \wedge p\}] \rightarrow p$.

In each of Exercises 9–14 determine whether the form (argument form) of the given argument is valid or fallacious.

9. If the experimental mouse is hungry, he will enter the food box on the next trial. If the mouse enters the food box on the next trial, then it is not the case that the hunger response was less on the previous trial. The hunger response on the previous trial was less. Therefore, the mouse is not hungry.

10. If the factory completes production in one day, the factory will receive an extra allocation of funds. The factory gets an extra allocation of funds, or the workers will not get a bonus. The factory completes the production in one day. Therefore, the workers get a bonus.
11. If a candidate running for office is poor, then he cannot spend much money on television. Moreover, if the candidate cannot spend much money on television, his chances for election are not good. But his chances for election are good; therefore, he is poor.
12. If interest rates decrease, consumers borrow more money; when consumers borrow more, more houses are built. Therefore, interest rates decrease if more houses are built.
13. If the population of China increased, then the Mississippi River flows north. China had increased population. Therefore, the Mississippi River flows north.
14. If the state of Vermont triples its population, she will get at least one extra seat in the House of Representatives. Vermont tripled her population or Mt. Mansfield erupted. Mt. Mansfield did not erupt. Therefore, Vermont will get an extra House seat.
15. Prove Theorem 18.
16. Prove Theorem 19.
17. Prove Theorem 20.
18. Prove Theorem 21.
19. Prove Theorem 22.
20. Prove Theorem 23.
21. Prove Theorem 24.
22. Determine whether the argument of Exercise 9 is sound or unsound or neither.
23. Determine whether the arguments of Exercises 11 and 13 are sound or unsound or neither.

NEW VOCABULARY

statement 2.1
statement form 2.1
instance of a statement form 2.1
connectives of statement forms 2.1
simple statement forms 2.1
compound statement form 2.1
simple statements 2.1
compound statements 2.1
component of a statement form 2.1
conjunction 2.1
disjunction 2.1
conditional 2.1
biconditional 2.1
algebra of statement forms 2.1
negation 2.1
value of a statement form 2.2
equivalent statement forms 2.3
contrapositive of $p \rightarrow q$ 2.3
tautology 2.3
implication 2.3
contradiction 2.3
argument form 2.5
premise 2.5
conclusion 2.5
valid argument form 2.5
fallacious argument form 2.5
converse 2.5
argument 2.5

3 Sets and the Algebra of Subsets

The purpose of this chapter is (1) to introduce a third mathematical structure in which elements other than numbers are combined by various operations, (2) to present material that will be prerequisite to the chapters on probability and linear programming, and (3) to unify the first three chapters by showing that each of the three new structures is an example of a single abstract structure known as a Boolean algebra.

Prerequisites: Only 1.1, 1.2, 2.1, and 2.2 are required if 3.3 and 3.4 are omitted; Section 2.3 is required for 3.3, and Sections 1.3, 1.4, and 2.4 are required for 3.4.
Suggested Sections for Moderate Emphasis: 3.1–3.3, 3.5, 3.6
Suggested Sections for Minimum Emphasis: 3.1, 3.2, 3.5

3.1 Basic Definitions

Although it is customary in mathematics to treat the words "set" and "element" as undefined terms, it is useful to have intuitive interpretations of these terms when undertaking an introductory study of set theory. Think of a well-defined collection of distinguishable objects; by a well-defined collection, we mean that it is perfectly clear whether an arbitrary object does or does not belong to the collection, and by distinguishable, we mean that we can distinguish one object from another. For example, we may think of a collection of postage stamps, or a collection of current U.S. Senators, or a collection of real numbers that satisfies the equation $x^2 = 1$. Such well-defined collections of distinguishable objects serve as intuitive interpretations of the word "set," while the objects themselves are examples of the elements of a set.

We usually denote sets by capital letters ($A, B, C, \ldots$) and the elements of a set by lowercase letters. The statement "b is an element of B" is expressed as $b \in B$.

There are several ways that a specific set can be described or defined. First, a defining statement can be used: "A is the set of the first three U.S. Presidents" defines a set A. Second, the same set can be defined by listing the elements within braces; for example,

$$A = \{\text{Washington, Adams, Jefferson}\}.$$

Third, using a colon to represent the expression "such that," we can use the set-builder notation; for example,

$$A = \{x : x \text{ is one of the first three U.S. Presidents}\},$$

which is read "the set of all x such that x is one of the first three U.S. Presidents." The letter x in the set-builder notation is simply a placeholder, and other symbols could have been used.

Next, the concept of a **subset** of a set is introduced by means of an example. Consider the set C of the first two U.S. Presidents; that is,

$$C = \{\text{Washington, Adams}\}.$$

Each element of C is also an element of the set

$$A = \{\text{Washington, Adams, Jefferson}\},$$

and we choose to express this important relationship between A and C by saying that C is a subset of A.

▶ **Definition 1**

Set R is a **subset** of a set S if every element of R is also an element of S; "R is a subset of S" will be denoted

$$R \subseteq S.$$

Example 1

If P represents the set of all living people in the world and F represents the set of living people that are female, then we can say

$$F \subseteq P,$$

which is read "F is a subset of P." If there is at least one living nonfemale person, then we say that F is a **proper subset** of P, and that concept is denoted

$$F \subset P.$$

In other words, a proper subset does not include all the elements of the parent set; a subset may or may not include all the elements of the parent set. ▮

It is possible to describe a set in which there are no elements. For example, suppose we say: Let S be the set of U.S. Presidents who were born in Mongolia. Because there are no elements of this set, we say that the set is the **empty set** or the **null set** and denote such a set with the symbol ∅ (Scandinavian modified "oh"). When working with sets, experience has shown that it is useful to identify a set U, such that all sets in the discussion are subsets of U; such a set is called the **universal set.** If, for example, a discussion revolves around various sets of nations in the world, then we could choose the set of all nations of the world as the universal set, that is, the universe of the discussion.

We are now in a position to use the set-builder notation to establish some relationships between sets and statements. Consider the set

$$A = \{x : x \text{ was built by General Motors}\}.$$

The assertion

"x was built by General Motors"

is not really a statement because it is not capable of being evaluated as true or false until x is identified. However, once a universe U is established, then we may think of the assertion "x was built by General Motors" as producing a collection of statements; a single statement of the collection is obtained by substituting an element of U for x. Thus, if we let U be the set of cars found on a certain parking lot, say

$$U = \{\text{a 1959 Ford, a 1970 Chevrolet, a 1974 Pontiac, a 1965 Dodge}\},$$

then the given assertion produces the statements

1. A 1959 Ford was built by General Motors,
2. A 1970 Chevrolet was built by General Motors,
3. A 1974 Pontiac was built by General Motors,
4. A 1965 Dodge was built by General Motors,

of which the second and third are true. The set A consists of those elements of U which when substituted for x in the given assertion produce a *true* statement, namely,

$$\{\text{a 1970 Chevrolet, a 1974 Pontiac}\}.$$

In other words, the notation

$$A = \{x : x \text{ was built by General Motors}\}$$

means the subset of U for which the assertion "x was built by General Motors" becomes a true statement. Set A can be referred to as the **truth set** of the assertion "x was built by General Motors."

We describe a **finite set** as a set having n elements, where n is a non-negative integer. Although the sets that we have discussed so far have been finite sets, this need not be the case. For example, consider the set of all natural numbers or the set of all rational numbers. Neither of these sets consists of a finite number of elements; hence they are called **infinite sets.**

Often it is helpful to draw graphical representations of sets. First, we draw a large rectangle that represents the universal set as shown in Figure 3.1. Then we can represent a subset A of U by a circle (or other closed curve) within the rectangle as shown in Figure 3.2. We may think of points within the circle as representing elements x of U for which $x \in A$ becomes a true statement and points outside of the circle as representing elements x of U for which $x \in A$ becomes a false statement. Next, we can represent a subset of A, which we will call B, by another closed curve within A as shown in Figure 3.3; such diagrams are often called **Venn diagrams.**

Finally, we wish to use an illustration to make a distinction between "a subset of a set" and "an element of a set." Consider a set A of all states

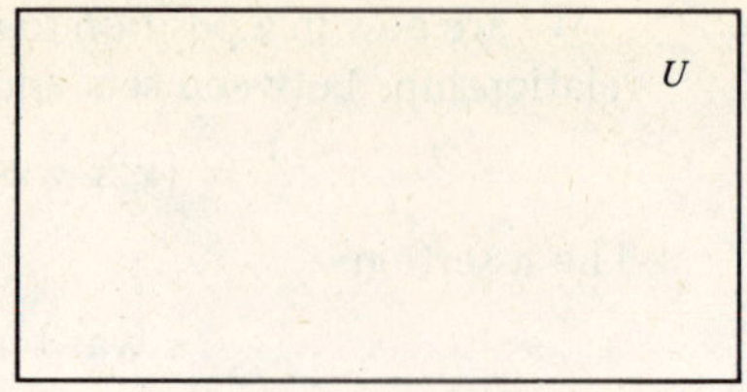

Figure 3.1. *Representation of universal set.*

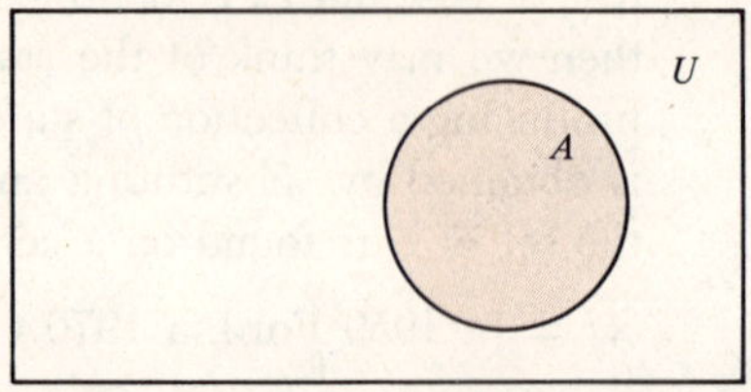

Figure 3.2. *Representation of set A.*

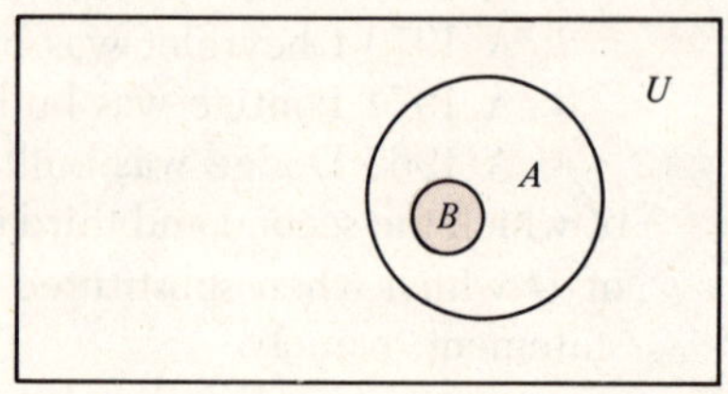

Figure 3.3. *Representation of a subset of A.*

in the United States. Consider a second set E of states that lie entirely east of the Mississippi River. We say that E is a subset of A; it is wrong to say that E is an element of A. Now consider the state of Maine, which we will call m. We say that m is an element of A, that is, $m \in A$; but it is wrong to say that m is a subset of A. We can say, however, that the set consisting of the single state of Maine is a subset of A; that is, we can say that $\{m\} \subseteq A$. Suppose that we let w represent the state of Wyoming; by the symbol $w \notin E$, we mean that Wyoming does *not* belong to set E. By the symbol $\{w\} \nsubseteq E$, we mean that the set consisting of the state of Wyoming is *not* a subset of E.

Example 2

Let $A = \{a, b, c\}$. The subsets of A are $\{a, b, c\}$, $\{a, b\}$, $\{a, c\}$, $\{b, c,\}$, $\{a\}$, $\{b\}$, $\{c\}$, and $\varnothing$; notice that both the empty set and the whole set are included in the list of subsets. The elements of A are a, b, and c. The following are *correct* statements:

$$\begin{array}{ll} \{a\} \subset A, & a \in A, \\ \{a, b\} \subset A, & \{a, b\} \subseteq A, \\ \{a, b, c\} \subseteq A, & \varnothing \subset A. \end{array}$$

The following are *incorrect* statements:

$$\{b\} \in A, \qquad b \subset A, \qquad \{a, b, c\} \subset A.$$

The first one is incorrect because $\{b\}$ represents a subset and not an element; the second is incorrect because b represents an element and not a subset; the third statement is incorrect because $\{a, b, c\}$ is *not* a *proper* subset of A. ▮

APPLICATIONS

Example 3

A student should have some familiarity with the notation and basic concepts of set theory because of the extent to which the notation is currently being used in the literature of the social, management, and life sciences. Articles and books abound in which the vocabulary of set theory is a prerequisite; for example, in political science, the book entitled *The Theory of Coalitions*, by Riker [37], depends heavily upon the basic ideas and notation of set theory. The student of political science should at least glance through this book to observe the relevance of set theory to the subject discussed in that book. ▮

We shall observe in the remainder of this book that the basic concepts of set theory are also fundamental in the study of probability, linear programming, and game theory.

EXERCISES

Suggested minimum assignment: 1, 3, 6, 7, 9, 13, 15, 17, 18, 19, 22, and 23.

In each of Exercises 1–8 express the given set in two additional ways.

1. A is the set of all states in the region of the United States that is called New England.
2. B is the set of living men who have been President of the United States.
3. C is the set of integers that satisfy the equation $x^2 - 1 = 0$.
4. D is the set of rational numbers that satisfy the equation

$$(2x + 3)(x - 2) = 0.$$

5. $E = \{x\colon (x - 3)(x + 2) = 0, \text{ and } x \text{ is a positive integer}\}$.
6. $F = \{x{:}x$ is the capital city of New York or New Jersey or South Carolina$\}$.
7. $G = \{1, 2, 3\}$.
8. $H = \{$Ford, General Motors, Chrysler, American Motors$\}$.

In each of Exercises 9–12 write a proper subset of the given set in symbolic notation. Also express an element of the given set using symbolic notation.

9. $A = \{3, 4, 5\}$.
10. $B = \{x, y, z\}$.

11. C is the set of all men who currently are President or Vice-President of the United States.
12. $D = \{x: 1 \leq x \leq 3, x \text{ is an integer}\}$.
13. If the universal set of discussion is the set of cities S = {Chicago, Cleveland, Pittsburgh}, list all the subsets of S.
14. If the universal set of discussion is the set of stocks B = {AT&T, GE, IBM, GM}, list all of the subsets of B.

In each of Exercises 15–22 state whether the given statement is true or false. If the statement is false, change it so that it becomes a true statement. Let

$$U = \{2, 3, 4, 5\}.$$

15. $2 \in U$.
16. $\varnothing \subset U$.
17. $\{2, 3\} \in U$.
18. $\{2, 3, 4, 5\} \subseteq U$.
19. $\varnothing \in U$.
20. $3 \subset U$.
21. $6 \notin U$.
22. $\{4, 5\} \not\subseteq U$.
23. If the universal set of discussion is the U.S. Senate, draw a Venn diagram showing the subset of all Democrats and the subset of all Republicans. List one member of each subset. Draw a second diagram in which another subset of all liberals (defined according to some index) is superimposed on a duplicate of the first diagram.
24. Use set notation to express the relationship between the set of all Republicans and the set of all liberal Republicans in the U.S. Senate.

3.2 Set Operations

We have begun the construction of still another mathematical structure; this time the building blocks of the structure are subsets of some designated universal set U. In this section we shall introduce various operations whereby pairs of subsets of U can be combined to produce a unique subset of U. We also shall present the concept of the complement of a single set.

First, however, we establish what is meant by the **equality** of two sets.

► **Definition 2**

Two sets A and B are said to be **equal** if their elements are the same.

The fact that the elements of two sets A and B are the same if $A \subseteq B$ and $B \subseteq A$ is frequently used to establish the equality of two sets.

Example 1

Let A = {Ford, Chrysler, General Motors}. Let B be the set of corporations that are the three largest manufacturers of automobiles in the United States. We can show that $A \subseteq B$ and we can show that $B \subseteq A$; therefore, $A = B$, that is, the elements of A and B are identical. ▮

We now turn our attention to ways of combining sets.

▶ **Definition 3**

Let A and B be arbitrary subsets of a universal set U. The set of elements that belong to both A and B is called the **intersection** of A and B and will be designated by

$$A \cap B.$$

Example 2

Let $A = \{a, b, c, d\}$ and let $B = \{c, d, e\}$. Then $A \cap B = \{c, d\}$. ▮

Example 3

Let F be the set of all Ford automobiles in this country. Let R be the set of all red automobiles in this country. Then $F \cap R$ is the set of all red Fords in this country. ▮

We now relate the intersection of two sets to the conjunction of two statements, evaluation tables, and Venn diagrams. For an arbitrary element x of U, the intersection, $A \cap B$, can be thought of as the set of elements of U which when substituted for x in

$$\text{“}x \in A \text{ and } x \in B\text{”}$$

produces a true statement. That is, for $x \in U$,

$$A \cap B = \{x : (x \in A) \wedge (x \in B)\}.$$

A variation of this alternative definition of $A \cap B$ may be given by means of the evaluation table shown in Table 3.1. With these ideas in mind we

Table 3.1. INTERSECTION OF A AND B.

	$x \in A$	$x \in B$	$x \in A \cap B$
	p	q	$p \wedge q$
(1)	T	T	T
(2)	T	F	F
(3)	F	T	F
(4)	F	F	F

have the Venn diagram representation of $A \cap B$ as shown by the colored region in Figure 3.4. In Figure 3.4 the number of each region corresponds to the number of the row of the evaluation table. The reader should realize that any of the numbered subsets shown in Figure 3.4 can be empty.

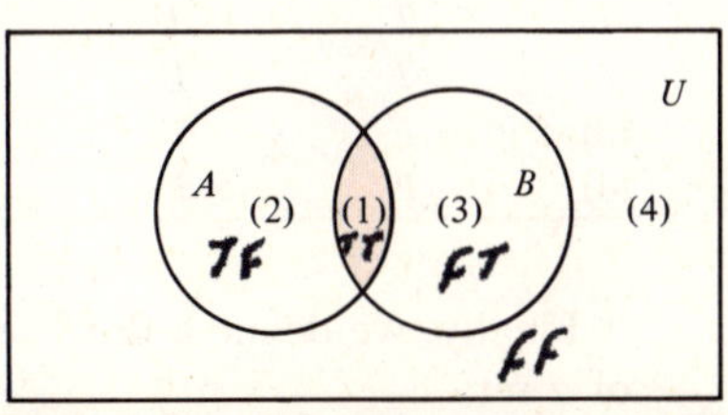

Figure 3.4. *Intersection of A and B.*

If the intersection of two sets is the empty set, then the sets are said to be **disjoint.** Next we define a second set operation known as the **union** of two sets.

▶ Definition 4

Let A and B be arbitrary subsets of a universal set U. The set of elements that belong to A or B (including those that belong to both) is called the **union** of A and B and will be designated by

$$A \cup B.$$

Example 4

Let $A = \{a, b, c, d\}$ and let $B = \{c, d, e\}$. Then $A \cup B = \{a, b, c, d, e\}$. ▮

Example 5

By using the same sets that were described in Example 3, $A \cup B$ represents the set of all automobiles in this country that are red or Fords or both. ▮

We now relate the union of two sets to the disjunction of two statements, evaluation tables, and Venn diagrams. For an arbitrary element x of U, the union, $A \cup B$, can be thought of as the set of elements of U which when substituted for x in

$$\text{“}x \in A \quad \text{or} \quad x \in B\text{”}$$

produces a true statement. That is, for $x \in U$,

$$A \cup B = \{x: (x \in A) \vee (x \in B)\}.$$

A variation of this alternative definition of $A \cup B$ may be given by means of the evaluation table shown in Table 3.2. With these ideas in mind we have the Venn diagram representation of $A \cup B$ as shown by the colored region in Figure 3.5. In Figure 3.5, the number of each region corresponds to the number of the row of the evaluation table. Again, it is pointed out that any of the numbered subsets shown in Figure 3.5 can be empty.

Table 3.2. UNION OF A AND B.

	$x \in A$	$x \in B$	$x \in A \cup B$
	p	q	$p \vee q$
(1)	T	T	T
(2)	T	F	T
(3)	F	T	T
(4)	F	F	F

Finally, we define a third set operation known as the **complementation** of a set.

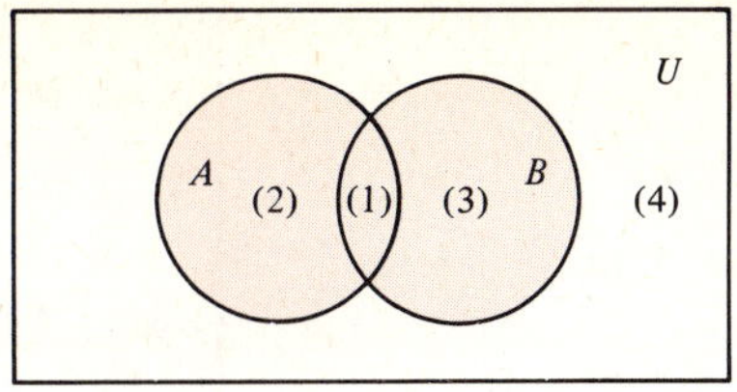

Figure 3.5. *Union of A and B.*

Definition 5

Let A be an arbitrary subset of a given universal set U. The set of all elements that belong to U but do not belong to A is called the **complement** of A and is denoted A'.

Example 6

Let U be the English alphabet, and let A be the set of consonants. Then A' is the set of vowels. ▮

Example 7

Let U be the set of U.S. Senators, and let R be the set of Republican U.S. Senators. Then R' represents the set of all non-Republican U.S. Senators. ▮

We now relate the complement of a set to the negation of a statement, evaluation tables, and Venn diagrams. For an arbitrary element x of U, the complement of A can be thought of as the set of elements of U which when substituted for x in

"it is not true that $x \in A$"

produces a true statement. That is, for $x \in U$,

$$A' = \{x : \sim(x \in A)\}.$$

A variation of this alternative definition of A' may be given by means of the evaluation table shown in Table 3.3. With these ideas in mind we have the Venn diagram representation of A' as shown by the colored region in Figure 3.6. In this figure the number of each region corresponds to the number of the row of the evaluation table.

Table 3.3. Complement of A.

	$x \in A$ p	$\sim(x \in A)$ $\sim p$
(1)	T	F
(2)	F	T

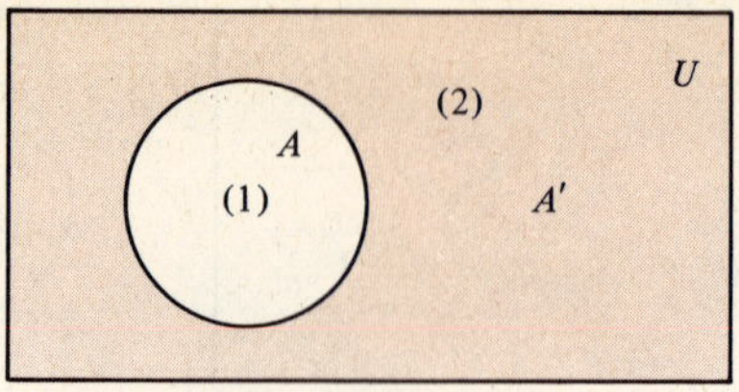

Figure 3.6. *Complement of A.*

Definitions 3 and 4 can be extended to define the intersection and union of three or more sets. Let $A_1, \ldots, A_n$ be subsets of a universal set U. By $A_1 \cap \cdots \cap A_n$ we mean the set of all elements of U which belong to all of the sets $A_1, \ldots, A_n$. By $A_1 \cup \cdots \cup A_n$ we mean the set of all elements of U which belong to at least one of the sets $A_1, \ldots, A_n$.

APPLICATIONS

Example 8

Prior to a blood transfusion certain tests are executed on both the recipient and the prospective donor in order to classify the blood of each and hence determine the suitability of the prospective donor's blood. Blood classification is accomplished by determining whether or not a person reacts to each of three tests. These tests and the corresponding classification are given in Table 3.4. If we let set A represent the set of all people who react to test A, let set B represent the set of all people who react to test B, and let set R represent the set of all people who react to the Rh test, then each of the classifications can be expressed using set notation as shown in the last column of Table 3.4.

Table 3.4. Set representations of blood classifications.

Blood classification	A test	B test	Rh test	Set notation
AB Rh positive	reacts	reacts	reacts	$A \cap B \cap R$
A Rh positive	reacts	no	reacts	$A \cap B' \cap R$
B Rh positive	no	reacts	reacts	$A' \cap B \cap R$
O Rh positive	no	no	reacts	$A' \cap B' \cap R$
AB Rh negative	reacts	reacts	no	$A \cap B \cap R'$
A Rh negative	reacts	no	no	$A \cap B' \cap R'$
B Rh negative	no	reacts	no	$A' \cap B \cap R'$
O Rh negative	no	no	no	$A' \cap B' \cap R'$

Now construct for each *donor* a set D of tests to which he *did* react, and for each *recipient* a set C of tests to which he *did* react. In order for a transfusion to be safe, $D \subseteq C$, and from this relationship, donors for particular recipients can be identified. For example, a donor in the set

$$(A' \cap B) \cap R,$$

for which $D = B \cap R$, can give blood only to those in the sets

$$(A' \cap B) \cap R \qquad \text{or} \qquad (A \cap B) \cap R.$$

On the other hand, a person in the set $(A' \cap B') \cap R'$ can give blood to anyone but can receive blood only from someone else in the same set. ▮

Example 9

The purpose of this example is to illustrate the use of sets and set operations in the analysis of legal documents. The Tenth Amendment to the U.S. Constitution states: "The powers not delegated to the United States by the constitution, nor prohibited by it to the states, are reserved to the states respectively, or to the people." Let U represent the universe of powers referred to by the Tenth Amendment. Under the assumption that these powers are well defined, let

D = set of all powers delegated to the United States by the Constitution;
P = set of all powers prohibited to the states by the Constitution;
S = set of powers of the states;
T = set of powers of the people.

Under the assumptions that $S \cap P = \emptyset$, $D \cup D' = U$, and $P \cup P' = U$, we have the Venn diagram of Figure 3.7.

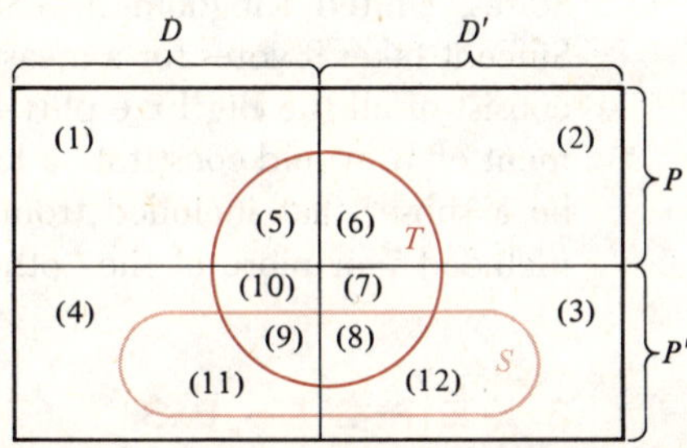

Figure 3.7. *Representation of Tenth Amendment.*

The various disjoint subsets defined by intersections of two or more of sets D, P, S, and T are numbered in Figure 3.7 for reference.

The Tenth Amendment, which says that

$$D' \cap P' \subseteq S \cup T,$$

requires that subset (3) be empty. Various arguments can be made that other of the numbered subsets of Figure 3.7 are empty. One point of view is that subsets (5), (7), (8), (9), (10), and (11) are all empty. Another point of view is that (2), (4), (5), and (6) are empty; many of these disagreements arise because of a violation of our first assumption that the powers are well defined. In other words, there is disagreement over membership in the sets D, P, S, and T, particularly the latter two. The interested reader may wish to list various powers and establish set membership of these powers according to predetermined criteria. Such an analysis using sets may be quite helpful in legal interpretations. ▮

Example 10

In our society we frequently encounter sets of people who have authority to make decisions: for example, boards of directors of corporations, state legislatures, and appeals courts. These sets of people usually operate under by-laws which specify the manner in which decisions can be accomplished. Often a simple majority of those voting is all that is required to effect a decision. In some cases, however, a majority of the members in the set, regardless of the number present, is required. There are also cases where a $\frac{2}{3}$ or $\frac{3}{4}$ majority is required. In the case of the stockholders of corporations, ordinarily a person's vote is weighted according to the number of shares of stock he owns.

Let the universal set U consist of the set of people charged with making certain decisions; a subset W of U is called a *winning coalition* if the voting power of W is sufficient to pass an issue. In such a case the subset W' is called a *losing coalition*. In some situations it is possible for a subset B of members to have sufficient voting strength to keep another coalition from passing a certain measure but not have sufficient voting strength for its own view to prevail; such a subset B is called a *blocking coalition*. An illustration of these concepts may be found in the 15-nation Security Council of the United Nations, where each of the Big Five nations (United States, United Kingdom, U.S.S.R, France, and China) has veto power. Since it takes 9 votes for a measure to pass, any winning coalition W must consist of all the Big Five plus 4 or more additional nations. The complement of W would constitute a losing coalition. A blocking coalition would be a subset that included from 1 to 5 of the Big Five or a subset that included 7 or more of the "other nations." ▮

EXERCISES

Suggested minimum assignment: Exercises 1, 3, 6, 7, 8, 9, 11, and 17.

1. Let $U = \{1, 2, 3, 4, 5, 6\}$, let $A = \{1, 2, 3, 4\}$, and let $B = \{4, 5, 6\}$. Find:
   (*a*) $A \cup B$; (*b*) $A \cap B$; (*c*) A'; (*d*) $A \cup (A' \cap B)$;
   (*e*) $A \cup \varnothing$; (*f*) $B \cap B'$; (*g*) $U \cup A$; (*h*) $\varnothing \cap B$.
2. Let $U = \{a, b, c, d, e\}$, let $A = \{a, b, c\}$, and let $B = \{b, c, d\}$. Find:
   (*a*) $A \cup B$; (*b*) $A \cap B$; (*c*) A'; (*d*) $B \cup (A \cap B')$;
   (*e*) $A \cap \varnothing$; (*f*) $B \cup B'$; (*g*) $U \cap A$; (*h*) $\varnothing \cup B$.
3. Let U be the set of all U.S. Senators, let P be the set of U.S. Senators from those states that are adjacent to the Pacific Ocean, and let R be the set of all Republican Senators. Use set-builder notation to represent
   (*a*) $P \cup R$, (*b*) $P \cap R$, (*c*) R',
   and draw Venn diagrams to represent these sets.
4. Let U be the set of all Justices on the U.S. Supreme Court, let A represent those who were appointed by a Democratic President, and

let E be those who resided in states east of the Mississippi at the time of their appointment. Use set-builder notation to represent
(*a*) $A \cup E$, (*b*) $A \cap E$, (*c*) A',
and draw Venn diagrams to represent these sets.

5. For a subset A of a universal set U write a set that is equal to the following:
(*a*) $A \cap U'$; (*b*) $A \cup \varnothing'$; (*c*) $\varnothing \cup U$; (*d*) $A \cup A$.
6. For a subset A of a universe U, write a set that is equal to the following:
(*a*) $A \cup U'$; (*b*) $A \cap \varnothing'$; (*c*) $\varnothing \cap U$; (*d*) $A \cap A$.

In Exercises 7–12 use the following set designations, assuming that none are empty:

Let U be the set of all members of a certain legislature.
Let D be the set of all Democrats.
Let $R = D'$ be the set of all Republicans.
Let C be the set of all conservatives.
Let L be the set of all liberals.
Let N be the set of all members that are neither conservatives nor liberals.

Use set notation to identify:

7. The set of liberal Democrats.
8. A conservative–Republican coalition.
9. The relation between liberal Democrats and all Democrats.
10. What is the intersection of N and L (using one term)?
11. What is the union of N and L (using one term)?
12. Draw a Venn diagram showing all sets. Draw this diagram in such a way that no null (empty) set is showing in the diagram.

13. Let U be the set of complex numbers.
Let R be the set of real numbers.
Let P be the set of imaginary numbers.
Let K be the set of rational numbers.
Let Z be the set of irrational numbers.
Let I be the set of integers.
Let N be the set of natural numbers.

Use set notation to identify the following:

(*a*) $R \cup P$; (*b*) $R \cap P$; (*c*) P'; (*d*) $I \cap N$;
(*e*) $Z \cup K$; (*f*) $Z \cap I$; (*g*) Is $N \subset P$? (*h*) Is $N \subset K$?

14. For subsets A and B of a universal set U, we define another operation by

$$A - B = A \cap B',$$

which is sometimes read "A minus B" or "the **relative complement of B in A**"; from the colored region of the Venn diagram shown in Figure 3.8 we see that $A - B$ represents the set of all elements in A that are not in B. For the sets defined in Exercises 1 and 2, find $A - B$ and $B - A$.

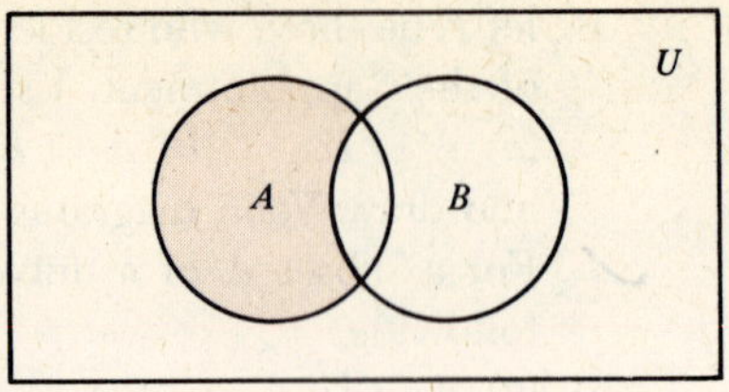

Figure 3.8. *Relative complement of B in A.*

15. For subsets A and B of a universal set U, we define another operation by

$$A \triangle B = (A \cup B) \cap (A \cap B)',$$

which is sometimes called the **symmetric difference** of A and B because it can be proved that

$$A \triangle B = (A - B) \cup (B - A).$$

See Figure 3.9 for the Venn diagram representation of $A \triangle B$. For the sets defined in Exercises 1 and 2, find $A \triangle B$.

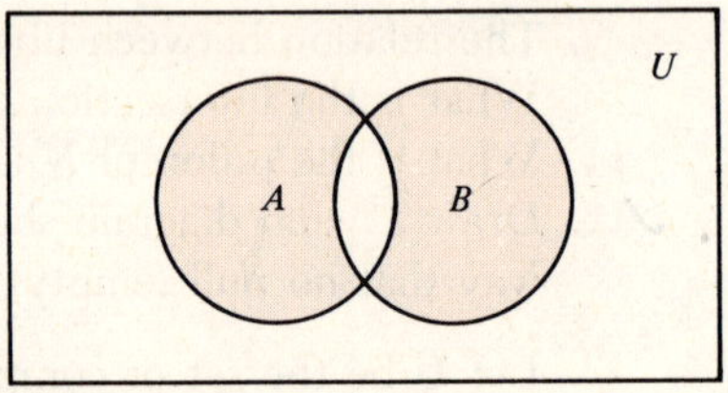

Figure 3.9. *Symmetric difference of A and B.*

16. Write an expression involving some or all of the symbols $\cap$, $\cup$, and $'$ for the colored region of the Venn diagram shown in Figure 3.10.

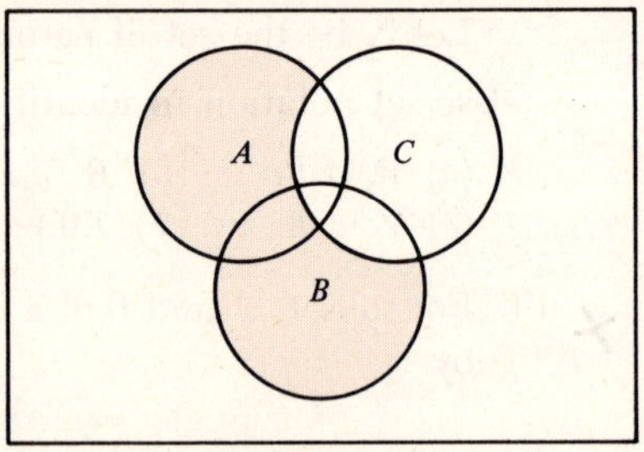

Figure 3.10

17. Write an expression involving some or all of the symbols, $\cap$, $\cup$, and $'$ for the colored region of the Venn diagram shown in Figure 3.11.

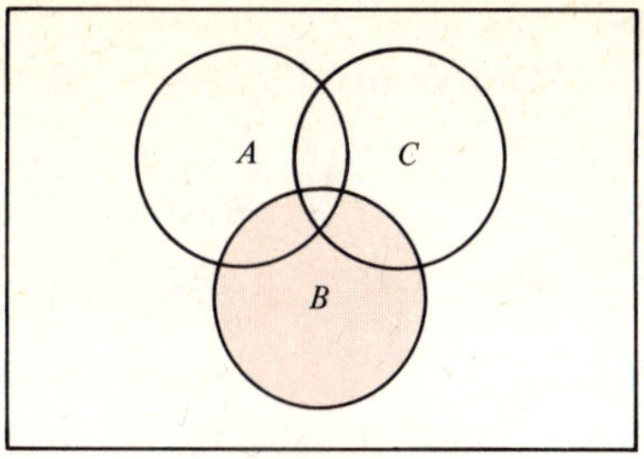

Figure 3.11

3.3 Properties of Set Operations

We have laid the foundation for a new mathematical structure called the algebra of sets. The elements of the structure are the subsets of a designated universal set U. Equality is a relation between the subsets of U, and the operations $\cup$, $\cap$, and $(')$ are defined over the elements of the structure. We also have the relations $\subseteq$ and $\subset$. As in the first two chapters we now build on this foundation. The resulting superstructure, presented as theorems, and fastened into place by means of the proofs, provides the reader with an increased capability to make use of set theory in realistic applications.

▶ Theorem 1

(Commutative Properties) If A and B are arbitrary subsets of a universal set U, then

$$(i) \quad A \cup B = B \cup A$$

and

$$(ii) \quad A \cap B = B \cap A.$$

Proof of (*i*): By the definition of the union of two sets we obtain the third and fourth columns of the evaluation table shown in Table 3.5. Then from the fifth column and the definition of set equality the conclusion follows.

The proof of (*ii*) is left as Exercise 20. ◀

Table 3.5. EVALUATION OF $(x \in A \cup B) \leftrightarrow (x \in B \cup A)$.

(1) $x \in A$	(2) $x \in B$	(3) $x \in A \cup B$	(4) $x \in B \cup A$	(5) $(x \in A \cup B) \leftrightarrow (x \in B \cup A)$
T	T	T	T	T
T	F	T	T	T
F	T	T	T	T
F	F	F	F	T

▶ **Theorem 2**

(Distributive Properties) If A, B, and C are arbitrary subsets of a universal set U, then

$$(i) \quad A \cap (B \cup C) = (A \cap B) \cup (A \cap C)$$

and

$$(ii) \quad A \cup (B \cap C) = (A \cup B) \cap (A \cup C).$$

Proof of (*i*): By the definitions of the union and intersection of sets we obtain the fourth and fifth columns of the evaluation table shown in Table 3.6. Then from the sixth column and the definition of set equality the conclusion follows.

The proof of (*ii*) is left as Exercise 21. ◀

Table 3.6. EVALUATION OF $[x \in A \cap (B \cup C)] \leftrightarrow [x \in (A \cap B) \cup (A \cap C)]$.

(1) $x \in A$	(2) $x \in B$	(3) $x \in C$	(4) $x \in A \cap (B \cup C)$	(5) $x \in (A \cap B) \cup (A \cap C)$	(6) $[x \in A \cap (B \cup C)] \leftrightarrow [x \in (A \cap B) \cup (A \cap C)]$
T	*T*	*T*	*T*	*T*	*T*
T	*T*	*F*	*T*	*T*	*T*
T	*F*	*T*	*T*	*T*	*T*
T	*F*	*F*	*F*	*F*	*T*
F	*T*	*T*	*F*	*F*	*T*
F	*T*	*F*	*F*	*F*	*T*
F	*F*	*T*	*F*	*F*	*T*
F	*F*	*F*	*F*	*F*	*T*

▶ **Theorem 3**

(Identity Properties) If A is an arbitrary subset of a universal set U, and if $\varnothing$ represents the empty set, then for every A

$$(i) \quad A \cup \varnothing = A$$

and

$$(ii) \quad A \cap U = A.$$

Proof of (*i*): By the definition of the union of two sets we obtain the third column of the evaluation table shown in Table 3.7. Then from the fourth column and the definition of set equality the conclusion follows.

The proof of (*ii*) is similar and is left as Exercise 22. ◀

Table 3.7. EVALUATION OF $(x \in A) \leftrightarrow (x \in A \cup \varnothing)$.

(1) $x \in A$	(2) $x \in \varnothing$	(3) $x \in A \cup \varnothing$	(4) $(x \in A) \leftrightarrow (x \in A \cup \varnothing)$
T	*F*	*T*	*T*
F	*F*	*F*	*T*

Theorem 4

(Complement Properties) If A is an arbitrary subset of a universal set U, and if $\varnothing$ represents the empty set, then for every A

$$(i) \quad A \cap A' = \varnothing$$

and

$$(ii) \quad A \cup A' = U.$$

Proof of (*i*): By the definition of the intersection of two sets we obtain the third column of the evaluation table as shown in Table 3.8. Because $x \in A \cap A'$ is a contradiction, then $A \cap A'$ must be the empty set.

The proof of (*ii*) is left as Exercise 23. ◀

Table 3.8. EVALUATION OF $x \in A \cap A'$.

(1) $x \in A$	(2) $x \in A'$	(3) $x \in A \cap A'$
T	F	F
F	T	F

As in the past chapters, the first four theorems have been proved using only the basic definitions. These four theorems were carefully chosen, however, so that the later theory can be based entirely on these four theorems. In other words, later theorems may be proved by using only preceding theorems, and evaluation tables are no longer required. Of course, one may choose to use an evaluation table for a proof if convenient. To illustrate these alternative methods of proof, we provide two proofs of the following theorem.

Theorem 5

If A is an arbitrary subset of a universal set U, and if $\varnothing$ represents the empty set, then for every A

$$(i) \quad A \cup U = U$$

and

$$(ii) \quad A \cap \varnothing = \varnothing.$$

Proof of (*i*)

Statement	*Reason*
1. $A \cup U = (A \cup U) \cap U$	1. Theorem 3.
2. $= U \cap (A \cup U)$	2. Theorem 1.
3. $= (A \cup A') \cap (A \cup U)$	3. Theorem 4.
4. $= A \cup (A' \cap U)$	4. Theorem 2.
5. $= A \cup A'$	5. Theorem 3.
6. $= U.$	6. Theorem 4.

Alternative proof of (*i*): By the definition of the union of two sets we obtain the third column of the evaluation table as shown in Table 3.9.

Table 3.9. EVALUATION OF $x \in A \cup U$.

(1) $x \in A$	(2) $x \in U$	(3) $x \in A \cup U$
T	T	T
F	T	T

Because $x \in A \cup U$ is a tautology, $A \cup U$ must be the universal set.

The proof of (*ii*) is left as Exercise 5. ◀

Further theorems are listed in Exercises 6–17.

By this time the similarities of the three systems constructed in Chapters 1, 2, and 3 should be apparent. To emphasize the fact that these three structures are simply variations of the same basic blueprint, we present a diagram in Table 3.10 showing the counterparts of the various structures.

Table 3.10. SUMMARY OF SYMBOLS.

Algebra of components p, q, etc.	Algebra of statement forms p, q, etc.	Algebra of subsets A, B, etc.
$\times$	$\wedge$	$\cap$
$+$	$\vee$	$\cup$
$\sim$	$\sim$	$(')$
U	U	U
Z	C	$\varnothing$
$\equiv$	$\equiv$	$=$

APPLICATIONS

Example 1

Consider the following passage from *Alice in Wonderland*, which was written by the mathematician Charles Dodgson (Lewis Carroll) in the 19th century.

"'Now let me see,' mused Alice, 'There are only Red Knights and White Knights, and not many White ones now, since the Duchess beheaded all the mounted White Knights for riding across the croquet lawn. That was the morning when the King confiscated the horses of any Knight who couldn't sing "Humpty-Dumpty" and that was a silly thing to do, because everybody knows that no Red Knight could ever sing a note.'

She looked up, and exclaimed, 'Look! There's a man on a horse! Now I wonder whether he's a Knight or not.'"

We can analyze the passage by the use of set notation as follows. Suppose that we let

$$U = \text{set of all knights},$$
$$R = \text{set of all red knights},$$

$$W = \text{set of all white knights},$$
$$M = \text{set of all mounted knights},$$
$$S = \text{set of all singing knights}.$$

From the quotation we make the following observations (in order):

1. $U = R \cup W, R \cap W = \varnothing$.
2. $M \cap W = \varnothing$.
3. $S' \cap M = \varnothing$.
4. $R \subseteq S'$.

If the intersection of two sets is the empty set (that is, the sets are disjoint), then each of the sets is a subset of the complement of the other. In other words,

$$(A \cap B = \varnothing) \Rightarrow (A \subseteq B').$$

Using this implication, we can say that

$$(2.\ M \cap W = \varnothing) \Rightarrow (M \subseteq W')$$

and

$$(3.\ S' \cap M = \varnothing) \Rightarrow (S' \subseteq M').$$

From observation 1 we have $R = W'$; hence 4 becomes $W' \subseteq S'$. Using these three inclusive relations, we have

$$M \subseteq W' \subseteq S' \subseteq M'.$$

Since M is a subset of its complement, then M must be an empty set. Thus we can conclude that there are no mounted knights, and hence if there is a mounted man he is *not* a knight. ∎

EXERCISES

Suggested minimum assignment: Exercises 1, 5, 7, 11, and 18.

In each of Exercises 1–4 illustrate the given theorem with Venn diagrams.

1. Theorem 2(*i*).
2. Theorem 2(*ii*).
3. Theorem 3.
4. Theorem 4.
5. Prove Theorem 5(*ii*).

Twelve theorems are stated in Exercises 6–17; their counterparts for switches and statement forms were stated as Theorems 6–17 of Sections 1.4 and 2.4, respectively. As Exercises 6–17 the reader is asked to prove these theorems by the use of previous theorems in this chapter or by evaluation tables.

Theorems 6–17. Let A, B, and C be arbitrary subsets of a universal set U, and let $\varnothing$ represent the empty set.

6. For every subset A, $A \cup A = A$.
7. For every subset A, $A \cap A = A$.
8. If X is a subset of U such that $A \cup X = A$ for every A, then $X = \varnothing$.

9. If X is a subset of U such that $A \cap X = A$ for every A, then $X = U$.
10. $A \cup (A \cap B) = A$ (Absorption Property).
11. $A \cap (A \cup B) = A$ (Absorption Property).
12. $A \cap B = A$ if and only if $A \cup B = B$.
13. If $A \cup B = A \cup C$ and if $A \cap B = A \cap C$, then $B = C$.
14. $A \cup (B \cup C) = (A \cup B) \cup C$ (Associative Property).
15. $A \cap (B \cap C) = (A \cap B) \cap C$ (Associative Property).
16. $(A \cup B)' = A' \cap B'$ (DeMorgan's Property).
17. $(A \cap B)' = A' \cup B'$ (DeMorgan's Property).

18. Illustrate Theorems 10, 14, and 16 with Venn diagrams.
19. Illustrate Theorems 11, 15, and 17 with Venn diagrams.

In Exercises 20–23 prove the indicated theorems using only the material that preceded the statement of the theorem.

20. Prove Theorem 1(*ii*).
21. Prove Theorem 2(*ii*).
22. Prove Theorem 3(*ii*).
23. Prove Theorem 4(*ii*).

3.4 Boolean Algebra (Optional)

In much the same way that biologists classify living organisms (structures) according to certain predetermined characteristics, mathematicians also classify their structures according to certain predetermined properties. The purpose of this section is to classify each of the three structures presented so far in this book, and in so doing to formally acknowledge that which the student probably has observed already—namely, that each of the three structures presented so far has four common properties, as stated in Theorems 1–4 of each chapter. It is customary to classify any mathematical structure that possesses these particular four properties as a *Boolean algebra*. In other words, one may think of a Boolean algebra as the basic blueprint, or more precisely the abstract algebraic system, having each of the three structures as a special case.

We now formally state the properties that are required in order for a particular structure to be classified as a Boolean algebra. Note that in the following definition, items 1–4 under *B* simply require that which Theorems 1–4 of Chapters 1–3 provide.

▶ Definition 6

A **Boolean algebra** is an algebraic system consisting of

A. A nonempty set S of elements.

B. A relation "=" (read "equals") and two binary operations $\cup$ and $\cap$ (read "cup" and "cap") over S, subject to the postulates:

1. (Commutative Postulate) For arbitrary elements x and y of S,

$$x \cup y = y \cup x, \text{ and } x \cap y = y \cap x.$$

2. (Distributive Postulate) For arbitrary elements x, y, and z of S,

$$x \cup (y \cap z) = (x \cup y) \cap (x \cup z)$$

and

$$x \cap (y \cup z) = (x \cap y) \cup (x \cap z).$$

3. (Identity Postulate) There are elements 0 and 1 of S such that for arbitrary x,

$$x \cup 0 = x \quad \text{and} \quad x \cap 1 = x.$$

4. (Complement Postulate) For each element x of S there is an element x' of S such that

$$x \cup x' = 1 \quad \text{and} \quad x \cap x' = 0.$$

We have proved in Sections 1.4, 2.4, and 3.3 that each of the three structures constructed so far in this book has the properties required in Definition 6. One advantage of this unifying observation is that the various abstract properties that follow from the basic postulates of Definition 6 will apply to each of our structures and to any other structure that is constructed and shown to be a Boolean algebra (that is, shown to have the properties required in Definition 6).

The following example illustrates a fourth structure that can be constructed and shown to be a Boolean algebra; Table 3.11 indicates the corresponding symbols used in the definition of a Boolean algebra and the four structures.

Table 3.11. SUMMARY OF SYMBOLS.

Boolean algebra x, y, etc.	Algebra of components p, q, etc.	Algebra of statement forms p, q, etc.	Algebra of subsets A, B, etc.	Binary Boolean algebra $\{0, 1\}$
$\cap$	$\times$	$\wedge$	$\cap$	$\odot$
$\cup$	$+$	$\vee$	$\cup$	$\oplus$
$(')$	$\sim$	$\sim$	$(')$	$(')$
1	U	U	U	1
0	Z	C	$\varnothing$	0
$=$	$\equiv$	$\equiv$	$=$	$=$

Example 1

In each of the first three chapters of this book we have developed a structure that satisfied the postulates of Definition 6, and hence each structure is an example of a Boolean algebra. In this example we will develop a much simpler structure that is also an example of a Boolean algebra. Let the elements or the building blocks of the structure be the symbols 0, 1. Define operations $\oplus$ and $\odot$ according to

$$\begin{aligned} &0 \oplus 0 = 0, && 0 \odot 0 = 0, \\ &0 \oplus 1 = 1 \oplus 0 = 1, && 0 \odot 1 = 1 \odot 0 = 0, \\ &1 \oplus 1 = 1, && 1 \odot 1 = 1. \end{aligned}$$

Also define $0' = 1$ and $1' = 0$. In Exercise 4 the reader is asked to show that the requirements of Definition 6 are met, and hence we have another illustration of a Boolean algebra. An interpretation of the simple structure just begun can be given by considering 0 to represent an open switch, 1 to represent a closed switch, $\oplus$ to represent the combination of *values* of two switches connected in parallel, and $\odot$ to represent the combination of *values* of two switches connected in series. Another interpretation can be given by considering 0 to represent the value assigned to a false statement, 1 to represent the value assigned to a true statement, $\oplus$ to represent the combination of *values* of two statements connected by the word "or," and $\odot$ to represent the combination of *values* of two statements connected by the word "and." It should be emphasized that in the latter interpretation, the operations $\oplus$ and $\odot$ represent a combination of values of statements rather than a combination of the statements themselves. ▮

APPLICATIONS

Example 2

The articles by Lawlor [30] and Kort [29] illustrate the use of Boolean algebra, logic, and set theory in the analysis of judicial decisions. For example, on page 152 of Kort's article, we find an equation associated with Lawlor's analysis of "the right to counsel rule" prior to its change by the Supreme Court in 1963. The equation is

$$D_{\text{pro}} \Leftrightarrow (f_{11} \vee f_{12}) \wedge f_{19} \wedge [L(1, S_a) \vee L(5, S_b)],$$

where the simple statements are

D_{pro} = court decision in favor of the petitioner,
f_{11} = petitioner had no assistance of counsel at time of arraignment,
f_{12} = petitioner had no assistance of counsel between the arraignment and the trial,
f_{19} = petitioner never waived explicitly the right to counsel,
$L(1, S_a)$ = at least 1 element in subset S_a is present,
$L(5, S_b)$ = at least 5 elements in subset S_b are present;

and where the subsets are

S_a = subset of facts (1) the petitioner was convicted of a crime subject to capital punishment, and (2) a jurisdictional issue or complicated charges were involved;
S_b = subset of 32 facts pertaining to the personal handicaps of the petitioner, to circumstances relating to a denial of representation by counsel, and to other irregularities.

With the aid of Boolean algebra, even much more complicated rules of law and statutory provisions can acquire a degree of clarity that cannot be obtained from the cumbersome and ambiguous verbal forms in which these rules and provisions often appear. ▮

EXERCISES

Suggested minimum assignment: Exercises 1, 3, and 5.

1. Let S be the set of all single and compound switches that can be constructed from the set of single switches $\{p_1, p_2, \ldots, p_n\}$ under the operations $\sim$, $+$, and $\times$. List the theorems of Chapter 1 that satisfy the required properties of a Boolean algebra and hence justify the statement that the set S under the operations $\sim$, $+$, and $\times$ forms a Boolean algebra.
2. Let S be the set of all simple and compound statement forms that can be constructed from the set of simple statement forms $\{p_1, p_2, \ldots, p_n\}$ under the operations $\sim$, $\vee$, and $\wedge$. List the theorems of Chapter 2 that satisfy the required properties of a Boolean algebra and hence justify the statement that the set S under the operations $\sim$, $\vee$, and $\wedge$ forms a Boolean algebra.
3. Let S be the set of all subsets of a universal set U. List the theorems of this chapter that satisfy the required properties of a Boolean algebra and hence justify the statement that the set S under the operations of $(')$, $\cup$, and $\cap$ forms a Boolean algebra.
4. Verify that the structure of Example 1 of this section is a Boolean algebra.
5. Using the notation of Definition 6, list 13 theorems of a Boolean algebra. *Hint:* See Theorems 5–17 of Sections 1.4, 2.4, and 3.3.
6. Prove any one of the theorems stated in Exercise 5.

3.5 The Number of Elements in a Set

Associated with a finite set A is a natural number $n(A)$ which simply represents the number of elements in the set. In this section we shall develop a few formulas concerning the number of elements of sets.

First we consider the union of two sets. It is easy to see that if two finite sets A and B are disjoint, then $n(A \cup B) = n(A) + n(B)$. Now consider the case where A and B are not disjoint. Suppose that A is the set of red automobiles on a certain parking lot and B is the set of all Fords on the same parking lot. Suppose that we let

$$n(A) = x + y,$$

where x is the number of red nonFords and y is the number of red Fords (see Figure 3.12). Suppose that we let

$$n(B) = z + y,$$

where z is the number of nonred Fords and y is the number of red Fords (see Figure 3.12). Then, since the sets A and B overlap by the amount

$$n(A \cap B) = y,$$

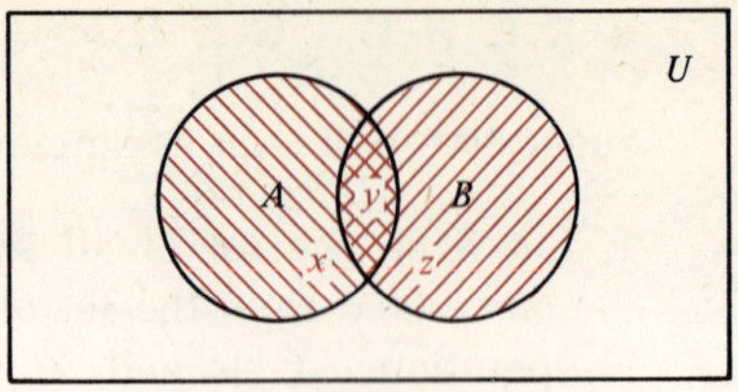

Figure 3.12

we have

$$\begin{aligned} n(A \cup B) &= x + z + y \\ &= (x + y) + (z + y) - y \\ &= n(A) + n(B) - n(A \cap B). \end{aligned}$$

This type of reasoning can be used to prove the following theorem.

▶ **Theorem 18**

If A and B are finite sets, then

$$n(A \cup B) = n(A) + n(B) - n(A \cap B).$$

Outline of Proof: Let $X = A \cap B'$, $Y = A \cap B$, and $Z = A' \cap B$.

1. Using the theorems of Section 3.3, it can be shown that X, Y, and Z are disjoint and that

$$\begin{aligned} A \cup B &= X \cup Y \cup Z, \\ A &= X \cup Y, \\ B &= Y \cup Z. \end{aligned}$$

2. Because of part 1,

$$\begin{aligned} n(A \cup B) &= n(X) + n(Y) + n(Z), \\ n(A) &= n(X) + n(Y), \\ n(B) &= n(Y) + n(Z). \end{aligned}$$

3. The last two equations of part 2 can be solved for $n(X)$ and $n(Z)$ and substituted into the first equation of part 2 to produce the desired conclusion. ◀

Example 1

Suppose that in a set of 100 patients, 20 exhibit symptom α, 30 exhibit symptom β, and 5 exhibit both symptoms. How many have symptom α or symptom β?

Let

$$\begin{aligned} A &= \text{set of patients with symptom } \alpha, \\ B &= \text{set of patients with symptom } \beta, \\ A \cap B &= \text{set of patients having both symptoms.} \end{aligned}$$

Then $A \cup B$ = set of patients having one or the other or both symptoms, and we are asking for $n(A \cup B)$. By Theorem 18,

$$\begin{aligned} n(A \cup B) &= n(A) + n(B) - n(A \cap B) \\ &= 20 + 30 - 5 \\ &= 45. \end{aligned}$$

This problem can be illustrated in Figure 3.13, where the set $A \cup B$ is thought of as being partitioned into three subsets, namely, $(A \cap B')$, $(A' \cap B)$, and $(A \cap B)$, having 15, 25, and 5 elements, respectively. The sum of these three numbers then represents the number of elements in $(A \cup B)$. ▮

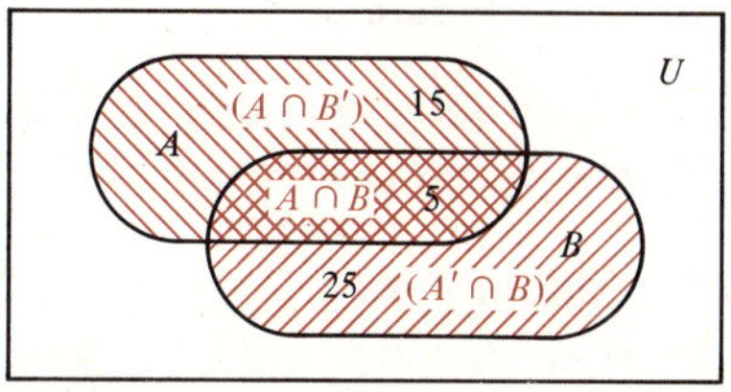

Figure 3.13. *$A \cup B$ partitioned into three disjoint subsets.*

In Exercise 22 the reader is asked to find a formula for $n(A \cap B')$, which in Example 1 would answer the question of how many patients have symptom α but not symptom β.

Next, suppose that in Example 1 we wanted to know the number of patients who did *not* have symptoms α or β. We could reason then that we are looking for

$$n[(A \cup B)'],$$

which is simply the total number of patients minus those who have symptoms α or β. That is,

$$n[(A \cup B)'] = n(U) - n(A \cup B).$$

We have here an illustration of a rather obvious theorem.

▶ **Theorem 19**

If A is a subset of a finite universal set U, then

$$n(A') = n(U) - n(A).$$

The proof of Theorem 19 is left as Exercise 21. ◀

Now suppose that we have three subsets A, B, and C of a universal set as shown in Figure 3.14, where

$$\begin{aligned} n(A) &= x + r + s + u, \\ n(B) &= z + s + t + u, \\ n(C) &= y + r + t + u. \end{aligned}$$

We see that

$$n(A \cup B \cup C) = x + y + z + r + s + t + u,$$

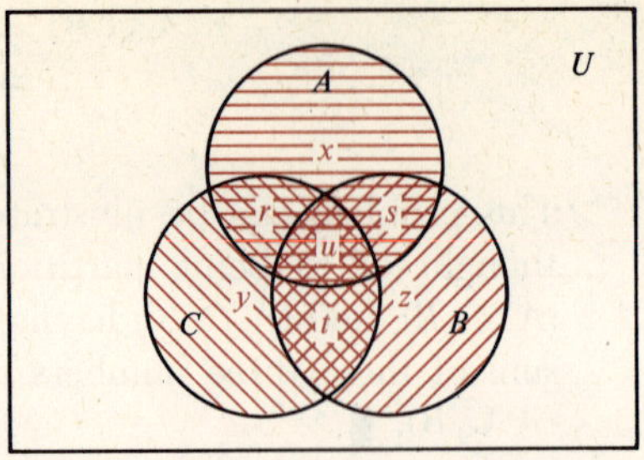

Figure 3.14. $A \cup (B \cup C)$ *partitioned into eight disjoint subsets.*

which, after considerable addition and subtraction of the natural numbers r, s, t, and u, agrees with the conclusion of the following theorem.

▶ **Theorem 20**

If A, B, and C are finite sets, then

$$n(A \cup B \cup C) = n(A) + n(B) + n(C) - n(A \cap B) - n(B \cap C) - n(A \cap C) + n(A \cap B \cap C).$$

A proof of Theorem 20 is similar to that outlined for Theorem 18. ◀

For an illustration of Theorem 20, see Example 2 in the "Applications."

APPLICATIONS

Example 2

Suppose that the following information is published in a survey.

68 countries receive foreign aid from United States,
40 countries receive foreign aid from U.S.S.R.,
20 countries receive foreign aid from France,
25 countries receive foreign aid from United States and U.S.S.R.,
10 countries receive foreign aid from United States and France,
4 countries receive foreign aid from U.S.S.R. and France,
2 countries receive foreign aid from United States, U.S.S.R., and France.

Suppose this survey goes on to say that there are 91 countries that obtain foreign aid from the U.S.S.R. or France or the United States and that there are 70 countries that receive aid from the United States or France but not from all three countries. The question arises: Are these last two assertions consistent with the original data? To analyze the problem, we let

S = set of countries receiving aid from United States,
R = set of countries receiving aid from U.S.S.R.,
F = set of countries receiving aid from France.

We are trying to verify that

$$n(S \cup R \cup F) = 91$$

and

$$n(S \cup F) - n(S \cap F \cap R) = 70.$$

From the information given at the outset of the example, we can write

$$n(S) = 68, \qquad n(R) = 40, \qquad n(F) = 20,$$
$$n(S \cap R) = 25, \qquad n(S \cap F) = 10, \qquad n(R \cap F) = 4,$$
$$n(S \cap R \cap F) = 2.$$

By Theorem 20,

$$n(S \cup R \cup F) = 68 + 40 + 20 - 25 - 10 - 4 + 2 = 91,$$

which is consistent with the report of the survey, but the second assertion of the report is not consistent with the data because

$$\begin{aligned} n(S \cup F) - n(S \cap F \cap R) &= n(S) + n(F) - n(S \cap F) - n(S \cap F \cap R) \\ &= 68 + 20 - 10 - 2 \\ &= 76 \\ &\neq 70. \end{aligned}$$

See Figure 3.15 for a Venn diagram with much of the above information given thereon. ▮

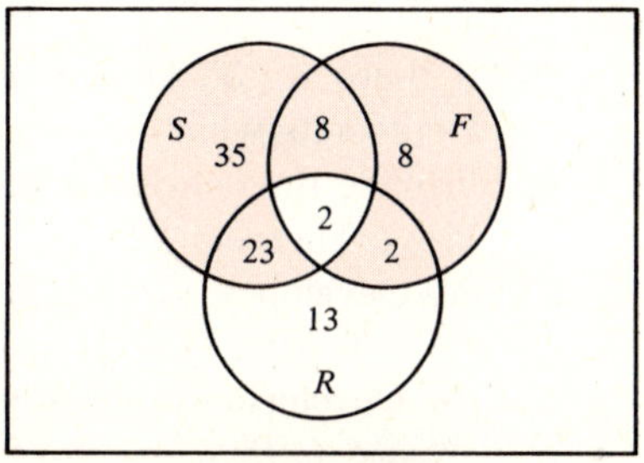

Figure 3.15

Example 3

In addition to the analysis of situations more complex but similar to Example 2, applications of the material of this section are indirectly obtained through the use of probability theory, as we shall demonstrate in Chapter 4. ▮

EXERCISES

Suggested minimum assignment: Exercises 1, 3, 5, 7, 8, 13, 19, and 22(a).

In each of Exercises 1–12 find the number of elements in the given sets if we let

$$n(A) = 12, \quad n(B) = 15, \quad n(C) = 30, \quad n(A \cap B) = 5,$$
$$n(A \cap C) = 6, \quad n(B \cap C) = 4, \quad n[A \cap (B \cap C)] = 2, \quad n(U) = 50.$$

Hint: Draw a Venn diagram similar to Figure 3.14, and first calculate the number of elements in each of the indicated disjoint subsets.

1. $n(A \cup B)$.
2. $n(A \cup C)$.
3. $n(B \cup C)$.
4. $n\{(A \cup B)'\}$.
5. $n[(A \cup C)']$.
6. $n\{(B \cup C)'\}$.
7. $n[A \cup (B \cup C)]$.
8. $n[A \cup (B \cap C')]$.
9. $n[A \cup (C \cap B')]$.
10. $n[A' \cup (B' \cup C')]$.
11. $n[A \cap (B' \cap C')]$.
12. $n[B \cap (A' \cap C')]$.

In a certain survey of 100 people it was found that 10 people buy *only* brand A, 30 people buy *only* brand *B*, 15 people buy *only* brand C, 8 people buy both brands *A* and *B*, 5 people buy brands *A* and *C*, 6 people buy brands *B* and *C*, and 4 people buy all three brands. In Exercises 13–18 answer the given questions based on these data.

13. How many people bought at least one of the three brands?
14. How many people bought none of the three brands?
15. How many people bought both brands *A* and *B*, but not *C*?
16. How many people bought brand *B* or *C*?
17. How many people bought brand *A*?
18. How many people bought brand *B*?

19. If there are 55 Democrats in the Senate, 50 liberals, and 30 liberal Democrats, what is the voting strength of a liberal–Democrat coalition?
20. Provide the details for the proof of Theorem 18.
21. Prove Theorem 19. *Hint*: Use Theorem 4.
22. (a) Develop a formula for finding $n(A \cap B')$ in terms of $n(A)$ and $n(A \cap B)$. *Hint*: $A = (A \cap B) \cup (A \cap B')$.
    (b) Develop a formula for finding $n(A \cup B')$ in terms of $n(U)$, $n(B)$, and $n(A \cap B)$.

3.6 Cartesian Product and Power Set

Prior to the national political conventions it is customary for various national pollsters to pit the leading Democratic candidates against the leading Republican candidates in hypothetical 2-man races. Suppose that the leading Democratic candidates are represented by the set $D = \{d_1, d_2, d_3\}$ and the leading Republican candidates are represented by the set $R = \{r_1, r_2\}$. The 2-man races that can be developed from these two sets can be represented by the set of ordered pairs

$$\{(d_1, r_1), (d_1, r_2), (d_2, r_1), (d_2, r_2), (d_3, r_1), (d_3, r_2)\},$$

where the first entry of each pair is an element of *D* and the second entry of each pair is an element of *R*; this set of ordered pairs is an example of a Cartesian product of *D* and *R* and is designated $D \times R$. It may be helpful to the reader to visualize the ordered pairs of $D \times R$ as entries of the rectangular array

	r_1	r_2
d_1	(d_1, r_1)	(d_1, r_2)
d_2	(d_2, r_1)	(d_2, r_2)
d_3	(d_3, r_1)	(d_3, r_2)

Definition 7

If a is an element of set A and if b is an element of set B, then the set of **all** ordered pairs (a, b) that can be constructed from A and B is called the **Cartesian product of A and B.**

Example 1

Let R be the set of all real numbers. The Cartesian product $R \times R$ is simply the set of all ordered pairs of real numbers. The elements (x, y) of $R \times R$ are used to identify points in the Cartesian plane. For example, the element (3, 2) graphed in Figure 3.16 is one member of the set $R \times R$; other points in the plane of Figure 3.16 may be thought of as a geometrical representation of other pairs belonging to $R \times R$. From this usage of the Cartesian plane, we get the name Cartesian product. ▮

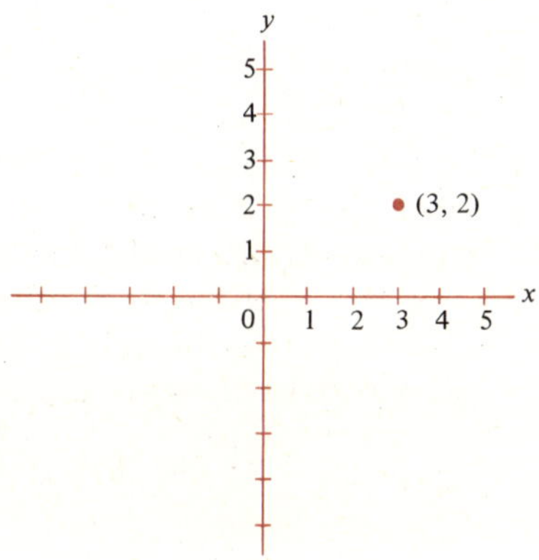

Figure 3.16

Reconsider the sets $D = \{d_1, d_2, d_3\}$, $R = \{r_1, r_2\}$, and $D \times R = \{(d_1, r_1), (d_1, r_2), (d_2, r_1), (d_2, r_2), (d_3, r_1), (d_3, r_2)\}$. Notice that associated with each of the 3 elements of D there are 2 elements of R, thus giving us $3 \cdot 2$, or 6, elements of $D \times R$. Generalizing this idea, we ascertain that for finite sets A and B,

$$n(A \times B) = n(A) \cdot n(B).$$

Example 2

If M is a set of 5 male rabbits and if F is a set of 6 female rabbits, then there are $5 \cdot 6$, or 30, different ways that the rabbits can be paired so that one is male and the other is female. The set of pairs can be represented by $M \times F$ and

$$n(M \times F) = n(M) \cdot n(F) = 5 \cdot 6 = 30. \blacksquare$$

The idea of a Cartesian product can be extended to apply to more than two sets, as the next example illustrates.

Example 3

A certain set of astronauts all have secondary training in some field. Suppose that there are 3 astronomers represented by the subset

$$A = \{a_1, a_2, a_3\},$$

4 meteorologists represented by the subset

$$M = \{m_1, m_2, m_3, m_4\},$$

and 2 engineers represented by the subset

$$E = \{e_1, e_2\}.$$

A space team is to be selected consisting of one astronaut from each of these subsets. One possible team is the triple

$$(a_1, m_1, e_1).$$

Another team is the triple

$$(a_1, m_1, e_2),$$

and of course there are many more that could be listed. The set of all such triples (potential teams) illustrates the concept of the Cartesian product $A \times M \times E$ of the three subsets. The number of elements (teams) in the Cartesian product can be determined by arranging the triples in a three-dimensional array shown in Figure 3.17. Thus we have the formula

$$n(A \times M \times E) = n(A) \cdot n(M) \cdot n(E). \blacksquare$$

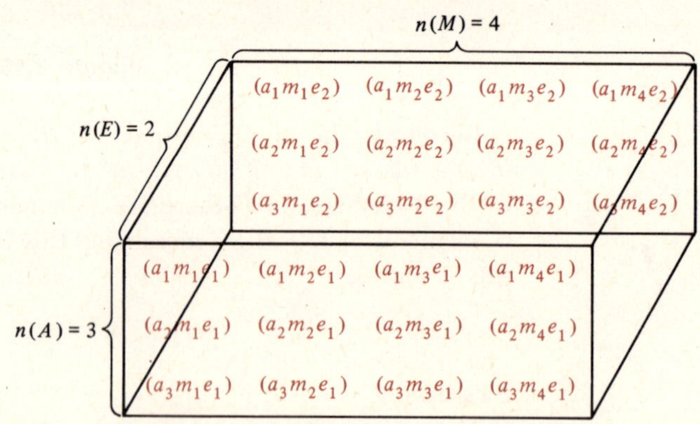

Figure 3.17. *Elements of $A \times M \times E$.*

The ideas presented in Example 3 are now formalized.

Definition 8

If a_i is an element of set A_i, where $i = 1, 2, \ldots, k$, then the set of all ordered k-tuples $(a_1, a_2, \ldots, a_k)$ is called the **Cartesian product**

$$A_1 \times A_2 \times \cdots \times A_k.$$

We state without proof that the number of elements in the Cartesian product $A_1 \times A_2 \times \cdots \times A_k$ of finite sets is given by the formula

$$n(A_1 \times A_2 \times \cdots \times A_k) = n(A_1)\, n(A_2) \cdots n(A_k).$$

A second concept that we wish to introduce in this section is that of the **power set** of a set.

Example 4

Consider the set $A = \{a, b, c\}$. Suppose that we list all possible subsets of A, namely,

$$\{a, b, c\},\ \{a, b\},\ \{a, c\},\ \{b, c\},\ \{a\},\ \{b\},\ \{c\},\ \varnothing.$$

The set of these subsets of A is known as the **power set of A.** ▮

Definition 9

The set of all subsets of a finite set A is called the **power set of A.** The power set of A will be designated P.

Notice that if $A = \{a\}$, then $P = \{\{a\}, \varnothing\}$, and $n(P) = 2^1$. If $A = \{a, b\}$, then $P = \{\{a, b\}, \{a\}, \{b\}, \varnothing\}$ and $n(P) = 2^2$. From Example 4 we can see that if $A = \{a, b, c\}$, then $n(P) = 2^3$.

If $A = \{a, b, c, d\}$, then the power set P of A has 2^4, or 16, elements. The latter result may be verified by studying the diagram shown in Figure 3.18; a dash in the figure means that the element of A in question is not an element of the subsets. We start with the element a, recognizing that there are two possibilities: Either a is included or it is not included. The same choice is faced for b and then c, and so forth. From Figure 3.18 we see that for each additional element of the set A, we double the number of subsets; hence the formula

$$n(P) = \underbrace{2 \cdot 2 \cdot \cdots \cdot 2 \cdot 2}_{n(A) \text{ factors}}, \qquad \text{or} \qquad n(P) = 2^{n(A)}$$

is plausible. (A proof of the formula may be found in Stoll [42], p. 11.)

APPLICATIONS

The concept of the Cartesian product of two or more sets will be used extensively in Chapter 4.

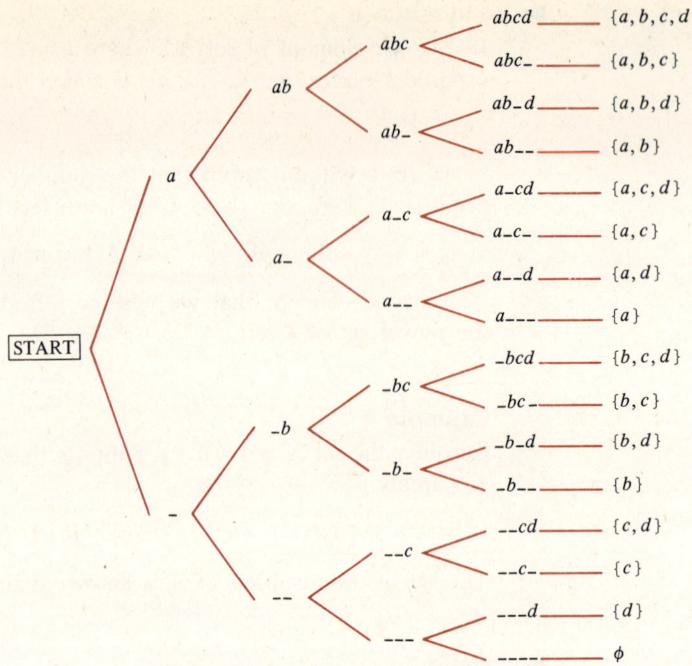

Figure 3.18. *Power set of* $\{a, b, c, d\}$.

EXERCISES

Suggested minimum assignment: Exercises 1, 3, 5, 8, 9, 11, and 12.

1. If $A = \{a, b, c\}$ and $B = \{x, y, z\}$, write $A \times B$.
2. For the sets in Exercise 1, write $B \times A$.
3. Let $A = \{1, 2, 3\}$, and $B = \{1, 2, 3, 4\}$. Plot the elements of $(A \times B)$ with respect to the rectangular coordinate system.
4. Repeat Exercise 3 for $B \times A$ rather than $A \times B$. Is $B \times A = A \times B$?
5. If $A = \{x, y\}$, write the power set P of A.
6. If $B = \{m, n, p, q\}$, write the power set P of B.
7. Use set notation to identify the set of all voting coalitions of a 20-man legislature (including a single legislator as a coalition). How many coalitions are there?
8. List all the committees consisting of at least 1 person that can be formed from 3 people. How many committees consisting of at least 1 person can be formed from 6 people (do *not* list the committees)?
9. Let $A = \{x, y, z\}$. Make a diagram similar to that shown in Figure 3.18 to illustrate the number of elements in the power set of A.
10. How many different matings of 7 females and 9 males are possible in a certain set of like animals? Express as set notation.
11. In Example 3 suppose that an additional member is to be added to the team, thus making it a 4-man team. Further, suppose that this fourth

person is to be selected from a set D of 3 medical doctors. State the set of teams as a Cartesian product and find the number of potential teams.

12. If there are 3 candidates for the Democratic nomination, 4 candidates for the Republican nomination, and 2 potential independent candidates, then for how many 3-man races in the general election must a poll-taker prepare? Express the set of potential 3-man races as a Cartesian product.

NEW VOCABULARY

set 3.1
subset 3.1
proper subset 3.1
empty set (null set) 3.1
truth set 3.1
universal set 3.1
finite set 3.1
infinite set 3.1
Venn diagrams 3.1
equal sets 3.2
intersection of sets 3.2
disjoint sets 3.2
union of sets 3.2
complement of a set 3.2
Boolean algebra 3.4
Cartesian product 3.6
power set of a set 3.6

4 Probability

The actions and behavior of a person often are results of conscious or subconscious decisions by the person. Many of these decisions are based on measurements of the likelihood of certain events. For example, people often make decisions based on a likelihood of bad weather at a future point in time. Of course, the significance of such a decision may range from the significance of scheduling a picnic to that of scheduling a space trip; and the methods of measurement of the likelihood of bad weather may vary from outright guesses to those based on sophisticated scientific analysis. As another illustration, consider a business decision based on the likelihood that the reader will be alive 10 years from now. Measurements of the likelihood of that event may vary from those that are based on little more than outright guesses to those that have been carefully obtained by the insurance industry.

Nevertheless, the decisions of men, however significant, and however they are based, do shape the course of events. It seems reasonable, therefore, that we concern ourselves with those measurements upon which decisions often can be based.

Prerequisites: 3.1, 3.2, 3.5; also 3.6 is prerequisite to 4.5–4.7
Suggested Sections for Moderate Emphasis: 4.1–4.6, 4.8, 4.10
Suggested Sections for Minimum Emphasis: 4.1–4.4, 4.10

4.1 Basic Terminology

The word **probability** is usually associated with the measurement of the likelihood of an event; this is illustrated by the television announcer who says, "the probability of rain tomorrow is 30 per cent." To gain some precision in the meaning of the word "probability," we begin with a universal set U, which we shall call the **sample space.** We then think of U as the set of outcomes of certain **experiments** or activities, such as flipping coins, or throwing dice, or swinging at a baseball, or buying stock, or trying to sell a piece of merchandise. Subsets of U are called **events,** or **event spaces;** in other words, an event or an event space is a subset of the set of outcomes.

Example 1

Consider a single die whose six faces appear as shown in Figure 4.1. We may think of casting the die on a level surface and reading the upper face as an experiment; each of the six faces can be considered as an outcome (result) of the experiment. Hence the sample space is the set shown in Figure 4.1. Of course, there are many events that can occur; the subset of odd numbers constitutes one event that we shall call E_1 (see Figure 4.2). The subset consisting of the outcomes shown in Figure 4.3 is another event, which we shall call E_2. If we form the ratio

$$\frac{n(E_1)}{n(U)} = \frac{3}{6} = .5,$$

we have a measure of the likelihood that an odd digit will appear on the upward face when the die is cast on a plane surface; we call this ratio the probability of event E_1. Likewise the probability of event E_2 is

$$\frac{n(E_2)}{n(U)} = \frac{2}{6} \approx .33. \blacksquare$$

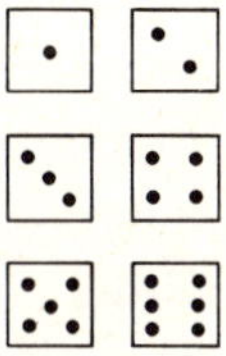

Figure 4.1. *Faces of a single die.*

Figure 4.2. *Odd-numbered faces of a die.*

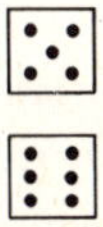

Figure 4.3. *Elements of E_2.*

In Example 1 we implicitly assumed that any one face was as likely to appear as any other face; such outcomes are said to be **equally likely.** Equally likely outcomes are not always the case, however. For example, it might *not* be the case that a magnetized die thrown on a metal table will produce equally likely outcomes.

▶ **Definition 1**

Let U be a finite set of equally likely outcomes of an experiment, and let E be a subset of U. The **probability $P(E)$ of event E** is defined as follows:

$$P(E) = \frac{n(E)}{n(U)}.$$

Example 2

Consider the probability of exactly two heads being thrown if a coin is flipped 3 times. First, we identify the sample space to be the set of outcomes

$$U = \{HHH, HHT, HTH, THH, HTT, THT, TTH, TTT\}.$$

Second, we identify the event space to be the subset of outcomes consisting of exactly two heads,

$$E = \{HHT, HTH, THH\}.$$

Hence, since the outcomes are equally likely,

$$P(E) = \frac{n(E)}{n(U)} = \frac{3}{8}. \blacksquare$$

Example 3

Suppose that an investor intends to buy stock in two companies. He has narrowed his interest to 3 industrials, which we shall label I_1, I_2, and I_3, and 2 utilities, which we shall label U_1 and U_2. If the investor chooses 2 of these 5 stocks at random, what is the probability of choosing 2 industrials? Here the sample space of equally likely outcomes is

$$U = \{I_1U_1, I_1U_2, I_1I_2, I_1I_3, I_2U_1, I_2U_2, I_2I_3, I_3U_1, I_3U_2, U_1U_2\},$$

where $n(U) = 10$, and the event space is

$$E = \{I_1I_2, I_1I_3, I_2I_3\},$$

where $n(E) = 3$. Therefore,

$$P(E) = \frac{n(E)}{n(U)} = \frac{3}{10}. \blacksquare$$

Now consider the case of finding the probability of an event where the outcomes are *not* equally likely. Recall that in Definition 1, $P(E)$ was defined to be $n(E)/n(U)$, where E was a subset of the set U of equally likely outcomes. If we examine the probability of each outcome e_i we find that each probability $P(\{e_i\})$ is $1/n(U)$; hence we can think of $P(E)$ as the sum of the probabilities of all the outcomes of the event space. That is,

$$P(E) = \frac{n(E)}{n(U)} = \underbrace{\frac{1}{n(U)} + \frac{1}{n(U)} + \cdots + \frac{1}{n(U)}}_{n(E)\text{ terms}}.$$

This idea is extended to sample spaces of outcomes that are not equally likely. We shall redefine the probability of an event to be the sum of the probabilities of obtaining the outcomes (equally likely or not) that comprise the event. The probabilities of obtaining the respective outcomes are often referred to as **weights** of the outcomes.

For example, consider the simple case of a jar containing 10 capsules, of which 3 are red, 6 are white, and 1 is blue. The experiment consists of drawing a capsule from the jar; the outcomes are red, white, and blue capsules, but these outcomes obviously are not equally likely. Therefore, we assign the respective weights $\frac{3}{10}$, $\frac{6}{10}$, and $\frac{1}{10}$ to the outcomes. Each weight is simply the probability of the event of obtaining each respective outcome. The probability of an event will be defined to be the sum of the weights of the respective outcomes that belong to the event; for example, the probability of drawing a nonred capsule is the sum of the weights of the nonred outcomes, namely,

$$\frac{6}{10} + \frac{1}{10} = \frac{7}{10}.$$

In Chapter 3 the reader was reminded of the conceptual difference between the element u_i of a set U and the subset $\{u_i\}$ of a set U; it is because of that distinction that we speak of the weight of an outcome (element) u_i, whereas we speak of the probability of an event (subset) $\{u_i\}$. We now state a generalization of Definition 1 to include nonequally likely outcomes.

Definition 2

Let U be a finite set of n outcomes $u_1, u_2, \ldots, u_n$ with nonnegative weights $w(u_1), w(u_2), \ldots, w(u_n)$, such that the sum of *all* the weights is 1. If an event E consists of k outcomes $e_1, e_2, \ldots, e_k$, then the **probability of E** is

$$P(E) = w(e_1) + w(e_2) + \cdots + w(e_k).$$

Example 4†

Consider the experiment of throwing two dice and adding the amounts shown on the two upward faces. What is the probability that a 7 or an 11 will occur? In Figure 4.4 we illustrate all 36 of the ways that the dice can appear.

We can reason that there are 11 outcomes, namely, the sums 2 through 12, and using Figure 4.4, we can assign weights as shown in Table 4.1.

Table 4.1. WEIGHTS FOR THE OUTCOMES OF A PAIR OF DICE.

Outcome	2	3	4	5	6	7	8	9	10	11	12
Weight	$\frac{1}{36}$	$\frac{2}{36}$	$\frac{3}{36}$	$\frac{4}{36}$	$\frac{5}{36}$	$\frac{6}{36}$	$\frac{5}{36}$	$\frac{4}{36}$	$\frac{3}{36}$	$\frac{2}{36}$	$\frac{1}{36}$

Hence by Definition 2 the probability of throwing a 7 or an 11 is the sum

$$P(E) = \frac{6}{36} + \frac{2}{36} = \frac{8}{36} \approx .22.$$

† Historically, the first recorded problem in probability is similar to this problem. See Example 6 on page 97.

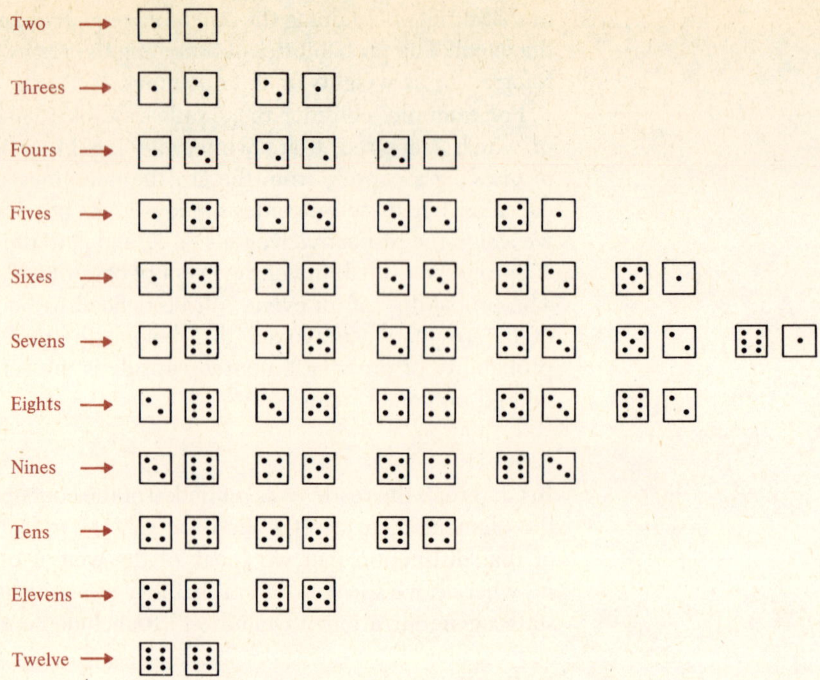

Figure 4.4. *Set of ways a pair of dice can appear.*

An alternative way of solving this problem is to think of each arrangement in Figure 4.4 as a separate outcome, thus producing a different sample space of 36 equally likely outcomes rather than a sample space of 11 nonequally likely outcomes. Then, using Definition 1, we have

$$P(E) = \frac{n(E)}{n(U)} = \frac{8}{36} \approx .22. \blacksquare$$

It is easy to show that the probability of an event E of a sample space U, as defined in Definition 2, satisfies the properties stated in the following theorem. Remember that sample spaces and events have been defined as sets.

Theorem 1

Let E be an event of a sample space U with a finite number of outcomes.

$$(i)\ 0 \leq P(E) \leq 1, \qquad (ii)\ P(U) = 1, \qquad (iii)\ P(\varnothing) = 0.$$

A proof of Theorem 1 is left as Exercises 15–17. ◀

At this point we wish to make a distinction between those probabilities that are obtained by a nonexperimental calculation and those that are extrapolated from some type of experiment. The flipping of a fair coin or the casting of a fair die are examples of **a priori probabilities** that can be calculated without experimentation. In the cases of the coin and die the calculations are based on the shape of the objects being thrown. On the other hand, the probability that a certain basketball player will make a free throw or the probability that the reader will die within 10 years are examples of **a posteriori probabilities**—probabilities obtained by past experimentation. The probability that a basketball player will make a free throw can be measured through consideration of his past performance, while the probability that a person will die within 10 years can be measured through a study of the experience of others of the same age. In any case, however, the reader should clearly recognize that a probability measurement is only the *measurement of the likelihood* of an event and in no way affects the event itself. For example, the probability of the event E of not flipping a head 6 times in succession on a fair coin is $\frac{63}{64}$, but if one tries the experiment, the large probability is no guarantee whatsoever that an outcome of the specified event E will occur.

Apparently the study of probability began through an interest in games of chance. It is recorded that Galileo (1564–1642) solved an isolated dice problem for the Grand Duke of Tuscany (see Example 6 of this section), but a serious study of probability did not begin until about 1654. This study was begun by Blaise Pascal (1623–1662) and Pierre de Fermat (1601–1665), and much of it is recorded in the correspondence between the two men. A French nobleman, Chevalier DeMéré, who had considerable interest in gambling, apparently initiated Pascal's interest by posing problems connected with games of chance.

APPLICATIONS

Example 5

In a certain genetic problem the assumption is made that there are two different alleles† A and a at a certain locus. Suppose that genotypes Aa and Aa are mated according to Figure 4.5, under the genetic rule that AA, Aa, aA, and aa are all equiprobable and that Aa and aA are biologically indistinguishable; then the outcomes AA, Aa, and aa are assigned weights

$$w(AA) = \tfrac{1}{4}, \qquad w(Aa) = \tfrac{1}{2}, \qquad w(aa) = \tfrac{1}{4}. \quad \blacksquare$$

Example 6

As with many subjects there is some difference of opinion concerning the origins of mathematical probability. There is, however, considerable agreement that these origins were a direct result of questions posed by gamblers. According to some, the first recorded result was due to Galileo (1564–1642) in response to a question by the Grand Duke of Tuscany. When 3 dice were thrown, it appeared to the Grand Duke that the sum of 10 occurred more often than did the sum of 9. This seemed strange to the Grand Duke because there were six ways of obtaining both the sum of

† An *allele* is one of two or more alternative hereditary units or genes at identical loci of homologous chromosomes, giving rise to contrasting Mendelian characters.

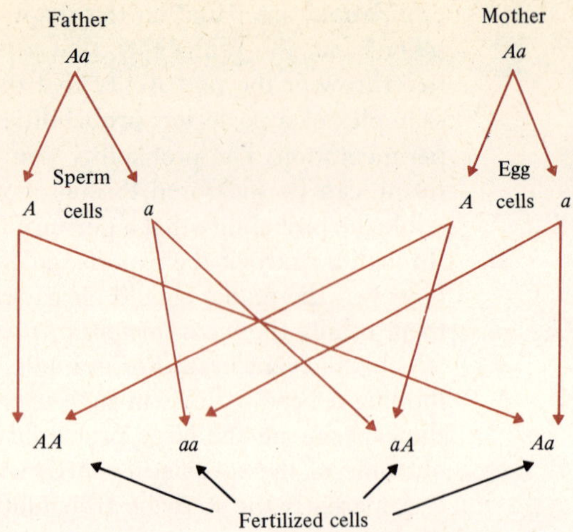

Figure 4.5. *Genotype matings.*

10 and the sum of 9, as shown in Figure 4.6. What the Grand Duke overlooked was that the outcomes shown in Figure 4.6 were not equally

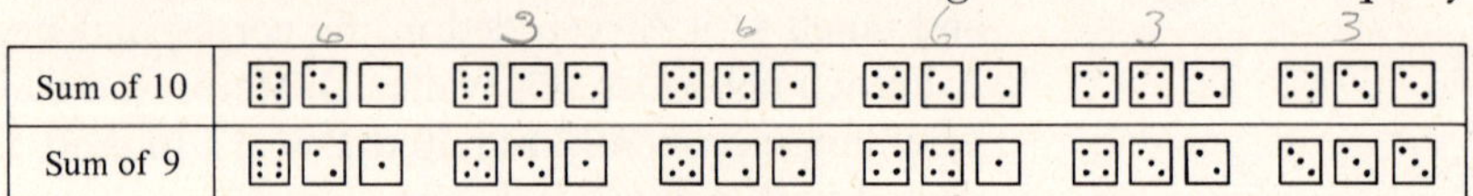

Figure 4.6

likely. For example, there are six ways that the sum 6 + 3 + 1 can be obtained:

$$6 + 3 + 1,$$
$$6 + 1 + 3,$$
$$3 + 1 + 6,$$
$$3 + 6 + 1,$$
$$1 + 3 + 6,$$
$$1 + 6 + 3.$$

Yet there are only three ways that the sum 6 + 2 + 2 can be obtained, namely,

$$6 + 2 + 2,$$
$$2 + 6 + 2,$$
$$2 + 2 + 6.$$

Galileo found that the probability of obtaining a sum of 10 is $\frac{27}{216}$, whereas the probability of obtaining a sum of 9 is $\frac{25}{216}$. In Exercise 18 the reader is asked to verify Galileo's results by assigning the proper weights to each of the outcomes of Figure 4.6. ∎

Example 7

In order to determine life insurance premiums and annuity payments, it is necessary for life insurance companies to be able to calculate the probability that a given person will be alive after a certain time interval. These calculations are based on mortality tables which give various types of information concerning deaths of humans. These tables are constantly revised by the insurance industry in order to reflect various influencing factors such as medical advances and living standards. According to one source, the probability of death of United States citizens at different ages is given in Table 4.2. If we let x_i be the outcome of death during the ith decade of life, then the second column of Table 4.2 can be thought of as a set of weights for the corresponding sample space of outcomes. Let E be the event of death during the first 3 decades after birth. Then

$$E = \{x_1, x_2, x_3\},$$

where from Table 4.2 we have

$$w(x_1) = .0323, \qquad w(x_2) = .0065, \qquad w(x_3) = .0121.$$

Therefore, according to Definition 2,

$$\begin{aligned} P(E) &= w(x_1) + w(x_2) + w(x_3) \\ &= .0323 + .0065 + .0121 \\ &= .0509. \blacksquare \end{aligned}$$

Table 4.2. PROBABILITY OF DEATH AT DIFFERENT AGES IN THE UNITED STATES.

Age (years)	Probability of death
0–10	.0323
10–20	.0065
20–30	.0121
30–40	.0184
40–50	.0431
50–60	.0969
60–70	.1821
70–80	.2728
80 and over	.3358
Total	1.0000

SOURCE: "United States Life Tables by Causes of Death: 1959–1961," Vol. 1, No. 6, 1967, p. 15, National Health Center for Health Statistics, Public Health Service, Washington, D.C.

EXERCISES

Suggested minimum assignment: Exercises 1, 3, 6, 7, 9, 11, and 13.

In solving probability problems, the reader may find it helpful to start by identifying both the sample space U and the event space E. If the elements of these sets are not too numerous, they can be listed; otherwise, the set can be described, or a sufficient number of elements can be listed, to give the reader a clear idea of the spaces with which he is working.

1. What is the probability that a telephone number picked at random will have a last digit which is divisible by 3?
2. What is the probability of drawing a face card (include aces) from an ordinary deck of 52 cards?

In Exercises 3–6 suppose that a pair of ordinary dice are cast.

3. What is the probability that an odd sum is cast?
4. What is the probability that an even sum is cast?
5. What is the probability that a 6 or 7 or 8 is cast?
6. What is the probability that a 2 or 3 or 4 or 10 or 11 or 12 is cast?

7. In a family of 3 children, what is the probability that exactly 2 of them are boys? Assume equally likely outcomes and state sample and event spaces.
8. A pair of animals are inoculated against a certain disease. There is an equally likely chance that the inoculation will take or will not take. What is the probability that the inoculation will take on at least one animal? State the sample space and the event space.
9. Let $U = \{u_1, u_2, u_3, u_4\}$ be a sample space of 4 outcomes. Suppose that the weights of these outcomes are $w(u_1) = \frac{1}{4}$, $w(u_2) = \frac{1}{2}$, $w(u_3) = \frac{1}{6}$, and $w(u_4) = \frac{1}{12}$. What is the probability of $E_1 = \{u_1, u_3\}$?
10. In Exercise 9 suppose that $w(u_1) = \frac{1}{3}$, $w(u_2) = \frac{1}{4}$, and $w(u_3) = \frac{1}{12}$. What must $w(u_4)$ be? Then calculate the probability of $E = \{u_3, u_4\}$.
11. Suppose that a jar contains 6 white capsules, 3 yellow capsules, and 2 orange capsules. What is the probability of picking a nonwhite capsule? List the sample space, the event space, and the weights of the outcomes.
12. Suppose that the probability for rain is 10 per cent, the probability for sleet is 20 per cent, and the probability for snow is 30 per cent; what is the probability that it will rain or snow? Assume that precipitation is classified in exactly one of the categories at any given time. List the sample space, the event space, and weights of the outcomes. Are the outcomes equally likely?
13. A pair of regular tetrahedrons (four-sided objects) are thrown on a flat surface in the same manner that one would throw a pair of dice. If the sides of the tetrahedrons are numbered from 1 to 4 (like dice), find the probability that the sum shown on one throw is 5.
14. In Exercise 13 find the probability that the sum shown on one throw is 7.
15. Prove Theorem 1(*i*).
16. Prove Theorem 1(*ii*).
17. Prove Theorem 1(*iii*).
18. In Example 6 verify the probabilities found by Galileo.

4.2 Probabilities of Complements, Unions, and Intersections of Events

Probability problems are often such that the event whose probability is sought can be treated as a union, or an intersection, or a complement of other events.

Example 1

Consider the experiment of drawing a single card from an ordinary deck of 52 cards. Suppose that we are requested to find the probabilities of 3 events: (1) the card drawn is a king *or* a diamond, (2) the card drawn is a king *and* a diamond, and (3) the card drawn is *not* a diamond. Let E_d represent the set of all diamonds. Let E_k represent the set of all kings; let the universal set be the sample space of all 52 cards. The three indicated probabilities that we seek are (1) $P(E_d \cup E_k)$, (2) $P(E_d \cap E_k)$, and (3) $P(E_d')$. Thus we see that event (1) can be thought of as a union, event (2) can be thought of as an intersection, and event (3) can be thought of as a complement. ▮

Before developing theorems that will give us formulas for solving Example 1, we state a second theorem, which together with Theorem 1, will be needed in the proofs of later theorems. In fact, in an axiomatic approach to probability theory, Theorems 1 and 2 can be considered as axioms from which later theorems follow. This axiomatic approach is, of course, similar to the approach described in Section 3.4, where it was mentioned that statements similar to Theorems 1–4 of each of the first three chapters could be accepted as axioms of a Boolean algebra; later theorems would then follow from the axioms. Of course, in this chapter we have chosen to define a specific probability function (Definition 2), and we shall now prove that it possesses the following property.

► **Theorem 2**

Let E_1 and E_2 be events within a sample space U with a finite number of outcomes. If $E_1 \cap E_2 = \varnothing$, then $P(E_1 \cup E_2) = P(E_1) + P(E_2)$.

Proof: Let $E_1 = \{e_1, \ldots, e_j\}$ and let $E_2 = \{e_k, \ldots, e_m\}$.

Statement	*Reason*
1. $E_1 \cup E_2 = \{e_1, \ldots, e_j, e_k, \ldots, e_m\}$.	1. Definition of $\cup$ and $E_1 \cap E_2 = \varnothing$.
2. $P(E_1 \cup E_2) = w(e_1) + \cdots + w(e_j) + w(e_k) + \cdots + w(e_m)$	2. Definition 2.
3. $= [w(e_1) + \cdots + w(e_j)] + [w(e_k) + \cdots + w(e_m)]$	3. Grouping terms.
4. $= P(E_1) + P(E_2)$.	4. Definition 2. ◄

Example 2

Consider the same experiment that we did in Example 1, that of drawing a single card from an ordinary deck of 52 cards. Let E_d be the event of drawing a diamond, and let E_s be the event of drawing a spade. Obviously these events are disjoint; no card is both a spade and a diamond. Thus the probability of drawing a spade or a diamond can be found by the formula of Theorem 2:

$$P(E_d \cup E_s) = P(E_d) + P(E_s).$$

Because there are 52 equally likely outcomes, 13 of which are diamonds and 13 of which are spades, then

$$P(E_d) + P(E_s) = \tfrac{13}{52} + \tfrac{13}{52} = \tfrac{1}{2}. \; \blacksquare$$

When two events are disjoint (that is, their intersection is the empty set) it is customary to say that the events are **mutually exclusive.** Events E_d and E_s of Example 2 are mutually exclusive.

A general formula for finding the probability of a union of two events (mutually exclusive or not) will be stated next.

▶ **Theorem 3**

Let E_1 and E_2 be events within a sample space U with a finite number of outcomes. Then

$$P(E_1 \cup E_2) = P(E_1) + P(E_2) - P(E_1 \cap E_2).$$

Proof: Let

$$E_1 = \{e_1, \ldots, e_j, e_k, \ldots, e_m\} \qquad \text{and} \qquad E_2 = \{e_k, \ldots, e_m, e_n, \ldots, e_p\},$$

where $E_1 \cap E_2 = \{e_k, \ldots, e_m\}$.

Statement	*Reason*
1. $E_1 \cup E_2 = (e_1, \ldots, e_j, e_k, \ldots, e_m, e_n, \ldots, e_p\}$.	1. Definition of the union of two sets.
2. $P(E_1 \cup E_2) = w(e_1) + \cdots + w(e_j) + w(e_k) + \cdots + w(e_m) + w(e_n) + \cdots + w(e_p)$	2. Definition 2.
3. $= [w(e_1) + \cdots + w(e_j) + w(e_k) + \cdots + w(e_m)] + [w(e_k) + \cdots + w(e_m) + w(e_n) + \cdots + w(e_p)] - [w(e_k) + \cdots + w(e_m)]$	3. Add and subtract $[w(e_k) + \cdots + w(e_m)]$ and regroup terms.
4. $= P(E_1) + P(E_2) - P(E_1 \cap E_2)$.	4. Definition 2. ◀

Example 3

Again returning to Example 1 with its 52 equally likely outcomes, 13 diamonds, 4 kings, and one card that is both a king and a diamond, we have

$$P(E_d) = \tfrac{13}{52},\ P(E_k) = \tfrac{4}{52},\ \text{and } P(E_d \cap E_k) = \tfrac{1}{52}.$$

Therefore, the probability of selecting a card that is a king or a diamond is given by the formula of Theorem 3.

$$\begin{aligned} P(E_d \cup E_k) &= P(E_d) + P(E_k) - P(E_d \cap E_k) \\ &= \tfrac{13}{52} + \tfrac{4}{52} - \tfrac{1}{52} \\ &= \tfrac{4}{13}. \end{aligned}$$ ∎

Notice that in the formula of Theorem 3 we need to know how to find $P(E_1 \cap E_2)$. Formulas for $P(E_1 \cap E_2)$ (which follow immediately from Definitions 1 and 2 and Theorem 1) are stated formally for future reference.

▶ **Theorem 4**

Let E_1 and E_2 be events (mutually exclusive or not) within a sample space U with a finite number of outcomes.

(*i*) If $E_1 \cap E_2 = \{e_1, \cdots, e_k\}$ with weights $w(e_1), \cdots, w(e_k)$, then

$$P(E_1 \cap E_2) = w(e_1) + \cdots + w(e_k).$$

If $E_1 \cap E_2 = \varnothing$, then $P(E_1 \cap E_2) = 0$.

(*ii*) If U consists only of equally likely outcomes, then

$$P(E_1 \cap E_2) = \frac{n(E_1 \cap E_2)}{n(U)}.$$ ◀

Additional formulas for $P(E_1 \cap E_2)$ will be given in Section 4.4.

Example 4

Consider the sample space of outcomes consisting of the sums of two dice when cast. The sample space together with their respective weights are given in Table 4.3.

Table 4.3. WEIGHTS FOR THE OUTCOMES OF A PAIR OF DICE.

u	2	3	4	5	6	7	8	9	10	11	12
$w(u)$	$\frac{1}{36}$	$\frac{2}{36}$	$\frac{3}{36}$	$\frac{4}{36}$	$\frac{5}{36}$	$\frac{6}{36}$	$\frac{5}{36}$	$\frac{4}{36}$	$\frac{3}{36}$	$\frac{2}{36}$	$\frac{1}{36}$

Let E_1 be the set of all sums less than 7; that is,

$$E_1 = \{2, 3, 4, 5, 6\}.$$

Let E_2 be the set of even sums; that is,

$$E_2 = \{2, 4, 6, 8, 10, 12\}.$$

Suppose that one wants the probability of the intersection

$$E_1 \cap E_2 = \{2, 4, 6\}.$$

By Theorem 4(i)

$$\begin{aligned} P(E_1 \cap E_2) &= w(2) + w(4) + w(6) \\ &= \tfrac{1}{36} + \tfrac{3}{36} + \tfrac{5}{36} \\ &= \tfrac{9}{36} = 0.25. \ \blacksquare \end{aligned}$$

Finally we consider the probability of the complement of an event.

▶ **Theorem 5**

Let E be an event within a sample space U with a finite number of outcomes, and let E' represent the complement of E. Then

$$P(E') = 1 - P(E).$$

Proof

Statement	*Reason*
1. $P(E \cup E') = P(U)$.	1. Given that $E \cup E' = U$.
2. $P(E) + P(E') = P(U)$.	2. Theorem 2.
3. $P(E) + P(E') = 1$.	3. Theorem 1.
4. $P(E') = 1 - P(E)$.	4. Subtracting $P(E)$ from both sides. ◀

Example 5

Again, from Example 1, an ordinary deck of 52 cards has 13 diamonds, and if E_d represents the event of selecting a diamond, then $P(E_d) = \frac{13}{52}$. Hence the event of *not* selecting a diamond is

$$\begin{aligned} P(E'_d) &= 1 - P(E_d) \\ &= 1 - \tfrac{13}{52} \\ &= \tfrac{39}{52} = .75. \ \blacksquare \end{aligned}$$

EXERCISES

Suggested minimum assignment: Exercises 1, 4, 5, and 7.

1. Consider the experiment in which one die is cast. Let E_1 be the event of casting an even number. Let E_2 be the event of casting a number greater than 2.
   (a) List the set of outcomes (sample space).
   (b) List the elements of E_1, E_2, $E_1 \cap E_2$, $E_1 \cup E_2$, and E'_2.
   (c) Calculate $P(E_1)$ and $P(E_2)$.
   (d) Calculate $P(E_1 \cap E_2)$ by Theorem 4.
   (e) Calculate $P(E_1 \cup E_2)$ by Theorem 3.
   (f) Calculate $P(E'_2)$ by Theorem 5.
2. Consider the experiment of flipping a coin 3 times. Let E_1 be the event

of achieving a head on the first flip, and let E_2 be the event of achieving a head on any 2 flips.

(*a*) List the set of outcomes (sample space).
(*b*) List the elements of E_1, E_2, $E_1 \cap E_2$, $E_1 \cup E_2$, and E_2'.
(*c*) Calculate $P(E_1)$ and $P(E_2)$.
(*d*) Calculate $P(E_1 \cap E_2)$ by Theorem 4.
(*e*) Calculate $P(E_1 \cup E_2)$ by Theorem 3.
(*f*) Calculate $P(E_2')$ by Theorem 5.

3. A pair of newlyweds plan a family of 3 children. Let E_1 be the event that the first child is a boy, and let E_2 be the event that all 3 children are the same sex. Assume that in a single birth a boy and a girl are equally likely.
(*a*) List the set of outcomes (sample space).
(*b*) List the elements of E_1, E_2, $E_1 \cap E_2$, $E_1 \cup E_2$, and E_2'.
(*c*) Calculate $P(E_1)$ and $P(E_2)$.
(*d*) Calculate $P(E_1 \cap E_2)$ by Theorem 4.
(*e*) Calculate $P(E_1 \cup E_2)$ by Theorem 3.
(*f*) Calculate $P(E_2')$ by Theorem 5.

4. Consider a deck of cards in which all the cards except the kings, queens, and jacks have been removed. Let E_1 be the event of drawing a queen from the deck. Let E_2 be the event of drawing a diamond or a heart from the deck.
(*a*) List the set of outcomes (sample space).
(*b*) List the elements of E_1, E_2, $E_1 \cap E_2$, $E_1 \cup E_2$, and E_2'.
(*c*) Calculate $P(E_1)$ and $P(E_2)$.
(*d*) Calculate $P(E_1 \cap E_2)$ by Theorem 4.
(*e*) Calculate $P(E_1 \cup E_2)$ by Theorem 3.
(*f*) Calculate $P(E_2')$ by Theorem 5.

5. Let the outcomes of an experiment be called O_1, O_2, and O_3, where $w(O_1) = \frac{1}{4}$, $w(O_2) = \frac{1}{3}$, and $w(O_3) = \frac{5}{12}$. Let E_1 consist of outcomes O_1 and O_2, and let E_2 consist of outcomes O_2 and O_3.
(*a*) List the set of outcomes (sample space).
(*b*) List the elements of E_1, E_2, $E_1 \cap E_2$, $E_1 \cup E_2$, and E_2'.
(*c*) Calculate $P(E_1)$ and $P(E_2)$.
(*d*) Calculate $P(E_1 \cap E_2)$.
(*e*) Calculate $P(E_1 \cup E_2)$.
(*f*) Calculate $P(E_2')$.

6. Let the outcomes of an experiment be called O_1, O_2, O_3, and O_4, where $w(O_1) = \frac{1}{4}$, $w(O_2) = \frac{1}{2}$, $w(O_3) = \frac{1}{6}$, and $w(O_4) = \frac{1}{12}$. Let E_1 consist of outcomes O_1 and O_2, and let E_2 consist of outcomes O_3 and O_4.
(*a*) List the set of outcomes (sample space).
(*b*) List the elements of E_1, E_2, $E_1 \cap E_2$, $E_1 \cup E_2$, and E_2'.
(*c*) Calculate $P(E_1)$ and $P(E_2)$.
(*d*) Calculate $P(E_1 \cap E_2)$.
(*e*) Calculate $P(E_1 \cup E_2)$.
(*f*) Calculate $P(E_2')$.

7. Let U be the sample space of outcomes consisting of the sum shown when two dice are cast. The nonequally likely outcomes are 2, 3, 4, 5, 6, 7, 8, 9, 10, 11, 12, and the weights (or probabilities of occurrence) of these outcomes are given in Table 4.3 (See page 103). Let E_1 be the event of achieving an even sum, and let E_2 be the event of achieving a sum of 7 or 11.
   (*a*) State the weight of each outcome.
   (*b*) List the elements of E_1, E_2, $E_1 \cap E_2$, $E_1 \cup E_2$, and E_2'.
   (*c*) Calculate $P(E_1)$ and $P(E_2)$.
   (*d*) Calculate $P(E_1 \cap E_2)$.
   (*e*) Calculate $P(E_1 \cup E_2)$.
   (*f*) Calculate $P(E_2')$.
8. Suppose that an enclosure contains 4 types of animals. There is 1 animal that has only disease A; there are 3 animals that have only disease B; and there are 4 animals that have only disease C. The enclosure also contains 2 animals that have all 3 diseases. Let E_1 be the event of selecting an animal with disease A or B (this includes those animals having all 3 diseases), and let E_2 be the event of selecting an animal that has disease C.
   (*a*) List the set of outcomes (sample space).
   (*b*) State the weight of each outcome.
   (*c*) List the elements of E_1, E_2, $E_1 \cap E_2$, $E_1 \cup E_2$, and E_2'.
   (*d*) Calculate $P(E_1)$ and $P(E_2)$.
   (*e*) Calculate $P(E_1 \cap E_2)$.
   (*f*) Calculate $P(E_1 \cup E_2)$.
   (*g*) Calculate $P(E_2')$.

4.3 Conditional Probability

In this section we shall be concerned with the probability of an event if it is known that another event has occurred. In other words, we shall revise our probability measure as additional information becomes known.

Example 1

Again consider the experiment of flipping a coin 3 times in succession. The sample space of *equally likely outcomes* is

$$U = \{HHH, HHT, HTH, THH, HTT, THT, TTH, TTT\}.$$

In Example 2 of Section 4.1 we found that the probability of obtaining exactly two heads was $\frac{3}{8}$. Recall that

$$E = \{HHT, HTH, THH\}.$$

Suppose that we now are given the condition that the first flip produces a head. Given this condition, now what is the probability of obtaining exactly 2 heads? We shall consider this question as a brand new problem. If we consider the fact that a head has been thrown the first time, then the reduced sample space is

$$F = \{HHH, HHT, HTH, HTT\},$$

and the reduced event space is

$$E \cap F = \{HHT, HTH\};$$

hence the probability that we seek is

$$\frac{n(E \cap F)}{n(F)} = \tfrac{2}{4} = \tfrac{1}{2}. \tag{4.1}$$

This new probability, subject to the condition of flipping a head on the first try, is called a **conditional probability;** notice that it differs from the probability of $\frac{3}{8}$ that was calculated without the condition. Since both numerator and denominator of (4.1) can be divided by $n(U)$, the conditional probability of E, given F, can be written as

$$\frac{n(E \cap F)/n(U)}{n(F)/n(U)}$$

or

$$\frac{P(E \cap F)}{P(F)}.$$

Visual assistance in understanding conditional probability may be obtained from the **tree diagram** of Figure 4.7; each complete sequence (from left to right) of connecting line segments in the tree diagram is called a **path** and represents an outcome of the experiment. In Figure 4.7 there are 3 of 8 equiprobable paths of the tree that produce 2 H's, hence the probability of $\frac{3}{8}$. However, once we are given that the first flip produces a head, then the tree diagram narrows to only the upper half of the tree; there we see that there are 2 of 4 equiprobable paths that produce 2 H's, hence the probability of $\frac{2}{4}$. ∎

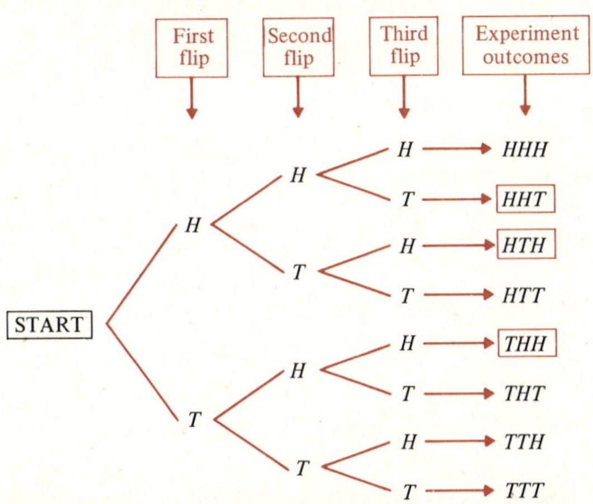

Figure 4.7. *Tree diagram of three flips of a coin.*

Although the last example illustrated a sample space of equally likely outcomes, the following definition is made for sample spaces with or without equally likely outcomes.

Definition 3
If E_1 and E_2 are two events of a finite sample space and if $P(E_1) \neq 0$, then the **conditional probability** of E_2, given E_1, is designated $P(E_2|E_1)$ and defined by

$$P(E_2|E_1) = \frac{P(E_2 \cap E_1)}{P(E_1)}.$$

Example 2
A biologist expects to obtain 2 offspring from a pair of animals. What is the probability that both will be male? Is this probability changed after the first born is a male? The sample space is

$$U = \{mm, mf, fm, ff\}.$$

Suppose that these outcomes are not equally likely and it is known that

$$w(mm) = .36,\ w(mf) = .24,\ w(fm) = .24,\ w(ff) = .16.$$

Before any birth the probability that both offspring are male is $P(\{mm\}) = w(mm) = .36$. To determine the probability that both are male after we know that the first is male, we use the formula of Definition 3. Let E_1 be the event of the first born being male; then

$$E_1 = \{mm, mf\}.$$

Let E_2 be the event of both being male; then

$$E_2 = \{mm\}.$$

Therefore,

$$E_2 \cap E_1 = \{mm\}.$$

To use the formula of Definition 3, we need

$$P(E_1) = w(mm) + w(mf) = .36 + .24 = .6,$$

and

$$P(E_2 \cap E_1) = w(mm) = .36.$$

Hence, by Definition 3,

$$P(E_2|E_1) = \frac{P(E_2 \cap E_1)}{P(E_1)} = \frac{.36}{.6} = .6.$$

Again, the formula just used can be illustrated by means of a tree diagram as shown in Figure 4.8. Before any births, the desired probability was the weight of the desired path relative to all 4 paths of the tree, but after know-

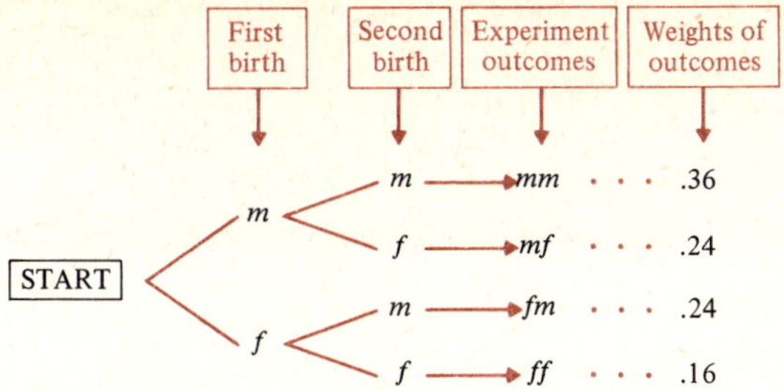

Figure 4.8. *Tree diagram of two births.*

ing that the first birth was male, we simply consider only the weight of the desired path relative to the top 2 paths. In other words, the weight of the desired path (outcome) is divided by the sum of the weights of the outcomes that are still possible after the condition is known. ∎

Example 3

Suppose that 55 per cent of the voters of a certain precinct voted for the Republican candidate in the last election, and suppose that 20 per cent of the voters voted for the Republican candidate *and* for the local bond issue. What is the probability that a voter who is selected at random voted for the bond issue if he tells us that he voted for the Republican candidate? Let R represent the set of all voters who voted Republican and B represent the set of all voters who voted for the bond issue. We are given that $P(R) = .55$ and that $P(B \cap R) = .20$; therefore,

$$P(B|R) = \frac{P(B \cap R)}{P(R)} = \frac{.20}{.55} \approx .36.$$

Figure 4.9 gives a tree diagram of the problem. ∎

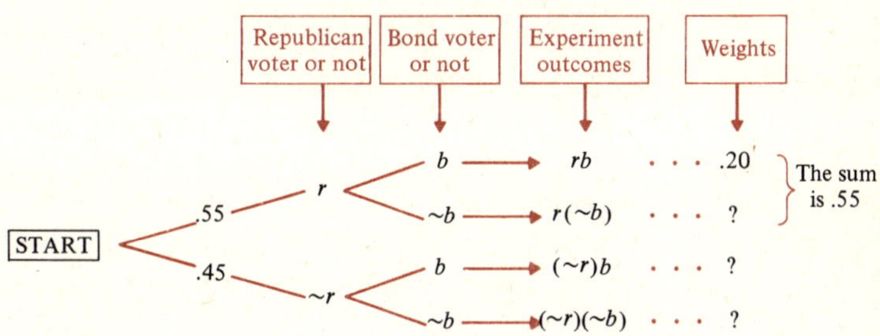

Figure 4.9. *Tree diagram of voter action.*

In Exercise 13 the reader will be asked to prove, using Definitions 1 and 3, that if E_1 and E_2 are events of *a finite sample space of equally likely outcomes* and if $n(E_1) \neq 0$, then

$$P(E_2|E_1) = \frac{n(E_2 \cap E_1)}{n(E_1)}.$$

APPLICATIONS

Example 4

The probability of death of U.S. citizens at different ages was given in Table 4.2 of Example 7 of Section 4.1. From that table, the probability of death during the fourth decade was listed as .0184. Suppose, however, that the probability of dying during the fourth decade is recalculated under the condition that the person survives the first three decades. Let A be the event that the person survives the first three decades. From Example 7 of Section 4.1, we found that the probability of dying during the first three decades was .0509; therefore, the probability of living three decades is $P(A) = 1 - .0509$ or $P(A) = .9491$. Let B be the event that the person dies during the fourth decade. The conditional probability of death during the fourth decade, given that death does not occur during the first three decades, is

$$\begin{aligned} P(B|A) &= \frac{P(B \cap A)}{P(A)} \\ &= \frac{P(\{x_4\})}{P(A)} \\ &= \frac{.0184}{.9491} \\ &\approx .0194. \end{aligned}$$

Conditional probability can be useful in calculating premiums for life insurance policies; for example, suppose a 30-year-old person wanted to take out a $10,000 term life insurance policy for 10 years and wanted to pay the premium in one lump sum at the outset. Excluding overhead costs, commissions, and so on, the premium, based on the information of this problem, would be

$$(.0194)(\$10{,}000) = \$194.$$

In reality, methods of calculating insurance premiums are more sophisticated, but this example does illustrate the basic idea. ▮

Example 5

A variety of applied problems can be solved in a manner illustrated by this example. Suppose that a company manufactures a product that includes a timer. The timer is purchased from several different subcontractors, including subcontractor A. The service department estimates that 5 per cent of all products have defective timers. They also estimate that, of the products serviced by them, 20 per cent of the defective timers and 60 per cent of the nondefective timers are purchased from A. What is the probability that a timer purchased from A is defective? Let

A = event that a timer was purchased from A,
D = event of a defective timer.

We have been given

$$P(D) = .05, \ P(A|D) = .20, \text{ and } P(A|D') = .60.$$

We wish to find $P(D|A)$:

$$P(D|A) = \frac{P(D \cap A)}{P(A)}.$$

Since $D \cap D' = \varnothing$ and $D \cup D' = U$,

$$P(D|A) = \frac{P(D \cap A)}{P(A \cap D) + P(A \cap D')}$$

and

$$\begin{aligned} P(D|A) &= \frac{P(A|D)P(D)}{P(A|D)P(D) + P(A|D')P(D')} \\ &= \frac{(.20)(.05)}{(.20)(.05) + (.60)(.95)} \\ &\approx .017. \end{aligned}$$

To illustrate how the method can be applied in other contexts, let

A = event that a person passed a certain test,
D = event that a person graduates from college.

The question would be to find the probability that a person would graduate from college if he passed the test. In psychology or medicine the model could be used to find the probability that a person actually had some physical or mental disorder, given that a certain test registered negative. Further discussion of the model presented in this example may be found in other books under the title of Bayes' Rule or Bayes' Theorem (for example, Freund [14], pp. 480–484). ▮

EXERCISES

Suggested minimum assignment: Exercises 1, 2, 3, 5, 7, 8, and 11.

1. Let the experiment consist of choosing a card from a deck in which all cards except the jacks, queens, and kings have been removed. Let E_1 be the event of choosing a king, and let E_2 be the event of choosing a diamond.
   (*a*) List the set of outcomes (sample space).
   (*b*) List the elements of E_1, E_2, and $(E_1 \cap E_2)$.
   (*c*) Find $P(E_1)$, $P(E_2)$, and $P(E_1 \cap E_2)$.
   (*d*) Find $P(E_1|E_2)$ and $P(E_2|E_1)$.
2. Let the experiment consist of casting one ordinary die. Let E_1 be the event of obtaining an odd number, and let E_2 be the event of obtaining a number less than 5.
   (*a*) List the set of outcomes (sample space).

(*b*) List the elements of E_1, E_2, and $(E_1 \cap E_2)$.
(*c*) Find $P(E_1)$, $P(E_2)$, and $P(E_1 \cap E_2)$.
(*d*) Find $P(E_1 | E_2)$ and $P(E_2 | E_1)$.

3. In a certain corporation the wages earned by union men vary directly with seniority. Suppose that 5 per cent of the union men have more than 10 years seniority and make more than \$4 per hour. If 10 per cent of the union men make more than \$4 per hour, what is the conditional probability that a man who makes \$4 per hour has more than 10 years of seniority? *Hint*: Start by identifying the sample and event spaces.
4. Suppose that 20 per cent of a seed mixture consists of seeds of red flowers and 5 per cent of the seed mixture consists of seeds of flowers that are both red and have diameters of less than 2 inches. What is the conditional probability that a plant selected at random has a flower with a diameter less than 2 inches if we know that the flower is red? *Hint*: Start by identifying the sample and event spaces.
5. Twenty per cent of the population of a certain area are both female and not eligible to vote (under 18). If 60 per cent of the population is female, what is the probability that a female who is selected at random is ineligible to vote?
6. Suppose that 30 per cent of the people in a certain precinct are over 18 and have registered to vote. If we know that 75 per cent of the people in the precinct are over 18, what is the probability that a 20-year-old is registered?
7. Suppose that there is a probability of $\frac{1}{2}$ that a ballistic missile will both survive an enemy's defenses and hit its target. Further suppose that there is a $\frac{3}{4}$ probability that a missile will survive the enemy's defenses. What is the probability that a missile which survives the defenses will hit the target?
8. Suppose that a specific drug can be administered to fight a certain disease. Studies have shown that 40 per cent of the victims of the disease have received the drug and 2 per cent of the victims of this disease have both received the drug and have died. What is the probability that a victim who receives the drug will die of the disease?
9. In Exercise 8 suppose that the studies further showed that 5 per cent of the victims of the disease both did *not* receive the drug and died. What is the probability that a victim who did not receive the drug will die of the disease?
10. A study shows that in one year 2 per cent of the population indulges in a certain activity and dies. If 20 per cent of the population indulges in that activity, what is the probability that a person who indulges will die?
11. A certain system consists of two components connected in series. If the probability that the system will perform satisfactorily is $\frac{3}{4}$, and if one component has a $\frac{7}{8}$ probability of performing satisfactorily, what is the probability that the other component performs satisfactorily?

12. There is a $\frac{1}{2}$ probability that a certain conjunction of two simple statement forms is true. If it is known that there is a $\frac{3}{4}$ probability that one of the simple statement forms is true, what is the probability that the other simple statement form is true?
13. Prove that if E_1 and E_2 are events of a finite sample space U of equally likely outcomes and if $n(E_1) \neq 0$, then

$$P(E_2 | E_1) = \frac{n(E_2 \cap E_1)}{n(E_1)}.$$

Hint: Use Definitions 1 and 3.

4.4 Independent Events

Example 1

Suppose that we toss a pair of dice; if the dice are fair, then certainly the upward face of one die in no way affects the upward face of the second die. Thus, if we let E_1 be the event of a 6 on the first die and let E_2 be the event of a 3 on the second die, then the occurrence of E_1 will not affect the probability of E_2; hence $P(E_2 | E_1) = P(E_2)$. We say that E_2 is **independent** of E_1. We now verify that $P(E_2) = P(E_2 | E_1)$ for this example.

$n(U) = 36;$

$E_1 =$ 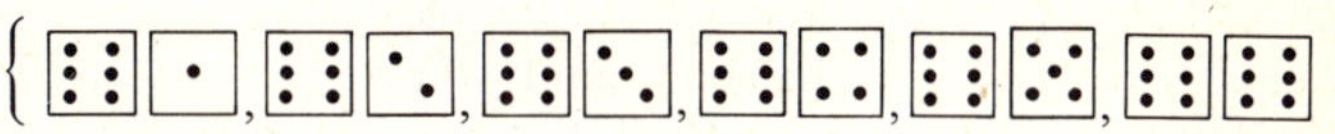;

$E_2 =$ 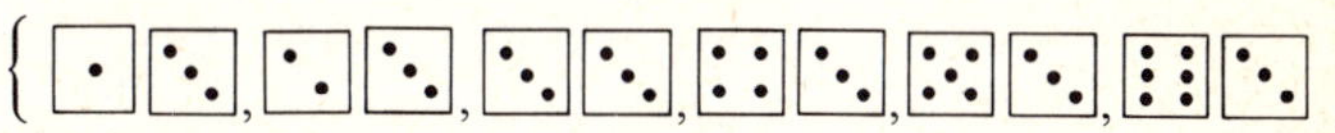;

$E_2 \cap E_1 =$ 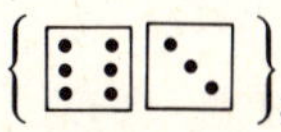;

$$P(E_2) = \frac{n(E_2)}{n(U)} = \frac{6}{36} = \frac{1}{6};$$

$$P(E_2 | E_1) = \frac{P(E_2 \cap E_1)}{P(E_1)} = \frac{\frac{1}{36}}{\frac{1}{6}} = \frac{1}{6}. \blacksquare$$

It can be shown (Exercise 13) that, for a given experiment, if event E_2 is independent of event E_1, then E_1 is independent of E_2.

▶ Definition 4

Two distinct events E_1 and E_2 [where $P(E_1) \neq 0$ and $P(E_2) \neq 0$] of a given finite sample space are said to be **independent** if

$$P(E_2 | E_1) = P(E_2) \qquad \text{or if} \qquad P(E_1 | E_2) = P(E_1).$$

Example 2

Consider 5 airlines that serve 3 cities according to the diagram shown in Figure 4.10. For the experiment of traveling from city 1 to city 3 the sample

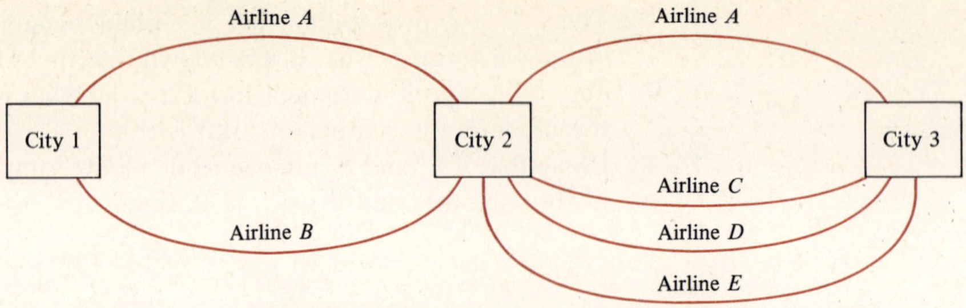

Figure 4.10. *Hypothetical airline service.*

space is

$$U = \{AA, AC, AD, AE, BA, BC, BD, BE\},$$

and we will assume that the outcomes are equally likely. Let E_1 be the event that a person chooses airline A from city 1 to city 2; hence

$$E_1 = \{AA, AC, AD, AE\}.$$

Let E_2 be the event that a person chooses airline D from city 2 to city 3; hence

$$E_2 = \{AD, BD\}.$$

We verify that the choice of airlines from city 1 to city 2 is independent of the choice from city 2 to city 3 by showing that $P(E_2 | E_1) = P(E_2)$; in other words, the probability of E_2 is the same regardless of knowledge of the condition E_1. First, since we are dealing with equally likely outcomes,

$$P(E_2) = \frac{n(E_2)}{n(U)} = \frac{2}{8} = \frac{1}{4}.$$

Second, since $E_1 \cap E_2 = \{AD\}$,

$$P(E_2 | E_1) = \frac{P(E_2 \cap E_1)}{P(E_1)} = \frac{n(E_2 \cap E_1)}{n(E_1)} = \frac{1}{4}. \blacksquare$$

We are now in a position to state still another pair of formulas for finding the probability of an intersection of two events.

▶ **Theorem 6**

Let E_1 and E_2 be events within a sample space with a finite number of outcomes, where $P(E_1) \neq 0$ and $P(E_2) \neq 0$. Then

(*i*) $P(E_1 \cap E_2) = P(E_2 | E_1)P(E_1) = P(E_1 | E_2)P(E_2)$.

(*ii*) $\{P(E_1 \cap E_2) = P(E_1)P(E_2)\} \Leftrightarrow \{E_1 \text{ and } E_2 \text{ are independent events}\}$.

The proof of Theorem 6 is left as Exercises 14 and 15. ◀

Example 3

This example illustrates Theorem 6(*i*). Suppose that a certain factory produces 3 products, labeled A, B, and C. What is the probability of selecting, at random, a defective product A if it is known that 30 per cent of the products produced at the factory are product A and 5 per cent of the A products are defective? Let A represent the event of selecting a product A, and we are given that $P(A) = \frac{3}{10}$. Let D represent the event of selecting a defective product, and we are given that $P(D|A) = \frac{1}{20}$. We are asked to find $P(D \cap A)$. By Theorem 6(*i*) we have

$$\begin{aligned} P(D \cap A) &= P(D|A)P(A) \\ &= \tfrac{1}{20}\tfrac{3}{10} = \tfrac{3}{200} = 1.5 \text{ per cent. } \blacksquare \end{aligned}$$

A visual illustration of Example 3 and Theorem 6(*i*) can be obtained from the tree diagram of Figure 4.11; the probability of $(D \cap A)$ is the product of the probabilities associated with the segments of the path. If, in Example 3, D had been independent of A, then $P(D|A)$ would have equaled $P(D)$.

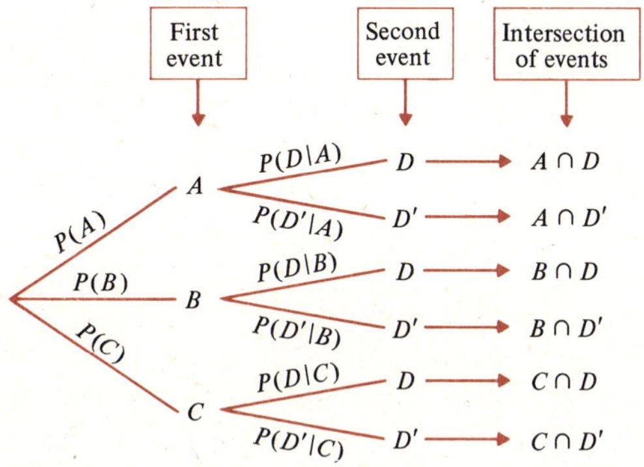

Figure 4.11. *Tree diagram of intersections of pairs of events.*

Example 4

To illustrate Theorem 6(*ii*), suppose that there is a 40 per cent probability that a missile will miss a given target and a 20 per cent chance that a bomber will miss the same target; assume that these two events are independent. (1) What is the probability that the target will be missed? (2) What is the probability that the target will be hit? Let E_1 be the event that a missile misses the target and E_2 be the event that a bomber will miss the target. We are asked to find $P(E_1 \cap E_2)$. The events are independent, hence by Theorem 6(*ii*) we have

$$\begin{aligned} P(E_1 \cap E_2) &= P(E_1)P(E_2) \\ &= \tfrac{4}{10}\tfrac{2}{10} = \tfrac{8}{100} = 8 \text{ per cent.} \end{aligned}$$

The probability that the target will be hit is determined by

$$P\{(E_1 \cap E_2)'\} = 1 - P(E_1 \cap E_2) = 1 - .08 = 92 \text{ per cent. } \blacksquare$$

There are some problems in which our intuition may fail us in determining whether or not a pair of events are independent; as illustrated in the next example, Theorem 6(*ii*) may be helpful in such cases.

Example 5
Let the experiment consist of the birth of three offspring of either male or female sex. Further suppose that the following outcomes are equally likely:

$$U = \{mmm, mmf, mfm, fmm, mff, fmf, ffm, fff\}.$$

Let E_1 be the event of obtaining at least two males, and let E_2 be the event of obtaining at least one of each sex. Are the events independent?

$$\begin{aligned} E_1 &= \{mmm, mmf, mfm, fmm\}; \\ E_2 &= \{mmf, mfm, fmm, mff, fmf, ffm\}; \\ E_1 \cap E_2 &= \{mmf, mfm, fmm\}. \end{aligned}$$

Hence

$$P(E_1) = \tfrac{1}{2}, \qquad P(E_2) = \tfrac{3}{4}, \qquad P(E_1 \cap E_2) = \tfrac{3}{8}.$$

Therefore, by Theorem 6(*ii*), since

$$P(E_1)P(E_2) = P(E_1 \cap E_2),$$

the events E_1 and E_2 are independent. ▮

APPLICATIONS

Example 6
In Section 1.1 (Example 2) the reliability of a system was defined as the probability that the system will perform satisfactorily for a specified time interval under specified environmental conditions. In a series system, such as the one shown in Figure 4.12, the event E that the system performs

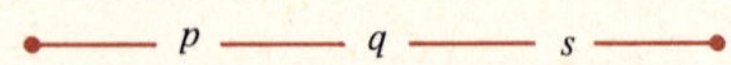

Figure 4.12. *Series system.*

satisfactorily is the intersection of the events $E_1, E_2, \ldots, E_n$ that each component functions satisfactorily. Thus the reliability R_s of the series system of Figure 4.12 is

$$R_s = P(E) = P(E_1 \cap E_2 \cap E_3).$$

We assume that the satisfactory or unsatisfactory performance of each component is independent of the performance of the other components. If r_1, r_2, and r_3 represent the reliabilities of the respective components p, q, and s, then it follows from Theorem 6(*ii*) that

$$\begin{aligned} R_s &= P(E_1)P(E_2)P(E_3) \\ &= r_1 r_2 r_3. \end{aligned}$$

In a parallel system, such as the one shown in Figure 4.13, the reliability R_p may be calculated in two ways. First, the event E that the system performs satisfactorily is the union of the events $E_1, E_2, \ldots, E_n$ that each

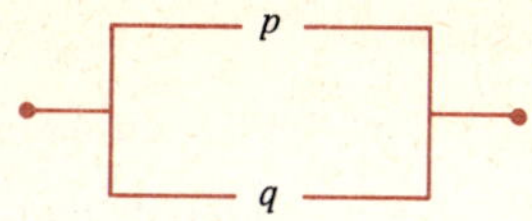

Figure 4.13. *Parallel system.*

component functions satisfactorily. Thus the reliability R_p of the parallel system of Figure 4.13 is

$$R_p = P(E) = P(E_1 \cup E_2).$$

By Theorem 3,

$$R_p = P(E_1) + P(E_2) - P(E_1 \cap E_2).$$

By Theorem 6(*ii*) and the assumption of independence,

$$R_p = r_1 + r_2 - r_1 r_2.$$

A second method of calculation of R_p recognizes that the event of satisfactory performance is the complement F' of the event F of failure to perform satisfactorily. Thus by Theorem 5,

$$R_p = P(F') = 1 - P(F).$$

But the event F of failure of the parallel system of Figure 4.13 is the intersection of the events F_1, F_2 of failure of each component; thus

$$P(F) = P(F_1)\,P(F_2)$$

and by Theorem 5,

$$\begin{aligned} P(F) &= [1 - P(F_1')]\,[1 - P(F_2')] \\ &= [1 - r_1]\,[1 - r_2]. \end{aligned}$$

Therefore, substituting for $P(F)$ in the formula $R_p = 1 - P(F)$, we get

$$R_p = 1 - [1 - r_1]\,[1 - r_2].$$

The two formulas for R_p are algebraically equivalent. The second formula, however, is more readily generalized for parallel systems of n components, namely,

$$R_p = 1 - [1 - r_1] \cdots [1 - r_n].$$

In Exercise 16 the reader will be asked to compute the reliability of the mixed system shown in Figure 4.14, where $r_1 = \frac{1}{3}$, $r_2 = \frac{2}{3}$, and $r_3 = \frac{3}{5}$. (r_1, r_2, and r_3 are the reliabilities of p, q, and s respectively.) ▮

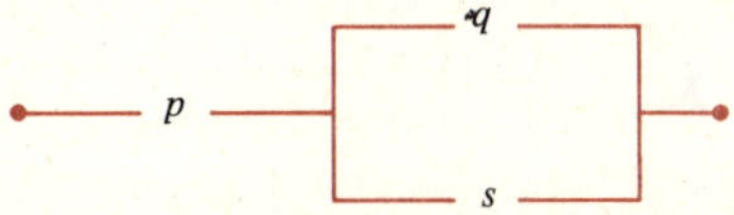

Figure 4.14

EXERCISES

Suggested minimum assignment: Exercises 1, 2, 3, 5, 8, 9, and 11.

1. Let A and B be events in a finite sample space U. If $P(A) = \frac{2}{3}$, $P(B) = \frac{1}{4}$, and $P(A \cap B) = \frac{1}{8}$, determine whether or not A and B are independent by means of Definition 4 and also by Theorem 6.
2. A pair of dice are cast. Let A be the event that the first die shows an odd number. Let B be the event that the sum of both dice is 7.
   (*a*) Find $P(A)$. (*b*) Find $P(B)$.
   (*c*) Find $P(B \cap A)$. (*d*) Find $P(B|A)$.
   (*e*) Are A and B independent events? Give two alternative reasons.

In each of Exercises 3–10 questions are raised about the probability of intersections of events. In some cases the events are independent and in some cases they are not; in either case state the formula used to solve each exercise and specifically identify each part of the formula.

3. Suppose that 30 per cent of the population has a certain vitamin deficiency because of their diet. Ten per cent of the people with this vitamin deficiency have a certain disease. What is the probability that a person selected at random will have the disease *and* the vitamin deficiency? *Hint*: Use Theorem 6(*i*).
4. Suppose that there is a $\frac{5}{8}$ probability that an intercontinental missile can survive an enemy's defense system and a $\frac{3}{4}$ probability that it can hit its target if it does survive. What is the probability that a missile will survive *and* hit its target? *Hint*: Use Theorem 6(*i*).
5. Suppose that a certain weed killer will become ineffective if it rains within 24 hours of application. Suppose that if it does not rain, there is an 80 per cent chance that the weed killer will be effective on a specified variety of weed. If the 24-hour weather forecast predicts that there is 10 per cent probability of rain, what is the probability of killing the specified weeds if the weed killer is administered now? That is, what is the probability of the weed killer's being effective and of having no rain?
6. If it rains, there is a 90 per cent probability that a certain type seed will germinate. If there is a 70 per cent chance of rain, what is the probability of both rain and germination?
7. In a deck of cards in which all the cards have been removed except the jacks, queens, and kings, what is the probability that a jack followed by a diamond will be drawn from the deck in 2 draws? Assume here that the first card is replaced before the second card is drawn. *Hint*: Use Theorem 6(*ii*).
8. Assume that the effect of one medicine is independent of the effect of another medicine. Suppose that experience has shown that the probability of recovery if one medicine is administered is 80 per cent, and if the other medicine is administered it is 70 per cent. What are the probabilities of *not* recovering if each single medicine is adminis-

tered? What is the probability of recovering if both medicines are administered?

9. Suppose that 2 switches are connected in series and that there is a 30 per cent chance that switch p is closed and a 20 per cent chance that switch q is closed; what is the probability that the compound switch $p \times q$ is closed? What is the probability that the compound switch is open? Assume that the events are independent.
10. Suppose that 2 statement forms are connected by the conjunction "and." If there is a 5 per cent probability that statement form p is false, and if there is 60 per cent probability that statement form q is false, what is the probability that the compound statement $p \wedge q$ is true? Assume independence.
11. Three fair coins are flipped: E_1 is the event that at least two tails appear, E_2 is the event that at most one tail appears, and E_3 is the event that all outcomes are heads or all are tails.
    (a) Are E_1 and E_2 independent events?
    (b) Are E_2 and E_3 independent events?
    (c) Are E_3 and E_1 independent events?
12. Two dice are cast: E_1 is the event that a 6 appears on *at least* one die, E_2 is the event that a 5 appears on *exactly* one die, and E_3 is the event that the same number appears on both dice.
    (a) Are E_1 and E_2 independent events?
    (b) Are E_2 and E_3 independent events?
    (c) Are E_3 and E_1 independent events?
13. Show that for a given experiment, if E_2 is independent of E_1, then E_1 is independent of E_2.
14. Prove Theorem 6(*i*).
15. Prove Theorem 6(*ii*).
16. In Example 6 determine the reliability of the system shown in Figure 4.14, where $r_1 = \frac{1}{3}$, $r_2 = \frac{2}{3}$, and $r_3 = \frac{3}{5}$.

4.5 Counting Techniques: Permutations†

The problems considered so far have been limited to those for which it has been easy to find the number of elements in the sample and event spaces. Many realistic problems, however, involve sets that require more sophisticated counting techniques than we have encountered so far. We shall introduce these techniques by means of examples.

Example 1

A certain manufacturer must decide on the design of a container that is to be used for his product. Suppose that this container consists of 3 components: the bottle, the cap, and the label. If there are 2 different bottle designs, 3 different cap designs, and 5 different labels, how many alternatives must he consider? We exhibit two ways of approaching this problem.

† Section 3.6 is a prerequisite for Sections 4.5–4.7.

1. We may use a tree diagram, shown in Figure 4.15. In the diagram one may start by choosing any one of the 2 different bottle designs; then for each bottle design there are 3 different cap designs, thus producing 2 times 3, or 6, different bottle and cap alternatives. Finally, for each of these bottle and cap alternatives there are 5 different labels. Each path of the tree represents a container that differs in some way from the others. We see that there are $2 \cdot 3 \cdot 5 = 30$ different alternatives.

2. Let B represent the set of bottle designs, let C represent the set of cap designs, and let L represent the set of label designs. The Cartesian product $B \times C \times L$ represents the set of all possible arrangements of the

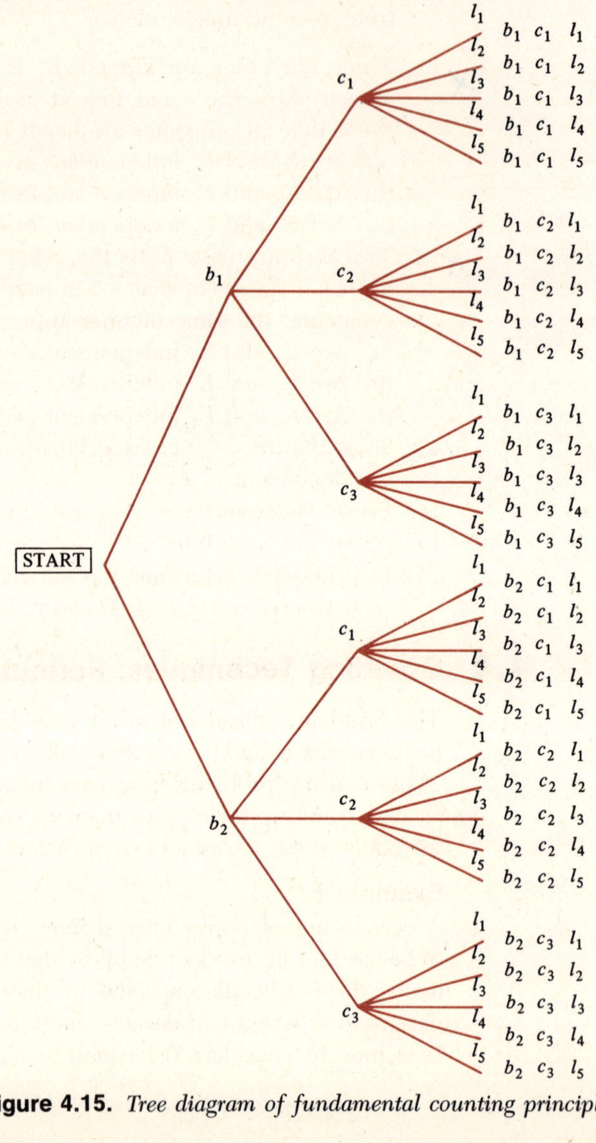

Figure 4.15. *Tree diagram of fundamental counting principle.*

elements of the respective sets. Of course each element of $(B \times C \times L)$ represents a path of the tree of Figure 4.15 and is listed there. From Section 3.6 we know that

$$\begin{aligned} n(B \times C \times L) &= n(B) \cdot n(C) \cdot n(L) \\ &= 2 \cdot 3 \cdot 5 \\ &= 30. \ \blacksquare \end{aligned}$$

The ideas presented in Example 1 can be generalized to produce the useful counting technique called the **fundamental counting principle:** *If one event can occur in n_1 ways, if a second event can occur in n_2 ways, if a third event can occur in n_3 ways, . . . , and if a kth event can occur in n_k ways, then the number of ways the k events can occur is the product* $n_1 \cdot n_2 \cdot n_3 \cdot \cdots \cdot n_k$.

Example 2

How many license tags can be constructed if the tag numbers cannot begin with 0, if each tag number has 4 digits, and if the digits may be repeated? There are 9 ways in which the first digit can occur, there are 10 ways in which the second digit can occur, 10 ways in which the third digit can occur, and 10 ways in which the fourth digit can occur. Hence by the fundamental counting principle, there are

$$9 \cdot 10 \cdot 10 \cdot 10 \quad \text{or} \quad 9000$$

license tags. Reconsider this solution in the following way: Let A be the set of digits $\{1, 2, 3, \ldots, 9\}$ and let B be the set of digits $\{0, 1, 2, 3, \ldots, 9\}$; then a 4-digit number with repeated digits allowed and a nonzero first digit can be considered as an element of

$$A \times B \times B \times B,$$

and, by the formulas of Section 3.6,

$$\begin{aligned} n(A \times B \times B \times B) &= n(A) \cdot n(B) \cdot n(B) \cdot n(B) \\ &= 9 \cdot 10 \cdot 10 \cdot 10 \\ &= 9000. \ \blacksquare \end{aligned}$$

A very important special case of the fundamental counting principle is illustrated in the next two examples.

Example 3

In order to obtain a priority list for foreign aid, suppose that experts in the State Department are asked to rank 4 eligible nations. How many priority lists are possible? Think of a priority list as a set of boxes.

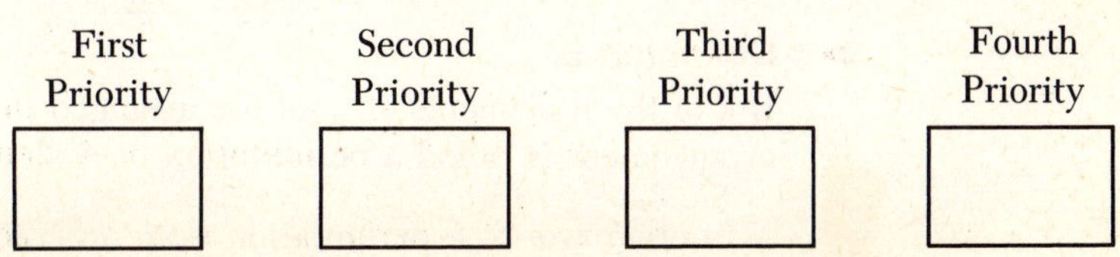

There are 4 nations that can be placed in the first box. After 1 nation has been placed there, 3 nations are left for placement in the second box. After the top 2 nations have been ranked, there are 2 nations remaining for placement in the third box. Finally, after the top 3 nations are chosen, there is no choice but to place the remaining nation in the fourth box. By the fundamental counting principle we have the product,

$$\boxed{4}\cdot\boxed{3}\cdot\boxed{2}\cdot\boxed{1} = 24.$$

A product of the form $n(n-1)(n-2)\cdots 3\cdot 2\cdot 1$ is ordinarily represented by the notation $n!$, which is read "n factorial"; hence the product $4\cdot 3\cdot 2\cdot 1$ could be represented by $4!$. As in Example 1, one may draw a tree diagram as shown in Figure 4.16 for this problem, and each of the 24 paths represents a priority list. In Figure 4.16, let a, b, c, and d represent the nations. ▮

Example 4

Suppose the last example is modified so that the experts are asked to rank only 3 nations from a set of 7 nations. Again think of the priority list as a set of boxes.

First Priority	Second Priority	Third Priority

There are 7 nations that can be placed in the first box. After 1 is chosen, there are 6 left that can be placed in the second box. Finally, after the first 2 have been chosen, there are 5 left that can be placed in the third box. By the fundamental counting principle we have the product

$$\boxed{7}\cdot\boxed{6}\cdot\boxed{5} = 210.$$

Such a product can be expressed using the factorial notation, namely,

$$7\cdot 6\cdot 5 = \frac{7\cdot 6\cdot 5\cdot 4\cdot 3\cdot 2\cdot 1}{4\cdot 3\cdot 2\cdot 1} = \frac{7!}{4!}. \ \blacksquare$$

Each priority list of Examples 3 and 4 is an example of what is commonly called a **permutation.** Each priority list is an *ordered* arrangement of a certain number of elements of a set. The ideas presented in Examples 3 and 4 are formalized in the following definition.

▶ **Definition 5**

If k of the n elements of a set are arranged in a specified order, then the arrangement is called a **permutation** of n elements taken k at a time.

In other words, a permutation is an ordered arrangement of elements;

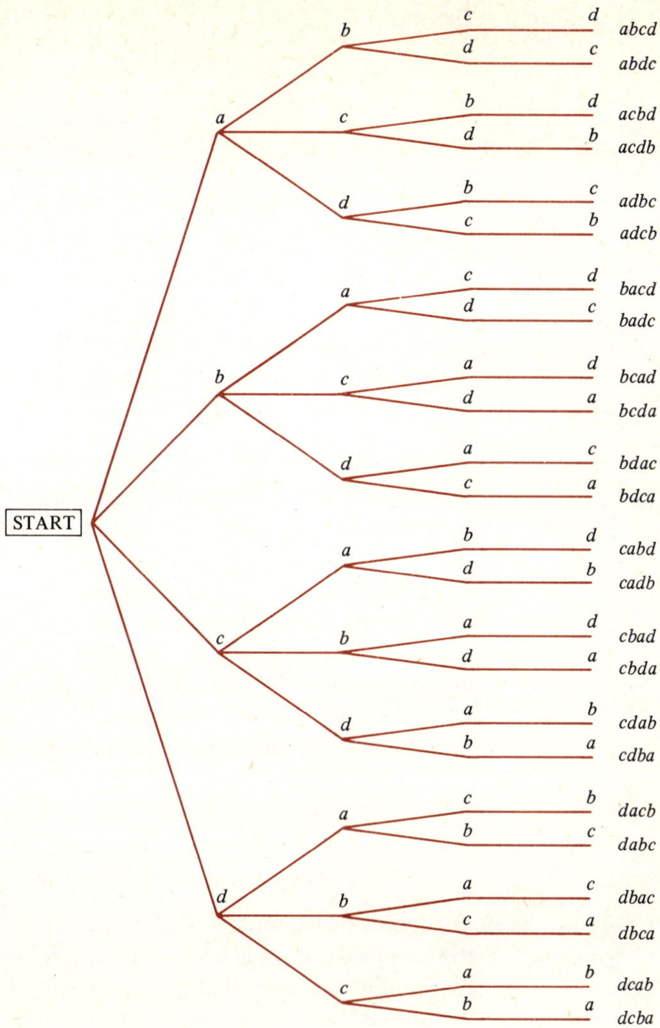

Figure 4.16. *Tree diagram of fundamental counting principle.*

if either the elements or the order of the elements is changed, a different permutation results.

The symbols

$$P_k^n, \quad {}_nP_k, \quad \text{and} \quad P(n, k)$$

are all used to designate the *number* of permutations of n elements or objects taken k at a time, and the formula for calculating this number is

$$P_k^n = \frac{n!}{(n-k)!}.$$

Verification of this formula involves a generalization of the reasoning illustrated in Examples 3 and 4.

Example 5

$$\begin{aligned} P_3^9 &= \frac{9!}{(9-3)!} \\ &= \frac{9!}{6!} \\ &= \frac{(9 \cdot 8 \cdot 7)\,6 \cdot 5 \cdot 4 \cdot 3 \cdot 2 \cdot 1}{6 \cdot 5 \cdot 4 \cdot 3 \cdot 2 \cdot 1} \\ &= 9 \cdot 8 \cdot 7 \\ &= 504. \ \blacksquare \end{aligned}$$

When using the formula for P_k^n, one will occasionally encounter the notation 0!, which is defined to be 1.

As one might expect, one of the common uses of the counting techniques presented in this section is in the calculation of probabilities, as we illustrate in the next two examples.

Example 6

In Example 4 what is the probability that nation a is first *or* that nation b is second on the priority list that is eventually chosen (assuming that all outcomes are equally likely)? Let E_1 be the event that nation a is first, and let E_2 be the event that nation b is second. We then seek $P(E_1 \cup E_2)$. Because the outcomes are assumed to be equally likely,

$$\begin{aligned} P(E_1 \cup E_2) &= P(E_1) + P(E_2) - P(E_1 \cap E_2) \\ &= \frac{n(E_1)}{n(U)} + \frac{n(E_2)}{n(U)} - \frac{n(E_1 \cap E_2)}{n(U)}. \end{aligned}$$

We found in Example 4 that $n(U) = P_3^7$. To find $n(E_1)$ we reason that if the specified nation is first, then the problem becomes a matter of finding the number of ways that the remaining 6 nations can be ordered in the 2 positions that are left on the list. Hence $n(E_1) = P_2^6$. Similarly, $n(E_2) = P_2^6$. If nation a is first *and* nation b is second, then there are 5 remaining nations to fill one position; thus $n(E_1 \cap E_2) = P_1^5$.

$$P(E_1 \cup E_2) = \frac{P_2^6}{P_3^7} + \frac{P_2^6}{P_3^7} - \frac{P_1^5}{P_3^7} = \frac{6!/4! + 6!/4! - 5!/4!}{7!/4!} = \frac{55}{210} = \frac{11}{42}. \ \blacksquare$$

Example 7

If 5 people sit down in a row of 5 chairs, what is the probability that a specified pair will be seated side by side? (Assume that the outcomes are equally likely.) The sample space U will consist of the set of permutations of the 5 people taken 5 at a time; hence $n(U) = P_5^5 = 5! = 120$. The event space E can be considered as the union of the mutually exclusive events $E_1, E_2, E_3, \ldots, E_8$, where each of the 8 events can be thought of as sets of elements illustrated in Figure 4.17. Because the events $E_1, E_2, \ldots, E_8$

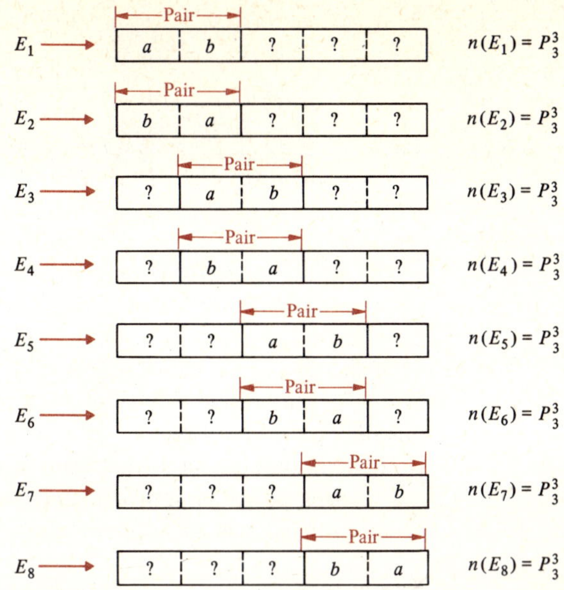

Figure 4.17. *Events in which two people are side by side.*

are mutually exclusive and because of the assumption of equally likely outcomes,

$$P(E) = P(E_1) + P(E_2) + \cdots + P(E_8)$$
$$= 8\frac{P_3^3}{P_5^5}$$
$$= 8\frac{3!}{5!} = \frac{2}{5}. \blacksquare$$

APPLICATIONS

Example 8

An interesting measure of voting power was developed in an article by Shapley and Shubik [40]. The basic idea is presented in the following illustration. Suppose there is a four-man committee that votes secretly. In case of a 2 : 2 tie the chairman has the power to break the tie. If we assume that the chairman votes the same way on both of his votes, then this method of voting amounts to 2 votes for the chairman and 1 vote for the other three members. We shall designate the members of the committee by a, b, c, d, where c is the chairman. The committee members are ordered according to their likelihood of voting yes on a certain motion. One such order might be ($c \quad d \quad b \quad a$); in such a permutation d is the crucial or *pivot* member because c has 2 votes, and hence d's vote could assure passage regardless of what b and a do. There are 4!, or 24, such permutations; they are listed next, with the pivot circled.

a b ⓒ *d*	*b a* ⓒ *d*	*c* ⓐ *b d*	*d a* ⓑ *c*
a b ⓓ *c*	*b a* ⓓ *c*	*c* ⓐ *d b*	*d a* ⓒ *b*
a ⓒ *b d*	*b* ⓒ *a d*	*c* ⓑ *a d*	*d b* ⓐ *c*
a ⓒ *d b*	*b* ⓒ *d a*	*c* ⓑ *d a*	*d b* ⓒ *a*
a d ⓑ *c*	*b d* ⓐ *c*	*c* ⓓ *a b*	*d* ⓒ *a b*
a d ⓒ *b*	*b d* ⓒ *a*	*c* ⓓ *b a*	*d* ⓒ *b a*

The voting power of a member is defined to be the ratio of the number of permutations in which the member is pivoted, to the total number of permutations.

Of the 24 permutations, c is pivotal in 12; therefore, the voting power of c is $\frac{1}{2}$. The voting power of each of the other three members is $\frac{1}{6}$. If this definition of voting power is applied to the Security Council of the United Nations, one finds that each of the Big Five with their veto power has a voting power of approximately .196, and each of the 10 other nations has a voting power of approximately .002. ▮

Example 9

This example offers a simple illustration of the fundamental counting principle. In International Morse code, dots and dashes are ordered in various ways to produce a code for the letters of the English alphabet. For example, the letter a is represented by the arrangement (·–), the letter b is represented by the arrangement (–···), and so on. Repetitions are allowed. This particular code is particularly adaptable to a telegraph key. Four characters or less will produce 30 arrangements of dots and dashes as shown in Table 4.4; hence the 26 letters of our alphabet can be coded by use of these 30 arrangements. The 4 leftover arrangements not used in the Morse code are (––––), (–––·), (·–––), and (·–·–). ▮

Table 4.4

No. of characters	No. of possible arrangements with repetitions allowed
1	2
2	4
3	8
4	16
Total	30

EXERCISES

Suggested minimum assignment: Exercises 1, 3, 5, 6, 7, 9, 11, and 14.

1. (*a*) Suppose that a certain student has 5 different reasons for failing calculus and 4 different reasons for failing English; using 1 reason for each failure, how many stories can the student give the Dean?

(b) Unknown to the student, the Dean is a reasonable man, and 3 of the reasons for failing calculus and 2 of the reasons for failing English would be acceptable to the Dean. Assume equally likely outcomes and calculate the probability that the student will come up with an acceptable story.

2. (a) If there are 10 different airlines serving route A, 8 different airlines serving route B, and 6 different airlines serving route C, how many different ways can one person make a trip covering all three routes?
   (b) If 6 of the airlines serving route A, 2 of those serving route B, and 4 of those serving route C use jets, find the probability that a person making a trip using all three routes will travel the entire way on jets. Assume equally likely outcomes in the choice of airlines.
3. Calculate P_2^4, P_5^7, P_3^{22}, P_8^8.
4. Calculate P_3^5, P_4^6, P_2^{19}, P_5^5.
5. In the game of Scrabble a player tries to make words from certain letters that are drawn. Suppose that a player draws the letters C, S, O, T, R, and E. In how many different ways may any 3 of the letters be arranged?
6. Suppose that a certain city decides to have automobile license tags with numbers that do not have repetitive digits. How many tags can be made if the tag number has exactly 4 digits? (Assume that the numbers beginning with 0 are permitted.)
7. In Exercise 6, if the city anticipates the need for 30,000 tags, how many digits must the numbers have in order to have enough tags to meet the demand? Assume that all tag numbers must have the same number of digits.
8. How many digits will be required in a telephone number if no digit is to be repeated, if the use of 0 is prohibited, and if 10,000 telephones are planned?
9. The ordinary telephone number has 4 digits plus a 3-digit exchange number. Digits can be repeated, but the number 0000 is prohibited. How many telephones can one exchange accommodate?
10. In relation to Exercise 9, the country is partitioned into areas and each area is assigned a 3-digit area code number. If the number 000 is prohibited from both the exchange number and the area code number, what is the maximum number of exchanges within an area, and what is the maximum number of areas?
11. (a) How many 4-letter "words" can be made from the letters *ABCDEF* if no letters are repeated?
    (b) In part (a), what is the probability that a "word" will not have a vowel in it?
    (c) In part (a), what is the probability that a "word" will have a vowel in it?
12. (a) How many 9-man batting orders can a coach obtain from 10 men?
    (b) What is the probability that a particular man will not be in the batting order?

(c) What is the probability that a particular man will be in the batting order?

13. What is the probability that a telephone number of 4 digits will have at least 1 repeated digit? Assume that the number 0000 is not permitted.
14. What is the probability that a license tag with 6 digits will not have any repeated digits? Assume that numbers beginning with 0 are not permitted.
15. In a 4-digit telephone number *without* repeated digits, what is the probability that a 3 will immediately follow a 2 in the number? Assume that numbers beginning with 0 are allowed.
16. In a 6-digit license number *without* repeated digits, what is the probability that all the digits 1, 2, and 3 will be in the number and adjacent to each other in some order? Assume that numbers beginning with 0 are allowed.

4.6 Counting Techniques: Combinations

In the last section we were concerned with counting the number of ways of selecting k elements from a set of n elements and arranging those k elements in some specified order; in this section we shall be concerned with only the selection of the k elements *without* regard to order. The next example will illustrate the method.

Example 1

In how many ways can a committee of 3 Senators be selected from a set of 4 Senators? The order in which the Senators are selected is immaterial as far as the makeup of the committee is concerned. To explain the method of calculating the number of unordered selections, however, we shall list all permutations of 4 Senators taken 3 at a time. We shall designate the Senators with the letters a, b, c, and d.

abc	*abd*	*acd*	*bcd*
acb	*adb*	*adc*	*bdc*
bac	*bad*	*cad*	*cbd*
bca	*bda*	*cda*	*cdb*
cab	*dab*	*dac*	*dbc*
cba	*dba*	*dca*	*dcb*

By the permutation formula there are P_3^4, or 24, ways of selecting the Senators in order. Notice, however, that each column of the list of permutations consists of 3! ways of arranging the same 3 Senators. Thus if we divide the number of ways (P_3^4) that 3 Senators can be selected in order by the number of ways (3!) that each threesome can be ordered, then we have simply unordered each threesome and consequently have the number of ways that 3 Senators can be selected from a set of 4 Senators without regard to order. In other words, we have the number of 3-man subsets of the original set of 4 Senators; there are 4 such subsets. ▮

Definition 6

A subset of k elements formed from a set of n elements is called a **combination** of n elements taken k at a time.

The symbols

$$C_k^n, \quad {}_nC_k, \quad C(n, k), \quad \text{and} \quad \binom{n}{k}$$

are all used to designate the *number* of combinations of n elements or objects taken k at a time, and the formula for calculating this number is

$$C_k^n = \frac{P_k^n}{k!} = \frac{n!}{k!(n-k)!}.$$

Verification of this formula can be obtained by generalizing the reasoning illustrated in Example 1.

Example 2

$$C_4^7 = \frac{P_4^7}{4!} = \frac{7!/3!}{4!} = \frac{7!}{4!3!} = \frac{7 \cdot 6 \cdot 5 \cdot 4 \cdot 3 \cdot 2 \cdot 1}{4 \cdot 3 \cdot 2 \cdot 1 \cdot 3 \cdot 2 \cdot 1} = 35. \quad ▮$$

As with permutations, one of the primary uses of combinations is their use in the calculation of probabilities.

Example 3

Ten men participate in the testing of a certain medicine. Five are given the medicine, and 5 are not given the medicine. What is the probability that a certain specified pair of men will both get the medicine? First, calculate the number of elements in the sample space U. The order of selection is not significant, hence

$$n(U) = C_5^{10} = \frac{P_5^{10}}{5!} = \frac{10!}{5!5!}.$$

An outcome of the event space E will consist of the 2 specified men plus 3 others selected from the remaining 8 men. Thus

$$n(E) = C_3^8 = \frac{P_3^8}{3!} = \frac{8!}{5!3!}.$$

Hence (assuming equally likely outcomes)

$$P(E) = \frac{n(E)}{n(U)} = \frac{8!/5!3!}{10!/5!5!} = \frac{8!5!5!}{10!5!3!} = \frac{2}{9}. \quad ▮$$

Example 4

Suppose in Example 3 that one seeks the probability of the 2 specified men *or* a specified third man getting the medicine. Let the event of Example 3 be designated E_1; let the event of the third man receiving the medicine be called E_2; and let the event $E_1 \cup E_2$ be designated F.

$$P(E_2) = \frac{n(E_2)}{n(U)} = \frac{C_4^9}{C_5^{10}} = \frac{9!/5!4!}{10!/5!5!} = \frac{1}{2}.$$

Now $P(F) = P(E_1 \cup E_2) = P(E_1) + P(E_2) - P(E_1 \cap E_2)$; hence we must find $P(E_1 \cap E_2)$. The event $E_1 \cap E_2$ consists of outcomes in which all 3 men get the medicine, and thus we are looking for the number of ways to fill the remaining 2 positions with the other 7 men. Hence $n(E_1 \cap E_2) = C_2$ and

$$P(E_2 \cap E_1) = \frac{n(E_2 \cap E_1)}{n(U)} = \frac{C_2^7}{C_5^{10}} = \frac{7!/5!2!}{10!/5!5!} = \frac{1}{12}.$$

Therefore,

$$P(F) = \frac{2}{9} + \frac{1}{2} - \frac{1}{12} = \frac{23}{36}. \blacksquare$$

Example 5

What is the probability that in a 5-card poker hand, all 5 cards will be diamonds? The sample space U will consist of the various combinations that can be selected 5 at a time from 52 cards, that is,

$$n(U) = C_5^{52} = \frac{52!}{5!47!}.$$

The event space will consist of the various combinations that can be selected 5 at a time from 13 diamonds, that is,

$$n(E) = C_5^{13} = \frac{13!}{5!8!}.$$

Therefore,

$$P(E) = \frac{n(E)}{n(U)} = \frac{13!/5!8!}{52!/5!47!} = \frac{13!}{8!} \cdot \frac{47!}{52!} \approx .000495. \blacksquare$$

A variation of Example 5 would be to determine the probability of getting 3 aces and 2 kings in a 5-card poker hand. Of course, $n(U)$ remains the same. The event space E can be thought of as the Cartesian product of two sets A and K shown in Figure 4.18.

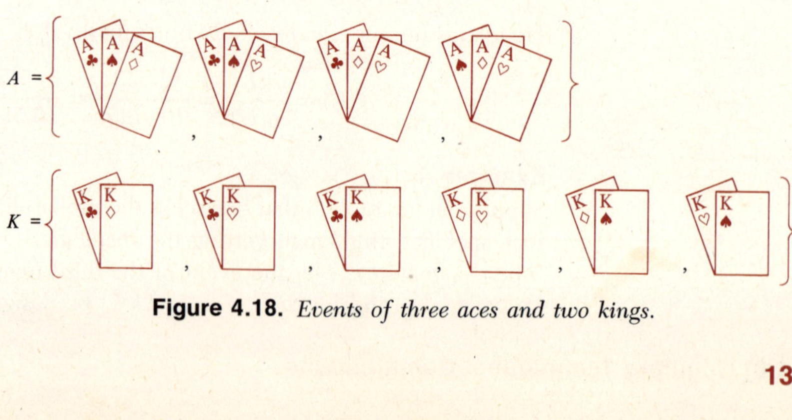

Figure 4.18. *Events of three aces and two kings.*

The number of elements of A is C_3^4 since we are selecting 3 aces from 4 in the deck, and $n(K) = C_2^4$ since we are selecting 2 kings from 4 in the deck. From Section 3.6 we know that

$$n(A \times K) = n(A) \cdot n(K),$$

hence

$$n(E) = n(A \times K) = 4 \cdot 6 = 24.$$

(The fact that $n(E)$ is the product of $n(A)$ and $n(K)$ can be viewed as a result of the fundamental counting principle.) Therefore,

$$P(E) = \frac{n(E)}{n(U)} = \frac{n(A \times K)}{n(U)} = \frac{24}{2{,}598{,}960} \approx .0000092.$$

Example 6

What is the probability that a committee of 3 Democrats and 2 Republicans will be selected from a group of 6 Democrats and 5 Republicans? (Assume equally likely outcomes.) Let R be the set of all possible pairs of Republicans and let D be the set of all possible triples of Democrats. Then

$$P(E) = \frac{n(E)}{n(U)} = \frac{n(R \times D)}{n(U)} = \frac{n(R) \cdot n(D)}{n(U)} = \frac{C_2^5 C_3^6}{C_5^{11}} \approx .43. \blacksquare$$

APPLICATIONS

Example 7

In one version of the game of poker, 5 cards are dealt to each player from an ordinary deck of 52 cards. Thus there is a sample space of C_5^{52} or 2,598,960 outcomes (5-card hands). The various hands that may be dealt are classified and ranked according to Table 4.5. After all bets are made, the winner of the game is the player who remains in the game and holds the highest ranking hand. The reader can observe from Table 4.5 that the classifications of hands are ranked on the basis of the relative size of their probabilities. Some of the methods of calculation given in Table 4.5 are not easy for the beginner; hence we give a discussion of the calculation of $n(E)$, where E represents the event of receiving 3 of a kind.

One way to think of the calculation of $n(E)$ is to use the fundamental counting principle:

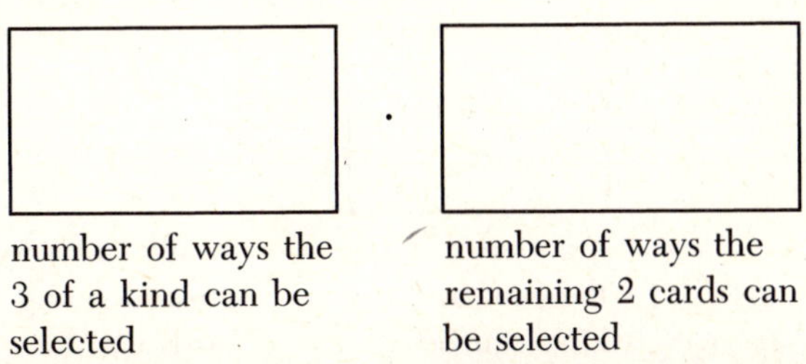

Table 4.5. Poker hand probabilities.

Classification of hand	Definition of the event space E	Method of calculation of the number of elements in E $n(E)$	$n(E)$	Probability of E $P(E)$†	Rank of hand
Royal flush	Ace, king, queen, jack, ten, in the same suit	C_1^4	4	.0000015	1
‡Straight flush	Any 5 cards in sequence in the same suit but not a royal flush	$(10 \cdot C_1^4) - 4$	36	.000014	2
Four of a kind	Any 4 cards all of the same kind (e.g., 4 kings)	$(C_1^{13}C_4^4)(C_1^{48})$	624	.00024	3
Full house	Any 3 cards of 1 kind and any 2 cards of another kind	$(C_1^{13}C_3^4)(C_1^{12}C_2^4)$	3,744	.0014	4
Flush	Any 5 cards of 1 suit but not a straight flush or a royal flush	$(C_1^4C_5^{13}) - 36 - 4$	5,108	.0020	5
‡Straight	Any 5 cards in a sequence but not all in the same suit (i.e., not a flush)	$(10[C_1^4]^5) - 36 - 4$	10,200	.0039	6
Three of a kind	Any 3 cards all of the same kind (e.g., 3 kings) but not a full house	$(C_1^{13}C_3^4)(C_2^{12}[C_1^4]^2)$ or $(C_1^{13}C_3^4)C_2^{48} - 3{,}744$	54,912	.0211	7
Two pairs	Any 2 cards of 1 kind and 2 more of another kind (e.g., 2 kings and 2 sixes); fifth card of a third kind	$(C_2^{13}[C_2^4]^2)(C_1^{11}C_1^4)$	123,552	.0475	8
One pair	Any 2 cards of 1 kind and each of the other cards of different kinds	$(C_1^{13}C_2^4)(C_3^{12}[C_1^4]^3)$	1,098,240	.4226	9

† $n(U) = C_5^{52} = 2{,}598{,}960$.

‡ In these calculations, an ace was considered to be high or low.

To find the entry of the first box we must select 1 kind from among 13 kinds in an ordinary deck (there are C_1^{13} such possible selections), and then we must select 3 of the 4 cards within that kind (there are C_3^4 ways of making this selection). Thus we have

$C_1^{13}C_3^4$	$\cdot$	
number of ways the 3 of a kind can be selected		number of ways the remaining 2 cards can be selected

To find the entry of the second box, we must select 2 different† kinds from the remaining 12 kinds (there are C_2^{12} such possible selections), and then we must select 1 of the 4 cards within each of the 2 kinds (there are $C_1^4 \cdot C_1^4$ ways of doing this). Thus the entry $C_2^{12}C_1^4C_1^4$ should go in the second box.

An alternative way of calculating $n(E)$ for 3 of a kind is to fill the first box as in the first method, but in filling the second box, simply select 2 cards from the remaining 48 (after the 4 cards of 1 kind have been eliminated from consideration); however, since this result, C_2^{48}, allows pairs of the same kind, and hence a full house, we must subtract the number of possible full houses from the product $(C_1^{13}C_3^4) \cdot (C_2^{48})$.

Because classifications are mutually exclusive, the probability of being dealt a ranked hand is the sum of the probabilities in the fifth column of Table 4.5; it is interesting to observe that there is a probability of approximately $\frac{1}{2}$ of receiving a ranked hand. Of course, in many versions of poker, cards may be replaced before the betting begins, and this will affect the probabilities discussed. ▮

Example 8

This example assumes an elementary knowledge of the game of bridge. In bridge, probabilities play a very important part in developing strategies for playing hands. In one of Goren's widely read books on bridge [17] there is a table of probabilities and a discussion of various uses of the table. One surprising result is that the probability of obtaining a 3–1 or 1–3 split when 4 cards of a certain suit are out is greater than the probability of an even 2–2 split. To illustrate the use of combinations in the calculation of Goren's table, we shall calculate the probability of a 2–2 split.

At the outset of play the contract winner cannot account for 26 of the 52 cards in the deck. Of these 26 cards, 4 belong to the suit in question, say hearts. Let E be the event that exactly 2 of the 4 hearts belong to the opponent on the left. We calculate $n(E)$ by multiplying the number of ways the left opponent can hold 2 of 4 hearts (that is, C_2^4) by the number of ways the left opponent can hold 11 of 22 nonhearts (that is, C_{11}^{22}). The total number of equally likely ways that 13 of the 26 cards can be held by the left opponent is C_{13}^{26}. Thus

$$P(E) = \frac{C_2^4 C_{11}^{22}}{C_{13}^{26}} \approx .407.$$

It is left to the reader as an exercise to show that the probability of a 3–1 or 1–3 split is approximately .497. ▮

EXERCISES

Suggested minimum assignment: Exercises 1, 3, 5, 8, 9, 11, and 14.

1. Calculate C_2^6, C_3^5, and C_{18}^{20}.
2. Calculate C_4^6, C_2^5, and C_2^{20}.

† They must be different to keep from having a full house.

3. How many ways can a team of 3 astronauts be selected from a group of 20 astronauts?
4. How many ways can a basketball team of 5 players be selected from a squad of 12 men?
5. In how many ways can a 5-man appellate court render a majority decision in favor of a motion?
6. How many different sums of money can be raised from the sale of a majority of 5 hotels that a certain man owns? Assume that each hotel has a different selling price and that each sum is different from every other sum. A majority consists of 3 or 4 or all 5 hotels.
7. What is the probability of getting 5 red cards (diamonds or hearts) in a 5-card poker hand? Express your answer in a form that uses the notation C_k^n.
8. In the game of blackjack 2 cards are dealt to each player at the outset of the game. What is the probability that a certain player will get 2 aces on the deal?
9. Suppose there are 60 Democrats and 40 Republicans in the 100-member U.S. Senate. What is the probability of selecting 3 Democrats and 2 Republicans on a 5-member committee if the outcomes of the sample space are equally likely? Express answer using the notation C_k^n.
10. A publisher decides to publish 6 college textbooks next year in the subjects of English and history. If he has 7 manuscripts in English and 8 manuscripts in history, what is the probability that he will select 3 of each? Assume that the outcomes of the sample space are equally likely.
11. An employer has 4 female applicants and 6 male applicants for 5 vacant positions.
    (*a*) In how many ways can he hire 3 females and 2 males?
    (*b*) In how many ways can he hire 4 females and 1 male?
    (*c*) In how many ways can he hire 5 females?
    (*d*) In how many ways can he hire 5 people regardless of sex?
    (*e*) What is the probability that the employer will hire a majority of females? (Assume equally likely hiring practices.)
12. Suppose that a certain city council has 5 Republicans and 6 Democrats.
    (*a*) In how many ways can one form a committee consisting of 3 Democrats and 1 Republican?
    (*b*) In how many ways can one form a committee consisting of 4 Democrats and 0 Republicans?
    (*c*) In how many ways can a 4-person committee be formed?
    (*d*) What is the probability that a majority of the committee members will be Democrats? (Assume equally likely assignment practices.)
13. A certain national committee of 6 people is to be composed to perform a certain task. A list of 12 names has been sent to those who must make the selection. There are 4 women among the 11 Democrats on the list, and the only Republican is a male. If the outcomes are equally likely, what is the probability that the committee will include all 4

women *or* the Republican? Leave your answer in a form using the notation C_k^n. *Hint*: Use Theorem 3.

14. A sample of 4 parts is to be selected from among 100 that have been manufactured in a certain period of time. If there are 2 defective, unusable parts and 2 defective but usable parts among the 100 parts, what is the probability that the 2 defective, unusable parts *or* the 2 defective, usable parts will show up in the sample? *Hint:* Use Theorem 3.
15. Verify the probability of obtaining a royal flush in a 5-card poker hand as shown in Example 7; try to determine $n(E)$ before looking at the third column of Table 4.5.
16. Verify the probability of obtaining a flush in a 5-card poker hand as shown in Example 7. [*Hint*: Do not forget to subtract the number of royal flushes in calculating $n(E)$.] Try to determine $n(E)$ before looking at the third column of Table 4.5.
17. Verify the probability of obtaining 4 of a kind in a 5-card poker hand as shown in Example 7. Try to determine $n(E)$ before looking at the third column of Table 4.5.
18. Verify the probability of obtaining a full house (3 of 1 kind and 2 of another kind) in a 5-card poker hand as shown in Example 7. Try to determine $n(E)$ before looking at the third column of Table 4.5.

4.7 Counting Techniques: Partitions (Optional)

In order to unify and extend the ideas presented in the last two sections, we shall find it very useful to use the concept of splitting or partitioning a set into nonempty cells or subsets such that every element of the given set belongs to exactly one (no more, no less) of these cells or subsets. For example, an ordinary deck of cards can be partitioned into 4 cells called spades, hearts, diamonds, and clubs; notice that every card in the deck belongs to exactly one cell. Of course, the same set of cards can be partitioned in other ways such as aces, face cards, and cards numbered 2 through 10.

▶ Definition 7

The set of nonempty subsets $A_1, A_2, \ldots, A_n$ of a set U is a **partition** of U if

$$(i)\ A_1 \cup A_2 \cup \cdots \cup A_n = U$$

and

$$(ii)\ A_i \cap A_j = \varnothing \qquad \text{for } i \neq j.$$

Example 1

A set of four objects $\{a, b, c, d\}$ can be partitioned into three cells, containing 1, 1, and 2 objects, in the following ways:

$$\{\{a\},\ \{b\},\ \{c, d\}\},$$
$$\{\{a\},\ \{c\},\ \{b, d\}\},$$

$$\{\{a\},\ \{d\},\ \{b, c\}\},$$
$$\{\{b\},\ \{c\},\ \{a, d\}\},$$
$$\{\{b\},\ \{d\},\ \{a, c\}\},$$
$$\{\{c\},\ \{d\},\ \{a, b\}\}.$$

The definition of a partition makes no mention of the order of the cells in the partition, and hence the sets

$$\{\{a\}, \{b\}, \{c, d\}\} \qquad \text{and} \qquad \{\{b\}, \{a\}, \{c, d\}\}$$

represent the same partition; to make that distinction, we define an **ordered partition.** ▮

Definition 8

An **ordered partition** is a partition in which the order of the cells is specified.

Example 2

Consider a set of 5 children that is to be partitioned into 2 cells. The 3 members of one cell will receive an experimental vaccine, while the 2 members of the second cell will serve as a control group and do not receive the vaccine. We may picture each cell as a box, and for illustrative purposes we list the ordered partitions of this problem. Let v, w, x, y, and z represent the children.

	Medicine	No Medicine
Partition 1	$v\ w\ x$	$y\ z$
Partition 2	$v\ w\ y$	$x\ z$
Partition 3	$v\ w\ z$	$x\ y$
Partition 4	$v\ x\ y$	$w\ z$
Partition 5	$v\ x\ z$	$w\ y$
Partition 6	$v\ y\ z$	$w\ x$
Partition 7	$w\ x\ y$	$v\ z$
Partition 8	$w\ x\ z$	$v\ y$
Partition 9	$w\ y\ z$	$v\ x$
Partition 10	$x\ y\ z$	$v\ w$

Notice that while the order of the cells is specified, the elements within each cell are *not* ordered. We now develop a formula for finding that there are 10 ordered partitions. Remember that there are 5! ways of ordering the 5 children. In partition 1, there were 3! ways that v, w, and x could have been ordered in the first cell, and there were 2! ways that y and z could have been ordered in the second cell. Thus if we divide the total number of ordered arrangements, 5!, by the number of ways the elements can be ordered within each cell, 3! and 2!, we get

$$\frac{5!}{3!2!} \quad \text{or} \quad 10$$

as the number of ordered partitions, each cell of which has unordered elements. Notice that this formula is simply the formula for finding the number of combinations of 5 elements selected 3 at a time. This should not be surprising because we can think of those elements *selected* in a combination as constituting one cell and those *not selected* as constituting a second cell. ▮

In the next example we consider an ordered partition with three cells.

Example 3

Suppose that 8 people apply for 2 jobs. If 3 people are to be accepted for job A, 2 are to be accepted for job B, and 3 are to be turned away, how many ordered partitions can be so constructed? Think of a partition with 3 cells,

A	B	N

First, we reason that the 8 people can be arranged in order in 8! ways. Those 3 who fall in cell A can be arranged in 3! ways, those 2 in cell B can be arranged in 2! ways, and those 3 in cell N can be arranged in 3! ways. Hence, to unorder the elements within each cell, we divide the total ordered arrangements, 8!, by the number of ordered arrangements in each cell, namely, 3!, 2!, and 3!. We get

$$\frac{8!}{3!2!3!} \quad \text{or} \quad 560$$

ordered partitions. ▮

The ideas presented in the last two examples can be generalized to produce the formula of the next theorem.

Theorem 7

If a finite set U of n elements is partitioned into k cells such that n_1 is the number of elements in the first cell, n_2 is the number of elements in the second cell, . . . , n_k is the number of elements in the kth cell, so that

$$n = n_1 + n_2 + \cdots + n_k,$$

then the number of such ordered partitions is given by the formula

$$\frac{n!}{n_1!\, n_2! \cdots n_k!}. \blacktriangleleft$$

Example 4

Consider a set of 4 animals that have a certain disease. In testing medicines for this disease, 1 of the animals is to receive a certain chemical A, 1 more is to receive chemical B, and the remaining 2 are to receive no chemical. The set of ordered partitions of the set of animals $\{a, b, c, d\}$ can be seen in Figure 4.19. The primary problem that we are concerned with is to determine the number of ordered partitions of the set $\{a, b, c, d\}$ of 4 animals. Of course, one way is to list and count the different ordered partitions as shown in Figure 4.19. As one can see, even for this small problem, the task of listing the ordered partitions is not simple. To facilitate matters, we use the formula of Theorem 7.

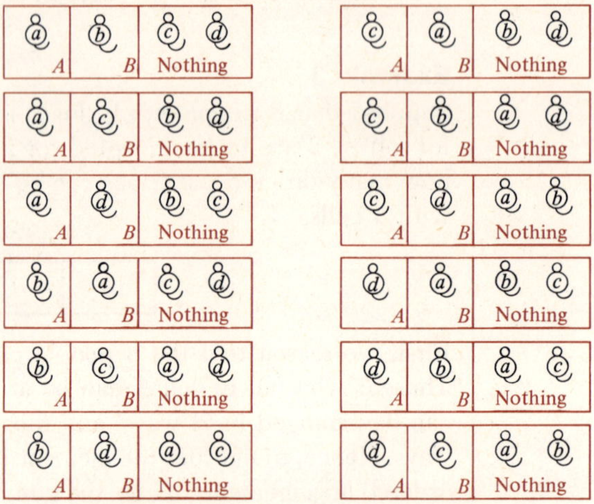

Figure 4.19. *Set of ordered partitions.*

First, however, in an ordered partition, we reemphasize that although the order of the cells is specified, the elements of the cells are not ordered. For example, in Figure 4.19 the interchange of the 2 animals in the last cell would not affect the ordered partition. In this problem $n = 4$, $n_1 = 1$, $n_2 = 1$, and $n_3 = 2$. Therefore, the number of ordered partitions is

$$\frac{4!}{(1!)(1!)(2!)} = \frac{(4 \cdot 3 \cdot 2 \cdot 1)}{(1) \cdot (1) \cdot (2 \cdot 1)} = 12. \blacksquare$$

Example 5

In Example 4 what is the probability that animal c, if selected at random, will receive medicine B? The event space is pictured in Figure 4.20. From Figure 4.20 we see that $n(E) = 3$, and in Example 4 we found the number of elements in the sample space to be 12; hence

$$P(E) = \frac{n(E)}{n(U)} = \frac{3}{12} = \frac{1}{4}.$$

It is not always feasible to list the elements of the event space in such a way that we can count the elements as we did in Figure 4.20. Therefore, we use Theorem 7 as an alternative approach in finding $n(E)$; if we specify that animal c is to receive medicine B, then it is a question of finding the number of ordered partitions of the remaining set of 3 elements into the two cells "A" and "NOTHING." Here we have $n = 3$, $n_1 = 1$, and $n_2 = 2$ and by the formula of Theorem 7

$$n(E) = \frac{3!}{1!2!} = 3. \blacksquare$$

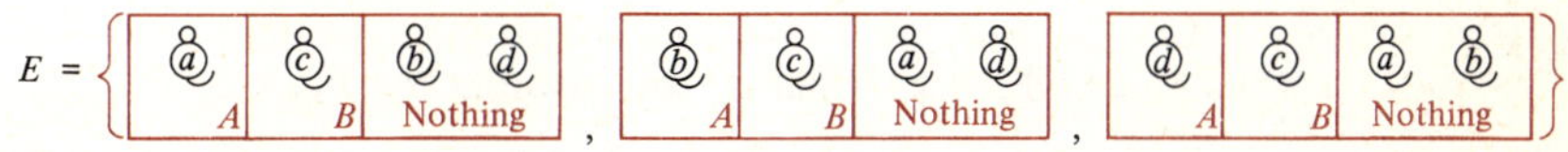

Figure 4.20

Example 6

A set of 6 workers is to be partitioned in such a way that 3 workers are assigned to task A, 2 workers are assigned to task B, and 1 worker is assigned to task C. Determine the total number of ways these assignments can be made. According to Theorem 7, $n = 6$, $n_1 = 3$, $n_2 = 2$, and $n_3 = 1$. Hence the number of ordered partitions subject to the stated requirements will be

$$\frac{6!}{3!2!1!} = \frac{6 \cdot 5 \cdot 4 \cdot 3 \cdot 2 \cdot 1}{(3 \cdot 2 \cdot 1)(2 \cdot 1)(1)} = 60. \blacksquare$$

Example 7

In Example 6 what is the probability that 2 specific workers will be assigned to task A? The event space is too large to list; hence we rely on the formula of Theorem 7. If 2 specific workers are assigned to task A, then we must simply calculate the number of ways the remaining 4 workers can be assigned in such a way that 1 worker is assigned as the third worker to task A, 2 workers are assigned to task B, and 1 worker is assigned to task C. Hence $n = 4$, $n_1 = 1$, $n_2 = 2$, $n_3 = 1$ and, by Theorem 7,

$$n(E) = \frac{4!}{1!2!1!} = \frac{4 \cdot 3 \cdot 2 \cdot 1}{(1)(2)(1)} = 12.$$

Therefore, since from Example 6, $n(U) = 60$, we have

$$P(E) = \frac{n(E)}{n(U)} = \frac{12}{60} = \frac{1}{5}. \blacksquare$$

We have observed that the formula for finding the number of combinations of n elements selected k at a time can be thought of as a special case of Theorem 7. It so happens that the formula for obtaining the number of permutations of n elements selected k at a time can also be thought of as a special case of Theorem 7; think of each of the k elements selected as k separate cells while the $n - k$ unselected elements fall into a single cell. Hence from Theorem 7 we have

$$\frac{n!}{1!1! \cdots 1!(n-k)!},$$

which is equal to the formula for P_k^n.

APPLICATIONS

Example 8

In many card games, j cards are dealt to each of k players; we are now in a position to resolve the question of the number of ways this can be done. Suppose that there are n different cards in the deck and after each player receives j cards there will be $(n - kj)$ cards left undealt. (In some games all the cards are dealt and $n - kj$ is 0.) We are in essence asking for the number of ordered partitions of n cards having $k + 1$ cells (k players plus the undealt cards). Using Theorem 7, we have the number of deals equal to

$$\frac{n!}{(j!)^k(n-kj)!}.$$

For example, in the game of bridge 13 cards, from a deck of 52 cards, are dealt to each of 4 players with none left over; hence the number of bridge deals is

$$\frac{52!}{(13!)^4 0!}.$$

In the game of poker, 5 cards from a deck of 52 cards are dealt to k people. Thus we partition the 52 cards into $k + 1$ cells with 5 cards in each of k cells and $(52 - 5k)$ cards in the remaining cell. The number of ways this can be accomplished is

$$\frac{52!}{(5!)^k(52-5k)!}. \blacksquare$$

EXERCISES

Suggested minimum assignment: Exercises 1, 3, 4, 6, 8, and 9.

1. List all of the ordered partitions of $\{a, b, c, d\}$, each partition having 2 cells with 2 elements in each cell.
2. List all of the ordered partitions of $\{a, b, c, d\}$, each partition having 3 cells with 1, 2, and 1 elements in the respective cells.
3. Suppose that there are 10 cases pending in the office of a district attorney. His personnel have time to bring 5 to trial immediately, defer 3 until next month, and defer 2 until the next court session. Find the number of ways that the 10 cases can be disposed.
4. In how many ways can an 11-game football season have a record of 7 wins, 3 losses, and 1 tie?
5. How many deals of 5-card poker hands can be dealt from an ordinary deck of 52 cards in a 4-player game? (Do not try to reduce your answer to a single real number.) *Hint*: Treat the undealt cards as a fifth hand.
6. How many ways can 9 animals be partitioned so that 3 of them are inoculated against disease A, 3 are inoculated against disease B, and 3 are selected as the control group?
7. In Exercise 5 what is the probability that
   (*a*) The first hand has 4 aces?
   (*b*) Any of the 4 players has 4 aces?
8. In Exercise 6 what is the probability of choosing a pair of the animals that have been inoculated against
   (*a*) Disease A?
   (*b*) Any single disease?
9. A corporation has a plan by which it sends 12 of its employees to 3 specific colleges. If colleges A, B, and C will be attended by 6, 4, and 2 employees, respectively, what is the probability that 2 specific friends among the 12 will both be assigned to college A? (Assume equally likely assignments.)
10. Four 3-man space flights are planned. Twelve of 14 astronauts are to be assigned to these flights. Two will not get chosen, and none will make more than 1 flight. What is the probability that 2 specific friends will be assigned to the first flight? (Assume equally likely assignments.)
11. (*a*) Find the number of ways that 4 bridge hands can be dealt.
    (*b*) Find the probability that the dealer of a bridge hand will receive all 4 aces.
    (*c*) Find the probability that any one of the 4 players of a bridge hand will receive all 4 aces.

4.8 Binomial Experiments

An important class of probability problems consists of those having n repeated trials of the same activity, such as successive tosses of a coin, successive casts of a die, repetitive mass production of some product, and so on. We define a **binomial experiment** to be an experiment consisting of n successive trials that satisfy the following three requirements: (1) the result of any trial is independent of the result of any other trial; (2) each trial has exactly 2 outcomes; and (3) the probability of either trial outcome will not change from one trial to another trial.

Example 1

Consider the experiment of flipping an ordinary coin 5 times. The trial outcomes are "heads" and "tails." The result of any trial is independent of the result of any other trial, and the probabilities of the 2 trial outcomes do not change for successive trials, namely,

$$P(\{\text{heads}\}) = \tfrac{1}{2} \text{ on every trial}$$

and

$$P(\{\text{tails}\}) = \tfrac{1}{2} \text{ on every trial;}$$

hence the experiment can be classified as a binomial experiment. ▌

The 2 outcomes of each trial are often viewed as a **success** or a **failure** to achieve a certain goal for each trial. The object then is to determine the probability of obtaining exactly k successes in n trials; as an illustration consider the following example.

Example 2

An assembly line production is sampled from time to time for purposes of quality control. If $\frac{9}{10}$ of the items are nondefective, what is the probability that a sample of 3 items will produce exactly 2 nondefective items? In a single trial consisting of drawing 1 item from the assembly line, there is a $\frac{9}{10}$ probability p of success (that is, drawing a nondefective part), and there is a $\frac{1}{10}$ probability q of failure (that is, drawing a defective part). The 3 successive draws can be illustrated on the tree diagram shown in Figure 4.21. Each path of the diagram represents an outcome of the total experiment and may be represented by a triple such as *ssf*; we notice that there are $C(3, 2)$ or 3 paths having exactly 2 successes. Because the experiment outcomes,

$$sss,\ ssf,\ sfs,\ sff,\ fss,\ fsf,\ ffs,\ fff,$$

are not necessarily equally likely, weights must be assigned to each. Consistent with Theorem 6, each weight is defined to be the product of the probabilities of the respective *trial* outcomes that make up each *experiment* outcome. (One can prove that each of these weights is nonnegative and that the sum of the weights of all the experiment outcomes is 1 as required in Definition 2.)

Thus if we let E be the event that exactly 2 successes are achieved, then from the tree diagram we see that

$$\begin{aligned} P(E) &= w(ssf) + w(sfs) + w(fss) \\ &= (\tfrac{9}{10}\cdot\tfrac{9}{10}\cdot\tfrac{1}{10}) + (\tfrac{9}{10}\cdot\tfrac{1}{10}\cdot\tfrac{9}{10}) + (\tfrac{1}{10}\cdot\tfrac{9}{10}\cdot\tfrac{9}{10}) \\ &= C(3,2)(\tfrac{9}{10})^2(\tfrac{1}{10}) = \tfrac{243}{1000} = .243. \ \blacksquare \end{aligned}$$

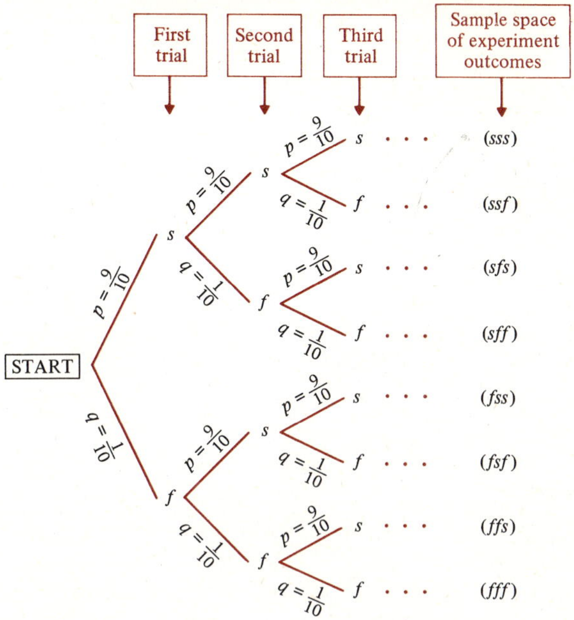

Figure 4.21. *Tree diagram of a binomial experiment.*

The method illustrated in Example 2 is generalized by the following theorem, which is stated without proof.

▶ **Theorem 8**

If p is the probability of achieving a specified outcome on each single trial of a binomial experiment, then the probability of event E of achieving the specified outcome exactly k times out of n trials is given by the formula

$$P(E) = C_k^n p^k (1-p)^{n-k}. \blacktriangleleft$$

For the case where $n = 4$, the validity of Theorem 8 can be demonstrated by a tree diagram as shown in Figure 4.22, where $1 - p = q$. Each distinct path from "start" to a terminal on the right of the tree represents an experiment outcome (we can count 2^4 experiment outcomes in the sample space). Consistent with Theorem 6, the weight of each outcome is defined to be the product of the probabilities of the respective trial outcomes that make up each experiment outcome; hence each experiment outcome has a weight, or a probability of occurring of $(p^k q^{n-k})$. Notice that there are C_k^4 ways in which an experiment outcome will contain exactly k specified trial outcomes; in other words, there are C_k^4 ways that one can succeed exactly k times in achieving a certain trial outcome in 4 trials.

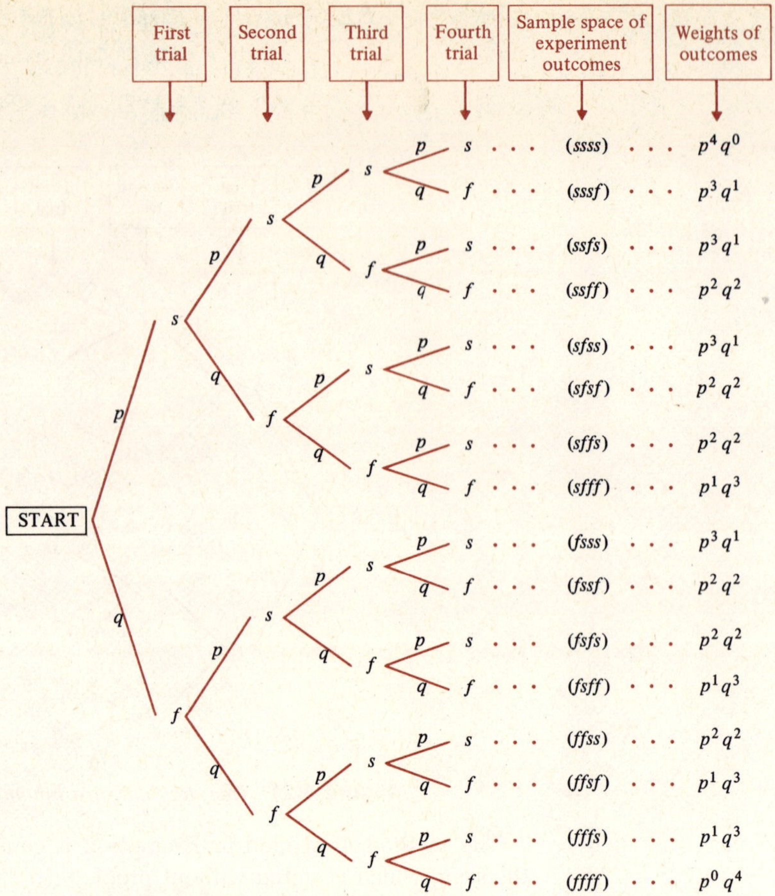

Figure 4.22. *Tree diagram of Theorem 8 for n = 4.*

Example 3

If 40 per cent of the population favors candidate A, what is the probability that 6 of 10 people polled will favor candidate A and thus give an erroneous impression of the preference of the population? By Theorem 8,

$$P(E) = C(10, 6)\left(\tfrac{4}{10}\right)^6\left(\tfrac{6}{10}\right)^4 \approx .11. \blacksquare$$

Example 4

In casting a fair die the probability of casting a 6 is $\frac{1}{6}$, while the probability of *not* casting a 6 is $1 - \frac{1}{6}$, or $\frac{5}{6}$. In 4 trials, the probability of event E_4 of achieving a 6 on all 4 trials is given by the formula of Theorem 8:

$$P(E_4) = C_4^4\left(\tfrac{1}{6}\right)^4\left(\tfrac{5}{6}\right)^0 = \tfrac{1}{1296}.$$

In 4 trials, the probability of event E_3 of achieving a 6 on exactly 3 trials is

$$P(E_3) = C_3^4\left(\tfrac{1}{6}\right)^3\left(\tfrac{5}{6}\right)^1 = \tfrac{20}{1296}.$$

In 4 trials, the probability of event E_2 of achieving a 6 on exactly 2 trials is

$$P(E_2) = C_2^4\left(\tfrac{1}{6}\right)^2\left(\tfrac{5}{6}\right)^2 = \tfrac{150}{1296}.$$

In 4 trials, the probability of event E_1 of achieving a 6 on exactly 1 trial is

$$P(E_1) = C_1^4\left(\tfrac{1}{6}\right)^1\left(\tfrac{5}{6}\right)^3 = \tfrac{500}{1296}.$$

In 4 trials, the probability of event E_0 of achieving a 6 on none of the trials is

$$P(E_0) = C_0^4\left(\tfrac{1}{6}\right)^0\left(\tfrac{5}{6}\right)^4 = \tfrac{625}{1296}. \blacksquare$$

The various events $E_0, \ldots, E_4$ listed in Example 4 are mutually exclusive and exhaust all possibilities (that is, they partition the sample space); therefore, the sum of these probabilities should equal 1. That is, in Example 4,

$$1 = P(E_4) + P(E_3) + P(E_2) + P(E_1) + P(E_0),$$

or

$$1 = C_4^4\left(\tfrac{1}{6}\right)^4\left(\tfrac{5}{6}\right)^0 + C_3^4\left(\tfrac{1}{6}\right)^3\left(\tfrac{5}{6}\right)^1 + C_2^4\left(\tfrac{1}{6}\right)^2\left(\tfrac{5}{6}\right)^2 + C_1^4\left(\tfrac{1}{6}\right)^1\left(\tfrac{5}{6}\right)^3 + C_0^4\left(\tfrac{1}{6}\right)^0\left(\tfrac{5}{6}\right)^4. \quad (4.2)$$

From elementary algebra recall the binomial expansion for $(p + q)^4$,

$$(p + q)^4 = p^4 + 4p^3q + 6p^2q^2 + 4pq^3 + q^4,$$

which is the same as

$$(p + q)^4 = C_4^4p^4q^0 + C_3^4p^3q^1 + C_2^4p^2q^2 + C_1^4p^1q^3 + C_0^4p^0q^4. \quad (4.3)$$

If we let p be the probability of success in achieving the trial outcome described in Example 4 and if q is $1 - p$ or the probability of failure in achieving that outcome, then we can see that the expression (4.2) can be thought of as a special case of the binomial expansion (4.3), where $p = \frac{1}{6}$ and $q = \frac{5}{6}$.

In fact, we can observe that, in general, each term of the binomial expansion of $(p + q)^n$ represents the probability of success in achieving a certain trial outcome a specific number of times in n trials, where the probability of success in a given trial is p and the probability of failure is q. The terms of the expansion represent the probabilities of mutually exclusive events, and hence the probabilities of various unions of these events can be found by simply adding the appropriate terms of the expansion.

Example 5

If a fair coin is flipped 10 times, what is the probability that a head will be flipped *at least* 8 times? By Theorem 8 the probability of getting exactly 8 heads is $C_8^{10}\left(\frac{1}{2}\right)^8\left(\frac{1}{2}\right)^2$, the probability of getting exactly 9 heads is $C_9^{10}\left(\frac{1}{2}\right)^9\left(\frac{1}{2}\right)^1$, and the probability of getting exactly 10 heads is $C_{10}^{10}\left(\frac{1}{2}\right)^{10}\left(\frac{1}{2}\right)^0$. Because these are probabilities of mutually exclusive events, the probability of the union of these events is the sum of the respective probabilities, or simply the sum of the last three terms of the binomial expansion of $(p + q)^{10}$, where $p = \frac{1}{2}$ and $q = \frac{1}{2}$.

$$P(E) = C_8^{10}(\tfrac{1}{2})^8(\tfrac{1}{2})^2 + C_9^{10}(\tfrac{1}{2})^9(\tfrac{1}{2})^1 + C_{10}^{10}(\tfrac{1}{2})^{10}(\tfrac{1}{2})^0 \approx .055. \blacksquare$$

APPLICATIONS

Example 6

An interesting physical illustration of the binomial expansion is the fountain shown in Figure 4.23. Water runs into the top basin at a rate of, say, 1 gallon per minute; the water overflows from each basin on two sides according to a fixed ratio for all basins. The ratio chosen for Figure 4.23 was 1:2, and 4 levels of basins are shown; this produces a physical analog of the binomial expansion for $(p + q)^n$, where $p = \frac{1}{3}$, $q = \frac{2}{3}$, and $n = 3$. ▮

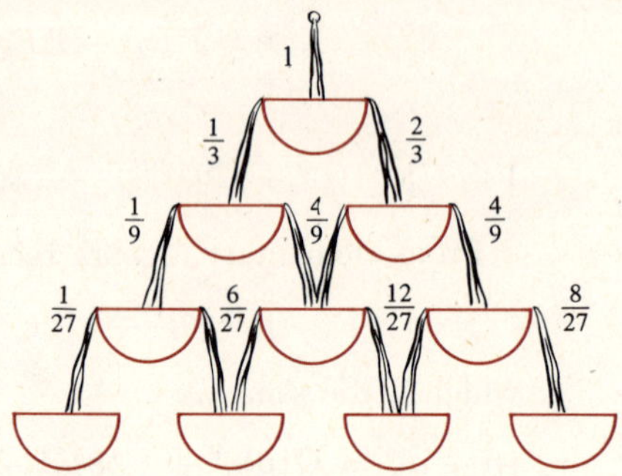

Figure 4.23. *Fountain analog of a binomial expansion.*

Example 7

Sampling is used in a variety of ways, such as political polls, opinion polls, market surveys, acceptance sampling, and so on. There are certain risks, however, in drawing conclusions based on samples; for example, suppose that a buyer makes a small sample of the items that he intends to purchase. One risk is the possibility that there *are* a very large number of defective items which will *not* appear in the sample in sufficient numbers to cause a rejection of the order. On the other hand, there may be very *few* defective items, and yet enough of them *will* appear in the sample to cause an unfortunate rejection of the entire lot.

As an illustration, suppose that a sample of 10 items is selected with the stipulation that the order will be accepted if no more than 2 members of the sample are defective. Let p be the unknown fraction of the lot that is actually defective, and let E be the event that 0, 1, or 2 defective items will appear in the sample. By the methods of this section,

$$\begin{aligned} P(E) &= C_0^{10}p^0(1 - p)^{10} + C_1^{10}p(1 - p)^9 + C_2^{10}p^2(1 - p)^8 \\ &= (1 - p)^{10} + 10p(1 - p)^9 + 45p^2(1 - p)^8. \end{aligned}$$

A graph relating the probability of acceptance $P(E)$ to the actual fraction p of defective parts is given in Figure 4.24.

Notice that there is a probability of approximately .26 that the lot will be accepted even though 35 per cent ($p = .35$) of the parts are actually defective. On the other hand, notice that when the actual percentage of defective parts is as low as 5 per cent ($p = .05$), there is a probability of approximately .01, $(1 - P(E))$, that the lot will be rejected. Of course, these risks can be reduced by taking larger samples, but the buyer may not be willing to pay that price. ▮

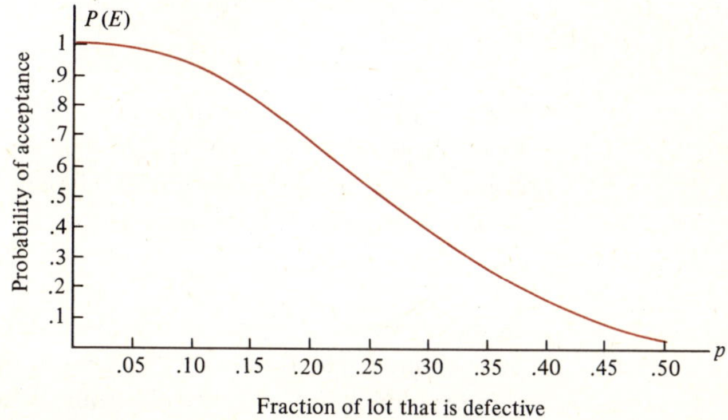

Figure 4.24. *Probability of acceptance based on sample size of 10 items.*

Example 8

Suppose that unknown to the investigator, a certain drug causes a very serious reaction in 5 per cent of the people who use the drug. What is the probability that the investigator will not detect this serious reaction in a sample of 10 people who are tested with the drug? If $p = .05$ is the probability that the reaction *will* appear in 1 person (1 trial), then $q = .95$ is the probability that the reaction will *not* appear in 1 person (1 trial). In $n = 10$ trials the probability of event E of no reactions is

$$\begin{aligned} P(E) &= C_0^{10} p^0 q^{10} \\ &= 1(.05)^0(.95)^{10} \\ &\approx .5987. \end{aligned}$$

If, however, a sample of 100 people were used (that is, $n = 100$ trials were used), the probability of event F of no reactions is

$$\begin{aligned} P(F) &= C_0^{100} p^0 q^{100} \\ &= 1(.05)^0(.95)^{100} \\ &\approx .0059. \end{aligned}$$

Thus it can be seen that an increase in the sample size from 10 to 100 can reduce the probability of nondetection of the reaction from about 60 per cent to about $\frac{3}{5}$ per cent. ▮

Example 9

Binomial experiments can be used to study the role of chance in bureaucratic decision making. For example, suppose a certain individual in a decision-making position makes correct decisions in at least 80 per cent of his chances; is this because of the individual's wisdom or is this a result of luck?

Under the assumption that each decision is independent of the others, that a distinction can be made between correct and incorrect decisions, and that the probability p of making a correct decision applies to all decisions and decision makers, then the probability $P(K)$ that an individual can make at least k correct decisions in n chances is

$$P(K) = C^n_k p^k (1 - p)^{n-k} + \cdots + C^n_n p^n (1 - p)^0.$$

If one considers m individuals as m components of a parallel system (see Example 6 of Section 4.4), then the probability $P(E)$ that at least one individual in an organization of m people will make at least k correct decisions is given by the formula

$$P(E) = 1 - (1 - P(K))^m.$$

In other words, $P(E)$ measures the reliability of the parallel system. (Of course, one could also consider series, or mixed, systems in connection with this problem.) We consider the role of luck in achieving an 80 per cent (or better) record of correct decisions out of 10 decisions by at least 1 of 30 people; suppose that every decision was based on the flip of a fair coin, which means that $p = \frac{1}{2}$. Let $k = 8$, $n = 10$, and $m = 30$. This information produces the result

$$P(E) = 1 - (1 - .0547)^{30} \approx .815.$$

Thus in this example, under the assumptions made, there is a probability of .815 that by simply flipping a coin for each decision, at least 1 of the 30 people would have 80 per cent, or better, correct decisions.

A study similar to this example may be found in Deutsch and Madow [12]. ▮

Example 10

In certain experiments in psychology there is a time interval, called the latency period, between a stimulus and corresponding response. Binomial experiments are used by McGill [33] to construct a model to study fluctuations in the latency period. Consider the binomial experiment shown in Figure 4.25 with outcomes, response r, and no response $\sim r$, having probabilities p and $1 - p$, respectively. Assume that the number of no responses occurring between 2 responses can be considered as a latency random variable, and suppose this variable is denoted x. From Figure 4.25 the probability of the event E that x is equal to k (where $k = 0, 1, 2, \cdots$) is given by the formula

$$P(E) = p(1 - p)^k. ▮$$

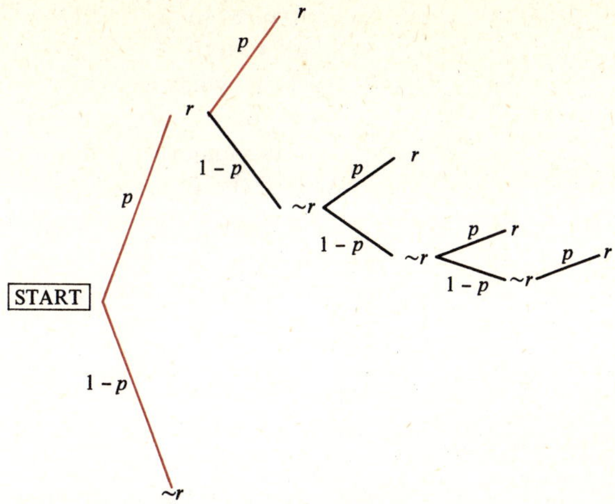

Figure 4.25

EXERCISES

Suggested minimum assignment: Exercises 1, 3, 4, 5, 7, 9, and 11.

In each of Exercises 1–4 draw a tree diagram representing the experiment, identify the sample space of experiment outcomes, state their respective weights, and calculate the requested probability from the diagram and also from the formula of Theorem 8.

1. If a fair coin is flipped 4 times, what is the probability of obtaining a head exactly 3 times? Are the experiment outcomes equally likely? Give reasons for your answer to the latter question.
2. If a die is cast 4 times, what is the probability of obtaining a number that is greater than 4 on exactly 1 throw?
3. Assume that the probability of a pair of animals having a male offspring is $\frac{4}{7}$ and the probability of having a female offspring is $\frac{3}{7}$. What is the probability of having 1 male and 3 females in 4 births?
4. Assume that the probability that a newborn member of a certain species will die before age 40 is $\frac{1}{3}$. What is the probability that exactly 2 of 4 randomly selected newborn members will live to be age 40?

5. In Exercise 3, what is the probability of obtaining at least 3 females in 4 births?
6. In Exercise 4, what is the probability that at least 2 newborn members will live to be age 40?
7. Assume that the reliability of a component of a certain system is $\frac{9}{10}$.
   (*a*) What is the probability that exactly 5 out of 6 of the same components are nondefective?
   (*b*) What is the probability that at least 5 out of 6 of the same components are nondefective?
8. (*a*) If 10 per cent of the parts produced by a certain factory are

defective, what is the probability that a sample of 10 items will produce exactly 1 defective part?

(*b*) In part (*a*), what is the probability that the sample will produce no more than 1 defective part?

9. Suppose that a majority of the people, say 60 per cent, favor a referendum. What is the probability that a poll of 5 people will indicate that the people oppose the referendum?

10. A multiple-choice test has four "answers" listed for each of 10 questions where only 1 answer is correct. What is the probability that a student who knows nothing about the material can guess at the answers and get

(*a*) 10 correct answers?

(*b*) 9 correct answers?

(*c*) At least 9 correct answers?

11. Suppose that a certain company is trying to determine whether or not to develop a certain piece of beachfront property. Certain predictable losses will occur if a hurricane strikes the property during the fall season. The company has calculated that during the first 3 years, the company can absorb the losses incurred by 1 hurricane but that 2 or more hurricanes would cause the venture to be unprofitable. If the probability of a hurricane hitting that property in 1 year is $\frac{1}{4}$, what is the probability of a profitable venture over the 3-year period?

If the restriction of 2 trial outcomes in Theorem 8 is relaxed to permit r trial outcomes with respective probabilities of $p_1, p_2, \ldots, p_r$, the formula of Theorem 8 can be generalized; the probability of event E of achieving exactly $n_1, n_2, \ldots, n_r$ of the respective trial outcomes is

$$P(E) = \frac{n!}{n_1!n_2! \cdots n_r!} p_1^{n_1} p_2^{n_2} \cdots p_r^{n_r},$$

where $n = n_1 + \cdots + n_r$. Apply this generalization to solve Exercises 12–14.

12. Assume that in a certain primitive country the probability of a live male offspring is 50 per cent, the probability of a live female offspring is 40 per cent, and the probability of a dead offspring is 10 per cent. What is the probability that in 6 births 3 will be male, 3 will be female, and 0 will be dead?

13. Suppose that in the manufacture of a certain product 10 per cent of the products will be defective and unusable, 20 per cent will be defective but usable, and 70 per cent will be nondefective. In a sample of 5 products selected at random, what is the probability that 3 products will be nondefective and 2 defective but usable?

14. Assume that the probability that a certain baseball player will get a hit is $\frac{3}{10}$, the probability that he will strike out is $\frac{2}{10}$, the probability that he will make a ground-ball out is $\frac{1}{10}$, and the probability that he will fly out is $\frac{4}{10}$. What is the probability that he will do all 4 in 4 at-bats?

4.9 Stochastic Experiments (Optional)

In this section we again consider a method of solving probability problems having a succession of n trials, but we remove the three restrictions of the last section. In other words, the trial results may not be independent of each other; each trial may have any finite number of outcomes, and the probabilities of these trial outcomes may vary from trial to trial. Such experiments, whose trial outcomes depend on some element of chance, are called **stochastic experiments.** The next two examples illustrate a method of solution of stochastic experiments.

Example 1

Consider 3 draws from an ordinary deck of cards. Each draw is made without replacement. What is the probability of the event E that a queen is drawn on exactly 2, including the first, of 3 draws? Let Q_1 be the event of drawing a queen on the first draw, let Q_2 be the event of drawing a queen on the second draw, let Q_3 be the event of drawing a queen on the third draw, and let N be the event of not drawing a queen on any 1 draw. The tree diagram of Figure 4.26 illustrates the problem. Each path of the diagram can be considered as an outcome of the total experiment. One should observe that these outcomes are not equally likely outcomes. The probability or weight of each of the experiment outcomes can be determined by taking the products of the probabilities along each path; the latter assignment of weights is consistent with a generalization of Theorem 6. The weights of the various outcomes are calculated and shown in Figure 4.27. In this example the event space of the total experiment is

$$E = \{(Q_1Q_2N), (Q_1NQ_3)\};$$

thus

$$\begin{aligned} P(E) &= w(Q_1Q_2N) + w(Q_1NQ_3) \\ &= \tfrac{4}{52} \cdot \tfrac{3}{51} \cdot \tfrac{48}{50} + \tfrac{4}{52} \cdot \tfrac{48}{51} \cdot \tfrac{3}{50} \\ &= \tfrac{1,152}{132,600}. \quad \blacksquare \end{aligned}$$

Example 2

A certain marketing report indicates that of the male shoppers, 10 per cent will purchase brand A, 30 per cent will purchase brand B, and 60 per cent will purchase brand C. The report also shows that of the female shoppers, 40 per cent will purchase brand A, 40 per cent will purchase brand B, and 20 per cent will purchase brand C. If 40 per cent of the shoppers are male and 60 per cent are female, what is the probability that an arbitrary shopper will purchase brand C? The tree diagram of Figure 4.28 illustrates the problem. Consistent with Theorem 6, the weight of each nonequally likely outcome of the event space is stated at the right of each path in the tree diagram. The event space of outcomes is

$$E = \{MC, FC\}.$$

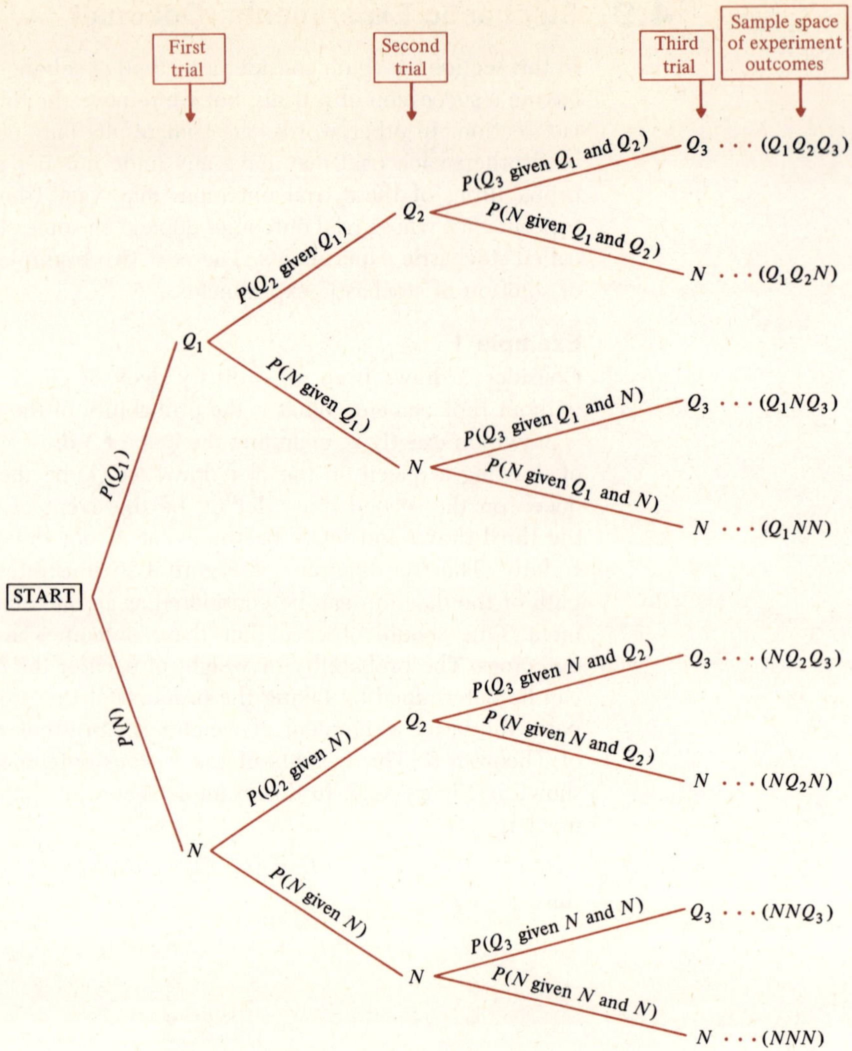

Figure 4.26. *Tree diagram of a stochastic experiment.*

Therefore,

$$\begin{aligned} P(E) &= w(MC) + w(FC) \\ &= .24 + .12 \\ &= .36. \ \blacksquare \end{aligned}$$

APPLICATIONS

Example 3

The purpose of this example is to illustrate the use of the methods of this section in analyzing a standard dice game. The player who is playing against the "house" wins if he rolls a 7 or 11 on the first throw of a standard pair

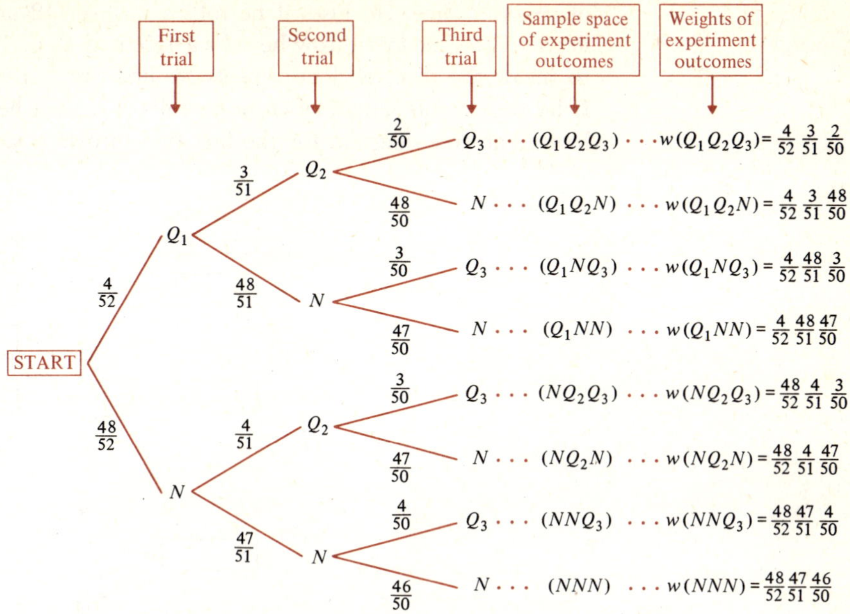

Figure 4.27. *Tree diagram of a stochastic experiment.*

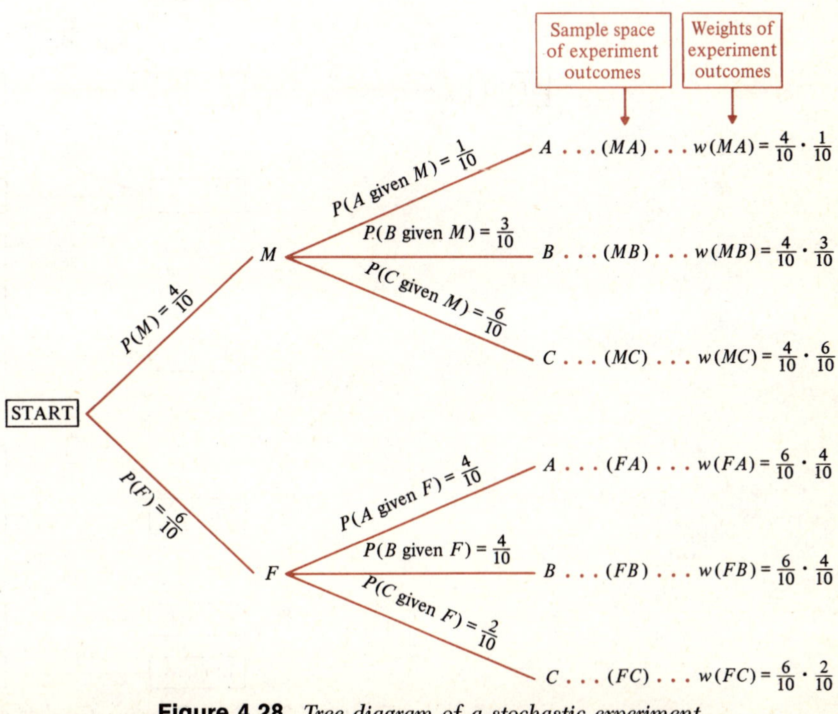

Figure 4.28. *Tree diagram of a stochastic experiment.*

of six-sided dice. He loses if he rolls a 2, 3, or 12 on the first throw. If, however, on the first throw he rolls a 4, 5, 6, 8, 9, or 10, then he continues to throw the dice until he repeats his first number or until he rolls a 7. If he repeats his number before he rolls a 7, then he wins; otherwise he loses. The tree diagram for the first two throws is shown in Figure 4.29.

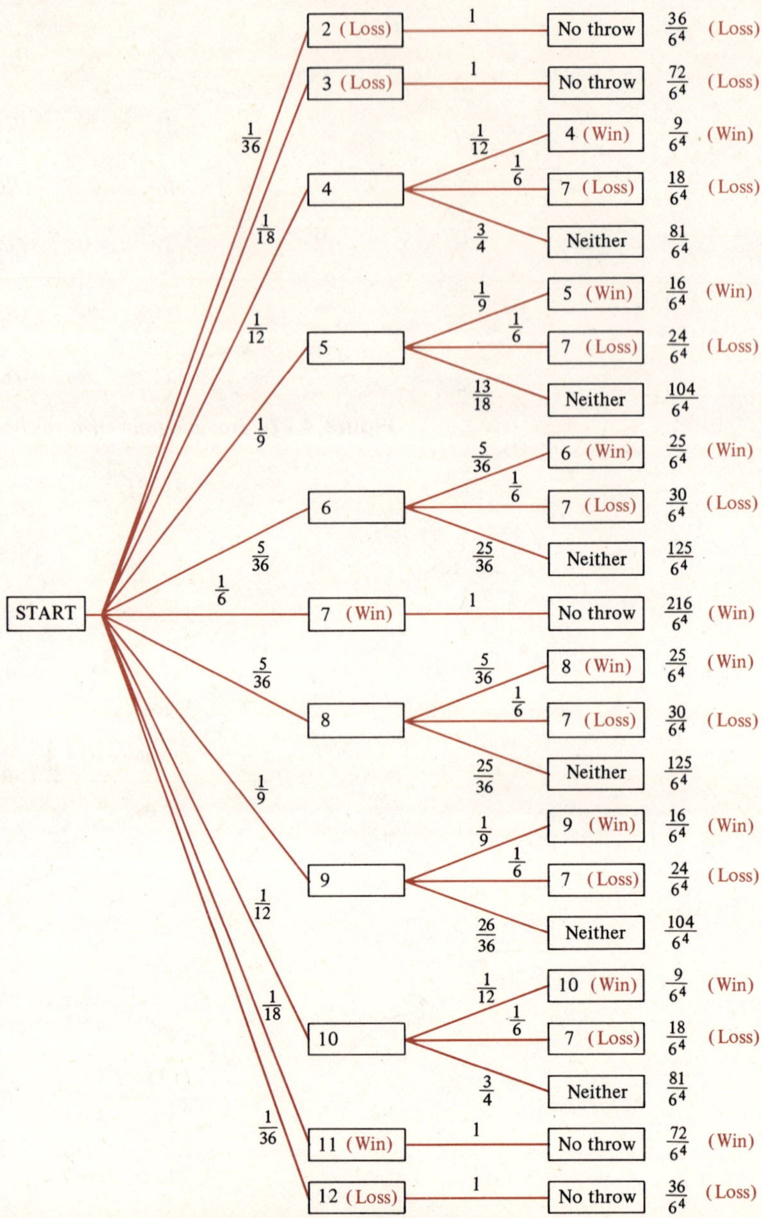

Figure 4.29. *Tree diagram of a dice game.*

If E is the event of winning in two throws or less, then

$$P(E) = \frac{9 + 16 + 25 + 216 + 25 + 16 + 9 + 72}{6^4}$$

$$= \tfrac{388}{1296} = \tfrac{97}{324} \approx .3.$$

Notice how this compares with the probability of winning after one throw, which is $\frac{1}{6} + \frac{1}{18} = \frac{2}{9} \approx .22$. The actual probability of winning, regardless of the number of throws, can be calculated to be $\frac{244}{495}$ if after the first throw one ignores those results which do not produce a win or loss. In other words, if the player throws a 4 on the first throw, then there are 3 ways that another 4 eventually can be thrown and 6 ways that a 7 eventually can be thrown; thus there is a $\dfrac{3}{3+6}$ probability that the player will eventually win and a $\dfrac{6}{3+6}$ probability of losing once he obtains a 4 on the first throw. A tree diagram similar to Figure 4.29 can then be drawn to analyze the problem and obtain the solution (Exercise 9). ▮

EXERCISES

Suggested minimum assignment: Exercises 1, 3, 4, 5, and 7.

In each of the following exercises draw a tree diagram and list the event space and the sample space of the total experiment.

1. On 3 unreplaced draws from an ordinary deck of cards, what is the probability of getting a diamond only on the second and third draws?
2. On 3 unreplaced draws from an ordinary deck of cards, what is the probability of getting a heart on the first draw and a spade on one or both of the last 2 draws?
3. A football coach estimates that the probabilities of winning the first 3 games of the season are $\frac{1}{2}$, $\frac{3}{4}$, and $\frac{1}{3}$, respectively. What is the probability that he will have 2 wins and 1 loss after the 3 games?
4. A political candidate for president estimates that the probabilities of winning states A, B, and C are $\frac{3}{4}$, $\frac{1}{4}$, and $\frac{1}{3}$, respectively. What is the probability that he will win exactly 2 of the 3 states?
5. Rework Exercise 3 if the coach states that winning the first 2 games will increase his chances in the third game to $\frac{1}{2}$.
6. Rework Exercise 4 if the candidate estimates that carrying state A will increase his chances in state B to $\frac{2}{3}$.
7. If in the U.S. Senate there are 50 Republicans, 45 Democrats, and 5 independents, and if 20 per cent of the Republicans, 60 per cent of the Democrats, and 40 per cent of the independents usually favor "welfare state" proposals, what is the probability that a Senator selected at random will support a "welfare state" proposal?
8. Suppose that a certain company has 3 suppliers of an item. The company buys 20 per cent of its items from company A, 30 per cent from B,

and 50 per cent from C. If 10 per cent of the items from A, 10 per cent from B, and 5 per cent from C are defective, what is the probability that an item selected at random from the imports will be defective?

9. In Example 3 verify that the probability of winning is $\frac{244}{495}$ if after one throw, one ignores those results that do not produce a win or loss.

4.10 Expected Value

Before we make a formal definition of **expected value,** we present an intuitive idea of the concept by means of several examples.

Example 1

One of the simplest gambling games is that of roulette. A wheel with 37 slots (some have 38) is spun, and a ball comes to rest in one of the slots numbered 0 through 36. The player may bet $1 on one or more numbers; if he wins on one number, he will be paid $35, and he will get his $1 stake back for the winning number.

If a player bet $1 on some number, then he has a $\frac{1}{37}$ probability of winning $35 and a $\frac{36}{37}$ probability of losing $1. If we designate the $1 loss by (−$1), we can calculate his expected winnings per game to be

$$\frac{1}{37}(\$35) + \frac{36}{37}(-\$1) = \$\frac{-1}{37}.$$

In other words, in the long run, each player can expect to *lose* at the rate of $\$\frac{1}{37}$ each time he bets $1. In *one* game, however, it is impossible to lose $\$\frac{1}{37}$; rather, the player will win $35 or lose $1. ▮

Example 2

A highway engineer knows from experience that his men and equipment can resurface a highway at the rate of 3 miles per day in rainy weather and 5 miles per day in nonrainy weather. The contractor has agreed to take on a large job in an area in which the Weather Bureau predicts a 20 per cent probability of rainfall for that time of year. Based on this information, the engineer can estimate the number of days that it will take him to do the job by calculating the expected mileage per day. There is a $\frac{1}{5}$ probability that in one day he will lay 3 miles of surface and a $\frac{4}{5}$ probability that in one day he will lay 5 miles of surface. The expected amount of surface to be laid in one day then is

$$\tfrac{1}{5}(3 \text{ miles}) + \tfrac{4}{5}(5 \text{ miles}) = \tfrac{23}{5} \text{ miles}.$$

It should be emphasized that $\frac{23}{5}$ miles per day is a long run expectation and not a literal expectation for one day. ▮

Example 3

If a single fair die is cast 3 times, what is the expected number of sixes? By methods previously studied in this chapter we can determine that there is a $(\frac{1}{6})^3$ or $\frac{1}{216}$ probability of casting 3 sixes, there is a $3(\frac{1}{6})^2(\frac{5}{6})$, or $\frac{15}{216}$, probability of casting 2 sixes, there is a $3(\frac{1}{6})(\frac{5}{6})^2$, or $\frac{75}{216}$, probability of casting 1 six, and $(\frac{5}{6})^3$, or $\frac{125}{216}$, probability of casting 0 sixes. Therefore, the expected number of sixes per experiment is

$$\frac{1}{216}(3) + \frac{15}{216}(2) + \frac{75}{216}(1) + \frac{125}{216}(0) = \frac{108}{216} = \frac{1}{2}.$$

In other words, in the long run, one could expect to get one six once every two times he played the game. Of course, in any one experiment he will not get $\frac{1}{2}$ sixes; he will get either 0 or 1 or 2 or 3 sixes. ∎

Betting \$1 on a number of the roulette wheel in Example 1 may be viewed as an experiment with two possible outcomes, namely, that of winning \$35 and losing \$1. These two outcomes had respective weights of $\frac{1}{37}$ and $\frac{36}{37}$. In Example 2 we could think of laying highway surface as an experiment with the two outcomes of 3 miles and 5 miles; the two outcomes had respective weights of $\frac{1}{5}$ and $\frac{4}{5}$. In Example 3 the casting of the die 3 times may be considered as an experiment with 4 outcomes of 3 sixes, 2 sixes, 1 six, and 0 sixes; the respective probabilities of achieving these outcomes were $\frac{1}{216}$, $\frac{15}{216}$, $\frac{75}{216}$, and $\frac{125}{216}$. Notice that each of the outcomes was represented by a real number; in Example 1 the real number measured monetary value, in Example 2 the real number measured the mileage of road surface laid per day, and in Example 3 the real number measured the number of sixes that appeared when a die was cast 3 times. In all cases we have multiplied the real number associated with each outcome by the probability of obtaining the outcome and then added the products. We now formalize this concept with the following definition.

▶ **Definition 9**

If the outcomes of an experiment are represented by the real numbers a_1, $a_2, \ldots, a_n$, and if the probabilities of obtaining the respective outcomes are $p_1, p_2, \ldots, p_n$, then the **expected value** V for the experiment is

$$V = p_1a_1 + p_2a_2 + \cdots + p_na_n.$$

Example 4

A hotel chain is considering the construction of a resort hotel at a certain beach. On clear sunny days in the summer season, the hotel can expect to fill its anticipated 200-room facility. On rainy days during the summer season, the experience of existing hotels shows that a 50 per cent capacity can be expected, while on threatening days a 70 per cent capacity can be expected. The Weather Bureau records indicate that during the summer season at that location, there is a 60 per cent probability of clear sunny weather, a 10 per cent probability of rainy days, and a 30 per cent probability of threatening days. Thus there is a $\frac{6}{10}$ probability of renting 200 rooms, there is a $\frac{1}{10}$ probability of renting 100 rooms, and there is a $\frac{3}{10}$

probability of renting 140 rooms. By Definition 9, the expected value for the experiment is

$$\begin{aligned} V &= \tfrac{6}{10}(200 \text{ rooms}) + \tfrac{1}{10}(100 \text{ rooms}) + \tfrac{3}{10}(140 \text{ rooms}) \\ &= 172 \text{ rooms.} \end{aligned}$$

As far as weather conditions are concerned, the hotel chain can expect in the long run to rent at the rate of 172 rooms per night during the summer season. ▮

Example 5

One of the most common ways of separating money from the uninformed (and in many cases the informed) is the "numbers game." There are many variations, but one version is that of choosing a three-digit number, which, with \$1 is delivered to the nearest "numbers runner." According to some previously announced method, a 3-digit number is selected by the operators of the game, and payments of some amount, say \$800, are awarded to the winners. From a player's point of view the outcomes of this experiment are (\$799) and (−\$1). By methods previously studied, the respective probabilities of these outcomes are $\frac{1}{1000}$ and $\frac{999}{1000}$ (assuming that numbers like 001 or 062 are 3-digit numbers). Therefore, the expected value for the experiment to a player is

$$V = \frac{1}{1000}(\$799) + \frac{999}{1000}(-\$1) = \$\frac{-1}{5}.$$

In other words, in the long run the player can expect to lose at the rate of \$0.20 per game. Of course, in any *one* game he will *not* lose exactly \$0.20. ▮

For an experiment with n equally likely real number outcomes $a_1, a_2, \ldots, a_n$, the expected value is

$$\begin{aligned} V &= \frac{1}{n}a_1 + \frac{1}{n}a_2 + \cdots + \frac{1}{n}a_n \\ &= \frac{a_1 + a_2 + \cdots + a_n}{n}, \end{aligned}$$

which is simply the *average* of the real numbers $a_1, a_2, \ldots, a_n$.

The reader has probably heard the expression "the odds in *favor* of a victory are 3 to 2." The statement simply means that the probability of winning is $\dfrac{3}{3+2}$ while the probability of not winning is $\dfrac{2}{3+2}$.

▶ **Definition 10**

If the probability of event E is $\dfrac{a}{a+b}$, then the **odds in favor of E** are a to b and the **odds against E** are b to a.

Example 6

In horse racing and dog racing one often hears the expression "the odds on a horse are 6 to 1." This means the odds are 6 to 1 against his winning; that is, there is a $\frac{6}{7}$ probability that he will *not* win. ▮

Example 7

Suppose that the final odds on a horse as determined by the bettors are 2 to 1 and that the race track operators are authorized by state rules to earn \$0.10 per \$2 bet. How much will they pay if the horse wins? (Show and Place are not considered here.) Let P represent the amount the operators will pay per \$2 ticket if the specified horse wins. The outcomes for a bettor are $-\$2$ and $+\$(P-2)$, with respective probabilities of $\frac{2}{3}$ and $\frac{1}{3}$; the expected value to the bettor is $-\$0.10$; hence we have

$$\frac{2}{3}(-2) + \frac{1}{3}(P-2) = -.10,$$

$$-\frac{4}{3} + \frac{P}{3} - \frac{2}{3} = -.10,$$

$$\frac{P}{3} = -\frac{1}{10} + \frac{6}{3},$$

$$P = -\frac{3}{10} + 6$$

$$= \$5.70. \ ▮$$

APPLICATIONS

Example 8

The concept of expected value can be very useful in certain types of decision making. For example, suppose that a group of actors is considering two different locations for their summer production. In one city there is an outdoor theater that will seat 500 people and an indoor theater that will serve 100 people. In the second city there is an outdoor theater that will serve 450 people and a 200-seat indoor theater. The probabilities of bad weather (sufficient to require an indoor performance) in the respective cities are 30 and 20 per cent. To which location should the drama group go for the summer if the expected audience size is the sole criteria for the decision? (Assume that the group will perform to a capacity audience each night.) A performance at each of the cities can be considered as an experiment. In the first experiment there is a $\frac{7}{10}$ probability of achieving an outcome of 500 people and a $\frac{3}{10}$ probability of achieving an outcome of 100 people; hence the expected value for the experiment is

$$V_1 = \tfrac{7}{10}\text{ (500 people)} + \tfrac{3}{10}\text{ (100 people)}$$

$$= 380 \text{ people.}$$

In the second experiment there is a $\frac{8}{10}$ probability of achieving an outcome of 450 people and a $\frac{2}{10}$ probability of achieving an outcome of 200 people; hence the expected value for the experiment is

$$V_2 = \tfrac{8}{10}\,(450 \text{ people}) + \tfrac{2}{10}\,(200 \text{ people})$$
$$= 400 \text{ people}.$$

Thus on the basis of the calculated expected values of audience size, the drama group should select the second city. ▮

Example 9

A fruit broker must decide how many trucks of watermelons to order for the summer season. Based on past seasons and other considerations, he estimates the probabilities of selling various numbers of truckloads as shown in Table 4.6. Each carload costs $3000 and he can make a profit of $4000

Table 4.6

No. truckloads	1	2	3	4
Probability of selling	$\frac{1}{6}$	$\frac{1}{4}$	$\frac{1}{3}$	$\frac{1}{4}$

on each carload that he can sell. Moreover, we assume that if potential sales exceed the amount he stocked, then the difference represents a loss for the broker. Because watermelons are perishable, he also loses whatever he overstocks. For the sake of simplicity we do not consider the sale of fractions of carloads. Table 4.7 provides the net profit for various numbers of carloads that he might buy and sell. (Each entry of the table is calculated by the formula [entry] = [sales profit] − [loss from overstocking or understocking].)

Table 4.7

Potential purchases (carloads)	Potential Sales (carloads)			
	1	2	3	4
1	4,000	0	−4,000	−8,000
2	1,000	8,000	4,000	0
3	−2,000	5,000	12,000	8,000
4	−5,000	2,000	9,000	16,000

From Tables 4.6 and 4.7 we calculate the expected value for each *purchase* possibility. We shall call the respective expected values E_1, E_2, E_3, and E_4.

$$E_1 = \tfrac{1}{6}(4{,}000) + \tfrac{1}{4}(0) + \tfrac{1}{3}(-4{,}000) + \tfrac{1}{4}(-8{,}000) = -2{,}667,$$
$$E_2 = \tfrac{1}{6}(1{,}000) + \tfrac{1}{4}(8{,}000) + \tfrac{1}{3}(4{,}000) + \tfrac{1}{4}(0) = 3{,}500,$$
$$E_3 = \tfrac{1}{6}(-2{,}000) + \tfrac{1}{4}(5{,}000) + \tfrac{1}{3}(12{,}000) + \tfrac{1}{4}(8{,}000) = 6{,}917,$$
$$E_4 = \tfrac{1}{6}(-5{,}000) + \tfrac{1}{4}(2{,}000) + \tfrac{1}{3}(9{,}000) + \tfrac{1}{4}(16{,}000) = 6{,}667.$$

Based on the assumptions made, the broker has a greatest expected value if he orders 3 truckloads of watermelons. If no information can be determined about the probabilities of sales, the problem can be approached through the use of game theory, which is to be studied in Chapter 7. ▌

Example 10

A very important application of the concept of expected value occurs in the study of game theory. Game theory is a relatively new branch of mathematics that permits the construction of models of adversary relationships; consequently, game theory has considerable potential in the social and managerial sciences and will be introduced in Chapter 7. ▌

EXERCISES

Suggested minimum assignment: Exercises 1, 3, 5, 7, 9, 10, 12, and 13.

1. In Example 1 what is the expected value of a player's making a $1 bet on a number of a roulette wheel with 38 slots numbered 00, 0, 1, 2, . . . , 36? The "house" will pay $35 for a winner plus the return of the $1 stake.
2. On a roulette wheel as described in Example 1 there are 18 black slots, 18 red slots, and 1 green slot (the number 0). A player may bet on the colors black or red rather than a number, and the "house" will pay $1 for a winner plus a return of the $1 stake. What is the expected value for the experiment of betting $1 on "red"?
3. A game is devised in which a single die is cast and a person agrees to pay in dollars the amount shown on the die if the number is odd and to collect the amount shown on the die if the number is even. What is the expected value of the game to the person?
4. A game is devised in which a single die is cast and a person agrees to collect $10 if a 1 or a 2 is cast and to pay the amount (in dollars) that is equal to the number shown on the die otherwise. What is the expected value of the game to the person?
5. In Exercise 3 suppose that the person has to pay $1 to play the game. What is the expected value of the game then?
6. In Exercise 4 suppose that the person has to pay $1 to play the game. What is the expected value of the game then?
7. A common money-raising device is that of the lottery. Suppose that 10,000 $1 tickets are sold in a lottery that promises a first prize of $1000 and 10 second prizes of $500 apiece. What is the expected value of 1 ticket? ~~What are the odds against winning a prize?~~
8. A merchant promises a free chance on an automobile to every person who purchases a special television model from him. If the television set cost the consumer $40 more than the same set sold by a competitor, if the automobile would cost the consumer $3000 (although it may not cost the merchant that much), and if the merchant sells 70 sets, what

is the expected value of a purchase to the consumer? What are the odds against winning?

9. A certain part takes 2 machine-hours to manufacture. All parts are tested and the defective parts are returned and repaired at an expense of 1 additional machine-hour per part. If there is a 20 per cent probability of manufacturing a defective part initially, what is the expected machine-hour expenditure per part?
10. A certain manufacturer produces a product that is graded according to the following categories: nondefective, defective but usable, and defective unusable; the respective values to the manufacturer are $30, $20, and $0 per item. What is the expected value of each item to the manufacturer if experience has shown that the probabilities of attaining items in these categories are $\frac{7}{10}$, $\frac{2}{10}$, and $\frac{1}{10}$, respectively?
11. Suppose that a person enters a game in which a pair of dice are thrown and a prize of $4 is paid if the numbers showing on the dice total 7. The game costs $1 to play. Find the odds against his winning and the expected value of the game for the player.
12. Three fair dimes are tossed simultaneously and a prize of $4 is paid if all three faces turn up to be the same; it costs $1 per game to play. Find the odds against winning and the expected value of the game.
13. On the day before election eve a candidate is faced with the decision of whether to go to city *A* or city *B* for an election-eve rally which is planned in each city. The candidate's staff has decided to base the decision on the expected value of the number of people the candidate will address. In city *A*, the rally has been scheduled for the 2000-seat city coliseum in case of rain and in the 5000-seat city stadium otherwise. In city *B*, the rally has been scheduled for a 1000-seat auditorium in case of rain and in the 10,000-seat baseball park otherwise. If there is a 20 per cent probability of rain in city *A* and a 40 per cent probability of rain in city *B*, and assuming that the candidate's appearance will guarantee a capacity crowd in any case, to which city should the candidate go?
14. Suppose that on the eve of a political rally the campaign manager of a certain candidate is faced with a decision as to whether to hold a rally for his candidate in the local outdoor stadium or the indoor auditorium. If he holds the rally in the outdoor stadium and it rains, he can expect a very small crowd of approximately 100 people, but if it does not rain he expects a crowd of 3000 people. If, however, he holds the rally indoors, he can expect the maximum capacity of 1000 people regardless of whether it rains or not. The latest weather report gives a 40 per cent probability for rain at the time of the rally; at which location should the manager schedule the rally if he bases his decision only on the expected value of the audience size?

NEW VOCABULARY

sample spaces 4.1
events or event spaces 4.1
equally likely outcomes 4.1
probability of an event 4.1
weights of outcomes 4.1
a priori probabilities 4.1
a posteriori probabilities 4.1
mutually exclusive events 4.2
conditional probability 4.3
tree diagram 4.3
path of a tree diagram 4.3
independent events 4.4
fundamental counting principle 4.5
permutation 4.5
combination 4.6
partition of a set 4.7
ordered partition of a set 4.7
binomial experiment 4.8
trial success 4.8
trial failure 4.8
stochastic experiment 4.9
expected value 4.10
odds in favor of E 4.10
odds against E 4.10

5 The Algebra of Matrices

In Chapters 1, 2, and 3, new types of useful mathematical structures were presented. The elements or building blocks of these structures were components, statement forms, and sets, respectively. In this chapter still another new structure will be presented; here the elements are simply rectangular arrays of numbers such as

$$\begin{bmatrix} 3 & 4 & 2 \\ \\ 1 & 6 & 5 \end{bmatrix} \quad \text{or} \quad \begin{bmatrix} 3 \\ \\ 4 \end{bmatrix}.$$

These rectangular arrays are called **matrices,** *and the new structure will be called the algebra of matrices.*

With the exception of the "Applications," the material in this chapter is not dependent upon the material of the previous chapters. The topics that are presented have been carefully selected for their current usefulness in the social, management, and biological sciences. For those readers who have access to a computer terminal that is capable of accepting the APL computer language, an appendix to this chapter is provided for instruction in the execution of matrix operations via the computer.

Prerequisites: high school algebra
Suggested Sections for Moderate Emphasis: 5.1–5.8
Suggested Sections for Minimum Emphasis: 5.1–5.3, 5.5

5.1 Matrices

In recent years there has been a dramatic increase in the use of rectangular arrays of numbers in applied problems. These arrays, such as

$$\begin{bmatrix} 2 & 3 & 4 \\ 6 & 9 & 2 \end{bmatrix},$$

are treated as single entities and are subject to various operations, some of which we shall define in this chapter. Evidence of the usefulness of arrays may be found by observing the very prominent position they occupy in some computer languages such as APL. In this section we shall define a matrix and a few operations involving matrices.

▶ **Definition 1**

Let m and n be positive integers; a rectangular array of real numbers arranged in m rows and n columns,

$$A = \begin{bmatrix} a_{11} & a_{12} & \cdots & a_{1n} \\ a_{21} & a_{22} & \cdots & a_{2n} \\ \vdots & \vdots & & \vdots \\ a_{m1} & a_{m2} & \cdots & a_{mn} \end{bmatrix}$$

is called an **m by n matrix.**

Example 1

The rectangular arrays

$$\begin{bmatrix} 2 & 4 \\ 3 & -2 \end{bmatrix}, \quad \begin{bmatrix} \sqrt{3} \\ 0 \\ 1 \end{bmatrix}, \quad \begin{bmatrix} 3 & \frac{2}{3} \\ 4 & 1 \\ 6 & 0 \end{bmatrix}, \quad [-2 \quad 0 \quad 1], \quad \begin{bmatrix} \sqrt{3} & 4 & 2 \\ 1 & \frac{7}{2} & 9 \\ 0 & 6 & 4 \end{bmatrix}$$

are all examples of matrices. ▮

The phrase "m by n" in Definition 1 is called the **order** of the matrix; for instance, in Example 1 the order of the first matrix is 2 by 2, while the order of the second matrix is 3 by 1. Notice that the number of rows is always listed first. The individual numbers in the array are called the **entries** of the matrix.

If the number of rows is the same as the number of columns of a matrix, then the matrix is said to be a **square matrix.** The **main diagonal** of a square matrix consists of those entries with the same row and column subscript; that is, $a_{11}, a_{22}, \ldots, a_{ii}$.

Example 2

The main diagonal entries of the following 3 by 3 matrix are enclosed in parentheses for emphasis.

$$A = \begin{bmatrix} (2) & 3 & 4 \\ 2 & (3) & 0 \\ 1 & 4 & (5) \end{bmatrix}. \quad ▮$$

In this book we shall consider matrices whose entries are real numbers, but the reader should be aware that studies of matrices can be made where the entries are other than real numbers. In this book when the word "matrix" is used, we shall mean a matrix with real number entries.

Rather than write out the array shown in Definition 1, it is customary to abbreviate the representation of a general m by n matrix by

$$A = [a_{ij}]_{(m,n)},$$

where it understood that the row subscript i assumes all integral values from 1 to m and the column subscript j assumes all integral values from 1 to n.

Example 3
Using abbreviated notation, we can represent the matrix

$$\begin{bmatrix} a_{11} & a_{12} & a_{13} & a_{14} \\ a_{21} & a_{22} & a_{23} & a_{24} \\ a_{31} & a_{32} & a_{33} & a_{34} \end{bmatrix} \quad \text{by} \quad [a_{ij}]_{(3,4)}. \blacksquare$$

We now use the abbreviated notation to define matrix equality, matrix addition, multiplication of a matrix by a real number, and matrix subtraction.

▶ **Definition 2**
Two matrices of the same order $A = [a_{ij}]_{(m,n)}$ and $B = [b_{ij}]_{(m,n)}$ are **equal** if $a_{ij} = b_{ij}$ for all i and j.

In other words, two matrices of the same order are equal when all their corresponding entries are equal. By the notation $A \neq B$ we mean that A is *not* equal to B.

Example 4

$$\begin{bmatrix} 2 & 3 & \frac{4}{2} \\ \sqrt{4} & 0 & -1 \end{bmatrix} = \begin{bmatrix} 2 & 3 & 2 \\ 2 & 0 & -1 \end{bmatrix},$$

$$\begin{bmatrix} 2 & 3 & 4 \\ 7 & 1 & 9 \end{bmatrix} \neq \begin{bmatrix} 2 & 3 & 4 \\ 7 & 1 & 4 \end{bmatrix}. \blacksquare$$

▶ **Definition 3**
The **sum** of two matrices of the same order $A = [a_{ij}]_{(m,n)}$ and $B = [b_{ij}]_{(m,n)}$ is defined to be

$$A + B = [a_{ij} + b_{ij}]_{(m,n)}.$$

In other words, two matrices of the same order can be added by adding corresponding entries. When two matrices are of the same order, they are said to be **conformable for addition.**

Example 5

$$\begin{bmatrix} 1 & 3 & -2 \\ \frac{2}{3} & -4 & 0 \end{bmatrix} + \begin{bmatrix} 2 & 0 & 4 \\ \frac{4}{3} & 1 & 6 \end{bmatrix} = \begin{bmatrix} (1+2) & (3+0) & (-2+4) \\ (\frac{2}{3}+\frac{4}{3}) & (-4+1) & (0+6) \end{bmatrix} = \begin{bmatrix} 3 & 3 & 2 \\ 2 & -3 & 6 \end{bmatrix}. \blacksquare$$

In Exercises 15 and 16 we will prove that matrix addition has the following properties.

▶ **Theorem 1**

(Commutative Property) If A and B are matrices of the same order, then

$$A + B = B + A. \blacktriangleleft$$

▶ **Theorem 2**

(Associative Property) If A, B, and C are matrices of the same order, then

$$(A + B) + C = A + (B + C). \blacktriangleleft$$

The reader will be asked to illustrate these properties in Exercises 13 and 14.

Now consider the sum $A + A + A + A$; the following definition is made in such a way that we can say that this sum is equal to $4A$.

▶ **Definition 4**

If $A = [a_{ij}]_{(m,n)}$ is a matrix and c is a real number, then

$$cA = [ca_{ij}]_{(m,n)} \qquad \text{and} \qquad Ac = [a_{ij}c]_{(m,n)}.$$

In other words, any matrix can be multiplied by a real number by multiplying each of the entries of the matrix by the real number.

Example 6

If $A = \begin{bmatrix} 2 & -3 & 4 \\ 6 & 9 & 0 \end{bmatrix}$ and $c = 2$, then

$$cA = (2)\begin{bmatrix} 2 & -3 & 4 \\ 6 & 9 & 0 \end{bmatrix} = \begin{bmatrix} 4 & -6 & 8 \\ 12 & 18 & 0 \end{bmatrix}. \blacksquare$$

Various properties of the operation defined in Definition 4 are given in Exercise 18. One such property is $cA = Ac$.

Now that the meaning of $(-1)A$ has been established by Definition 4, we can define **subtraction** of two matrices of the same order by

$$B - A = B + (-1)A.$$

APPLICATIONS

Example 7

Diagrams are often used to exhibit certain defined relations among persons, nations, animals, plants, genetic characteristics, or various other components of a system. As an illustration of one such relation, consider that of influence among 4 people. In Figure 5.1 let the circles represent the persons (components) involved and let an arrow from person i to person j indicate that person i has influence over person j; suppose that within the group of 4 people, influence can be exerted as shown in the figure. Figure 5.1 can be characterized by the matrix

$$A = \overset{\text{Influenced}}{\overbrace{\begin{matrix} 1 & 2 & 3 & 4 \end{matrix}}} \atop \begin{bmatrix} 0 & 1 & 0 & 0 \\ 1 & 0 & 0 & 1 \\ 0 & 0 & 0 & 1 \\ 1 & 1 & 0 & 0 \end{bmatrix} \left. \begin{matrix} 1 \\ 2 \\ 3 \\ 4 \end{matrix} \right\} \text{Influencer}$$

where $a_{ij} = 1$ if i influences j and $a_{ij} = 0$ if i does *not* influence j. We assume that no person influences himself, and hence the entries on the main diagonal are 0.

A very interesting use of this model was made in a study by Ullmer [43] of influence among the Justices of the Michigan Supreme Court. Similar models may apply to dominance of biological characteristics or sociological behavior, or to communication systems between nations or business associates. Of course, for large communication systems, diagrams like Figure 5.1 are very unwieldy, whereas the matrix representation of the system is a single entity that is subject to the various useful matrix operations that will be developed later. ∎

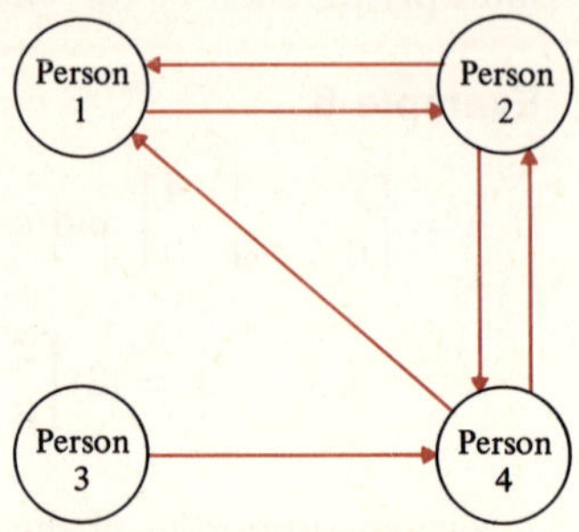

Figure 5.1. *Influence diagram.*

Example 8

The illustration that is begun in this example will be continued in later sections of this chapter as the reader attains more information about the algebra of matrices. The basic mathematical model that is developed has widespread use in a variety of applications. Suppose that at specified time intervals, a certain system can be in one of two states: it either operates satisfactorily or it does not operate satisfactorily (that is, has values of 1 or 0). Further suppose that this system has a self-correcting device that activates if needed at the specified time intervals. Experience has shown that when the system is operating satisfactorily at the beginning of one time interval, then 60 per cent of the time it will continue to operate satisfactorily and 40 per cent of the time it will not operate satisfactorily at the beginning of the next time interval. If, however, the system is not operating satisfactorily, then 20 per cent of the time it will remain in this state and 80 per cent of the time it will operate satisfactorily at the beginning of the next time interval. This information is given in Figure 5.2. As we shall see in future examples, it is advantageous to display these probabilities in matrix form.

$$
\begin{array}{c}
\overbrace{\begin{array}{cc}\text{satis-} & \text{unsatis-} \\ \text{factory} & \text{factory}\end{array}}^{\text{To}} \\
A = \left[\begin{array}{cc} .6 & .4 \\ .8 & .2 \end{array}\right]
\end{array}
\begin{array}{l} \\ \\ \left.\begin{array}{l}\text{satisfactory} \\ \text{unsatisfactory}\end{array}\right\}\text{From}\end{array}
$$

An n by n matrix whose ijth entry represents the probability that a system in state i will, after a specified time interval, be in state j, is called a **transition matrix** of the system. Other uses of the same basic model include (1) the transition matrix of consumer purchasing probabilities of brands on successive purchases, (2) the transition matrix of probabilities of a disease progressing from one state to another (for an illustration see Alling [2]), and (3) the transition matrix of probabilities of voting behavior. ▌

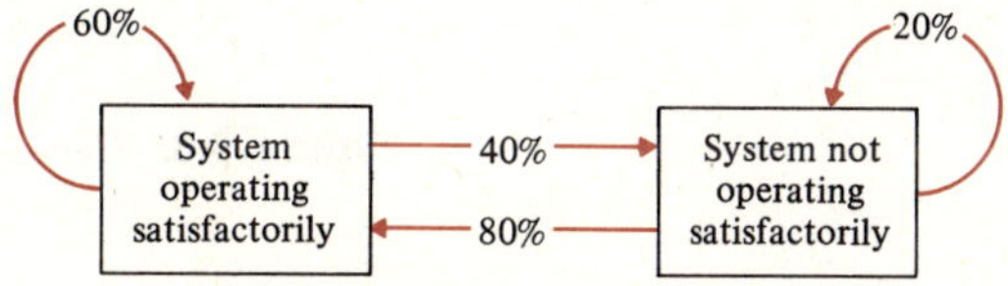

Figure 5.2. *Transition diagram.*

Example 9

This example is an illustration of the use of transition matrices in the field of genetics. For a particular gene locus with possible alleles A and a, assume that within a population, allele A is present with probability p and allele a is present with probability $1 - p$. Suppose that we consider a parent and his offspring; for each possible genotype of the parent, the probabilities of the genotype of the child are given in the tree diagrams of Figure 5.3 and also in the transition diagram of Figure 5.4. The same information can be displayed in the matrix

$$
\begin{array}{c}
\overbrace{\begin{array}{ccc} AA & Aa & aa \end{array}}^{\text{Genotype of child}} \\
A = \left[\begin{array}{ccc} p & 1-p & 0 \\ \frac{1}{2}p & \frac{1}{2} & \frac{1}{2}(1-p) \\ 0 & p & 1-p \end{array}\right]
\end{array}
\begin{array}{l} \\ \left.\begin{array}{l} AA \\ Aa \\ aa \end{array}\right\}\text{Genotype of parent}\end{array}
$$

One advantage of the matrix display over the displays of Figures 5.3 and 5.4 will be illustrated in Section 5.3 when the probabilities of the genotypes of later generations will be calculated by means of matrix multiplication. ▌

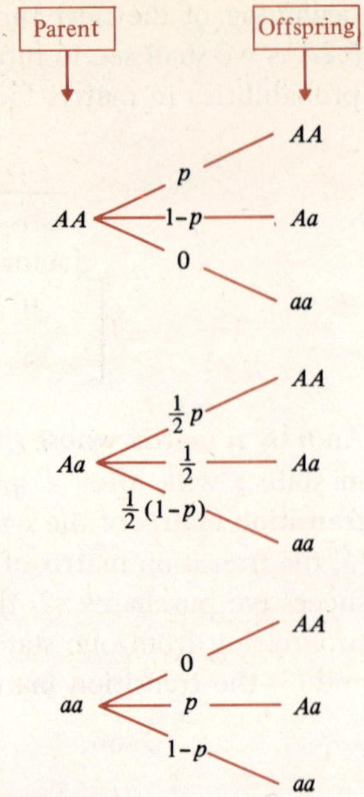

Figure 5.3. *Tree diagrams of transitions.*

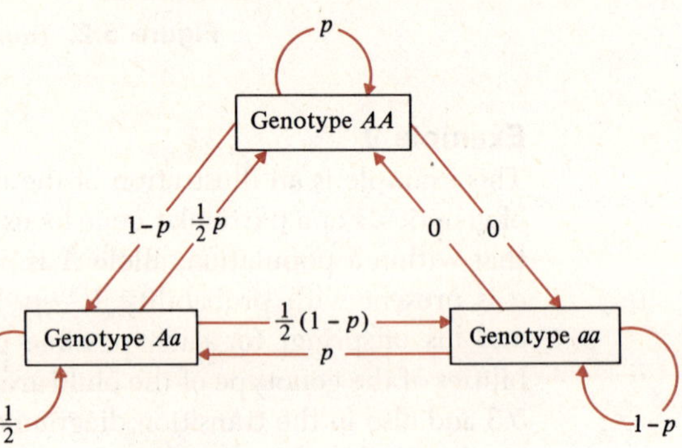

Figure 5.4. *Transition diagram.*

EXERCISES

Suggested minimum assignment: Exercises 1, 3, 5, 7, 9, 12, 14, and 18(a).

In Exercises 1–6 find the sum of each pair of matrices if possible.

1. $\begin{bmatrix} 1 & 2 \\ 4 & 0 \end{bmatrix}, \begin{bmatrix} 0 & 3 \\ 2 & 4 \end{bmatrix}.$

2. $\begin{bmatrix} 1 \\ 3 \end{bmatrix}, \begin{bmatrix} 2 \\ -4 \end{bmatrix}.$

3. $[2 \quad 3 \quad 4], \quad [7 \quad 9 \quad 1].$

4. $\begin{bmatrix} 4 & 3 \\ 2 & 1 \end{bmatrix}, \begin{bmatrix} 0 & 0 \\ 0 & 0 \end{bmatrix}.$

5. $\begin{bmatrix} 2 & 1 & 2 \\ 4 & 3 & 1 \end{bmatrix}, \begin{bmatrix} 2 & 4 \\ 2 & 1 \end{bmatrix}.$

6. $\begin{bmatrix} 2 & 1 \\ 3 & 4 \\ 6 & -2 \end{bmatrix}, \begin{bmatrix} 1 \\ 3 \\ 2 \end{bmatrix}.$

7. Let $A = \begin{bmatrix} 2 & 3 & -4 \\ 6 & 9 & 1 \end{bmatrix}$, $B = \begin{bmatrix} 0 & -2 & 4 \\ -1 & 3 & 1 \end{bmatrix}$, and $k = 4$.

(*a*) Find $A + kB$.
(*b*) What is the order of the result of part (*a*)?
(*c*) If $C = A - B$, find c_{12}.

8. Let $A = \begin{bmatrix} 2 & 8 & 1 \\ 1 & 0 & -2 \\ 2 & 1 & 0 \end{bmatrix}$, $B = \begin{bmatrix} 0 & -4 & 1 \\ 2 & 3 & 5 \\ 0 & 0 & 1 \end{bmatrix}$, and $k = 3$.

(*a*) Find $A + kB$.
(*b*) What is the order of the result of part (*a*)?
(*c*) If $C = A - B$, find c_{23}.

9. Express $A = \begin{bmatrix} a_{11} & a_{12} & a_{13} \\ a_{21} & a_{22} & a_{23} \end{bmatrix}$ in abbreviated notation.

10. Write the rectangular array for which

$$A = [a_{ij}]_{(4,2)}.$$

11. Two matrices can be added only when they have the same order; is there any restriction on when a matrix can be multiplied by a real number?

12. Are $A = [a_{ij}]_{(3,4)}$ and $B = [b_{ij}]_{(3,5)}$ conformable for addition? Give the reason for your answer.

13. Illustrate Theorem 1 with the matrices

$$A = \begin{bmatrix} 1 & 2 \\ 3 & 4 \end{bmatrix} \quad \text{and} \quad B = \begin{bmatrix} 0 & -1 \\ 4 & 2 \end{bmatrix}.$$

14. Illustrate Theorem 2 with the matrices

$$A = \begin{bmatrix} 2 & -4 \\ 6 & 9 \end{bmatrix}, \quad B = \begin{bmatrix} 2 & 4 \\ 0 & 3 \end{bmatrix}, \quad \text{and} \quad C = \begin{bmatrix} 2 & 0 \\ 7 & 1 \end{bmatrix}.$$

15. Prove Theorem 1. *Hint*: The statements are given, and the reader should complete the proof by supplying the reasons.

$$\begin{aligned} A + B &= [a_{ij}]_{(m,n)} + [b_{ij}]_{(m,n)} \\ &= [a_{ij} + b_{ij}]_{(m,n)} \\ &= [b_{ij} + a_{ij}]_{(m,n)} \\ &= [b_{ij}]_{(m,n)} + [a_{ij}]_{(m,n)} \\ &= B + A. \end{aligned}$$

16. Prove Theorem 2. *Hint*: See Exercise 15.
17. Prove: If A, B, and C have the same order, then

$$A + (B + C) = (B + A) + C.$$

18. For the operation defined in Definition 4, the following properties are valid. If A and B are matrices of the same order and if c and d are real numbers, then

$$cA = Ac,$$
$$(cd)A = c(dA),$$
$$c(A + B) = cA + cB,$$
$$(c + d)A = cA + dA.$$

(*a*) If $A = \begin{bmatrix} 2 & 1 \\ 4 & 0 \end{bmatrix}$, $B = \begin{bmatrix} 3 & 4 \\ 2 & 1 \end{bmatrix}$, $c = 2$, and $d = 3$, illustrate each property.
(*b*) Prove the first property. *Hint:* Let the first statement be $cA = c[a_{ij}]_{(m,n)}$.
(*c*) Prove the second property.
(*d*) Prove the third property.
(*e*) Prove the fourth property.

5.2 Vectors

There is often the need for expressing certain information by means of a list of numbers that are arranged in a specific order; for example, consider the alphabetical price list ($10, $20, $50, $25, $60) of 5 items in a store or the coordinates (3, 1, 4) of a point in 3-space. Such ordered lists of numbers are known as **vectors.** We shall limit our discussion of vectors to ordered lists of *real* numbers.

▶ Definition 5

An ordered list $(a_1, a_2, \ldots, a_n)$ of n real numbers is a **vector** of **dimension n.** A vector also may be displayed in the following two ways:

1. As a 1 by n matrix $[a_1 \quad a_2 \quad \cdots \quad a_n]$, which is called a **row vector,** and
2. As a n by 1 matrix $\begin{bmatrix} a_1 \\ a_2 \\ \vdots \\ a_n \end{bmatrix}$, which is called a **column vector.**

The numbers $a_1, a_2, \ldots, a_n$ are called the **components** of the vector $(a_1, a_2, \ldots, a_n)$. If all the components are 0, then the vector is called a **zero vector.** If at least one component is nonzero, then the vector is a **nonzero vector.** Ordinarily, we shall designate vectors by the Greek letters α, β, γ, and so on.

Example 1

The list of closing prices for the alphabetized stocks listed on the New York Stock Exchange is a vector. The dimension of the vector is the number of stocks listed. ▮

Example 2

The ordered list of coordinates (3, 6, 7) of a point in 3-space (see Figure 5.5) is a vector of dimension 3. The vector also can be displayed in either of the two forms

$$[3 \quad 6 \quad 7] \qquad \text{or} \qquad \begin{bmatrix} 3 \\ 6 \\ 7 \end{bmatrix}.$$

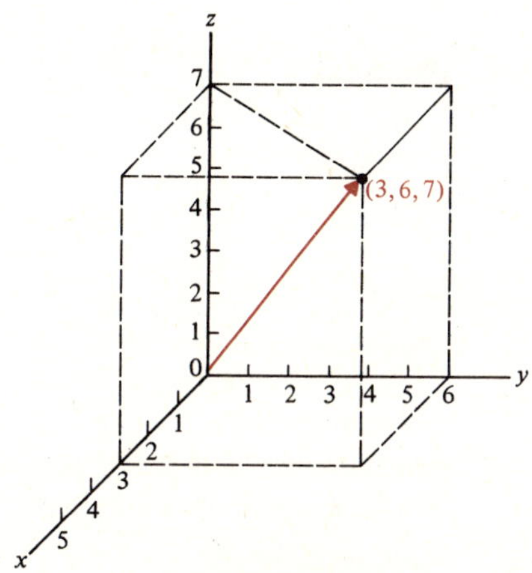

Figure 5.5

An arrow originating at the origin of coordinates and terminating at the point (3, 6, 7) (see Figure 5.5) may be used as a geometric representation of the row vector [3 6 7] or of the corresponding column vector. Such an interpretation is very useful because an arrow indicates both length and direction; as shown later in Example 12 of this section, these ideas lead to several applications. ▮

Example 3

Each column of a given m by n matrix is a column vector of dimension m, and each row of the given matrix is a row vector of dimension n. ▮

We now define equality of vectors and several operations.

▶ **Definition 6**

Two vectors of the same dimension are **equal** if their corresponding components are equal.

▶ **Definition 7**

The **sum** of two vectors of the same dimension, $\alpha = (a_1, a_2, \ldots, a_n)$ and $\beta = (b_1, b_2, \ldots, b_n)$, is defined by

$$\alpha + \beta = (a_1 + b_1, a_2 + b_2, \ldots, a_n + b_n).$$

In other words, two vectors can be added by adding corresponding components.

▶ **Definition 8**

If $\alpha = (a_1, a_2, \ldots, a_n)$ is a vector and c is a real number, then

$$c\alpha = c(a_1, a_2, \ldots, a_n) = (ca_1, ca_2, \ldots, ca_n).$$

Subtraction of vectors can be defined according to

$$\alpha - \beta = \alpha + (-1)\beta.$$

Example 4

Under what condition is $c\alpha - \beta = \gamma$ if $\alpha = (1, 2)$, $\beta = (4, 3)$, $c = 2$, and $\gamma = (x, y)$?

$$\begin{aligned} c\alpha - \beta &= 2(1, 2) + (-1)(4, 3) = (2, 4) + (-4, -3) \\ &= (-2, 1). \end{aligned}$$

Therefore, $\gamma = c\alpha - \beta$ if $x = -2$ and $y = 1$. ▮

At this point it is natural to wonder about the possibility of multiplying two vectors. Actually there are several types of products of vectors. The one that we shall define here is known as the **dot product.**

▶ **Definition 9**

If α and β are two vectors of the same dimension, where $\alpha = (a_1, a_2, \ldots, a_n)$ and $\beta = (b_1, b_2, \ldots, b_n)$, then the **dot product** $\alpha \cdot \beta$ is the real number determined by

$$\alpha \cdot \beta = a_1b_1 + a_2b_2 + \cdots + a_nb_n.$$

Example 5

Let $\alpha = (3, 2, 0)$ and $\beta = (2, 4, 6)$. Then according to Definition 9,

$$\alpha \cdot \beta = 3 \cdot 2 + 2 \cdot 4 + 0 \cdot 6 = 14. \text{ ▮}$$

Example 6

Let $\alpha = \begin{bmatrix} 2 \\ 6 \end{bmatrix}$ and $\beta = \begin{bmatrix} 3 \\ 4 \end{bmatrix}$. Then

$$\alpha \cdot \beta = 2 \cdot 3 + 6 \cdot 4 = 30. \text{ ▮}$$

Example 7

Suppose a stockbroker ordered shares of four different stocks in quantities revealed by the respective components of $\alpha = [100 \quad 200 \quad 200 \quad 300]$. The respective prices of these stocks are given by the components of

$$\beta = [\$60 \quad \$50 \quad \$20 \quad \$40].$$

The total value of the purchase is

$$\begin{aligned}\alpha \cdot \beta &= 100 \cdot 60 + 200 \cdot 50 + 200 \cdot 20 + 300 \cdot 40 \\ &= \$32{,}000. \blacksquare\end{aligned}$$

Example 8

A primary application of the dot product of vectors is its use in writing a linear expression. The dot product of $[a_1 \quad a_2 \quad \cdots \quad a_n]$ and $[x_1 \quad x_2 \quad \cdots \quad x_n]$ is the linear expression $a_1x_1 + a_2x_2 + \cdots + a_nx_n$. ▮

Definition 10

The **magnitude** (or **length**) of a vector α is denoted by $\|\alpha\|$ and is equal to the real number $\sqrt{\alpha \cdot \alpha}$. That is, if $\alpha = (a_1, a_2, \ldots, a_n)$, then

$$\|\alpha\| = \sqrt{a_1^2 + a_2^2 + \cdots + a_n^2}.$$

Example 9

Let $\alpha = (3, 6, 7)$. Then by Definition 10,

$$\begin{aligned}\|\alpha\| &= \sqrt{(3, 6, 7) \cdot (3, 6, 7)} \\ &= \sqrt{3^2 + 6^2 + 7^2} \\ &= \sqrt{94}.\end{aligned}$$

Geometrically $\|\alpha\|$ can be interpreted as the length of the arrow from the origin to the point (3, 6, 7) as shown in Figure 5.6. ▮

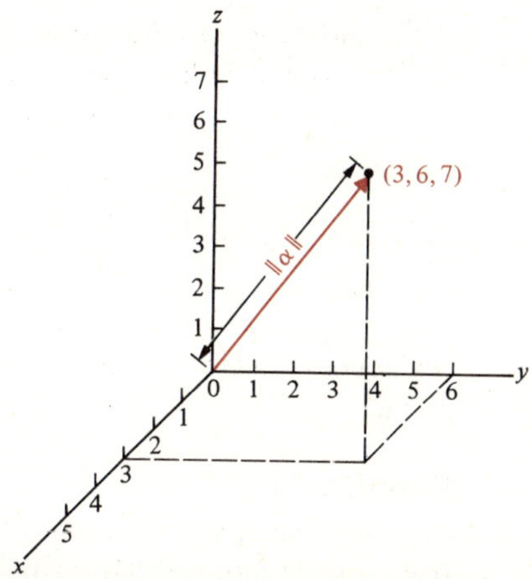

Figure 5.6. *Geometric interpretation of* $\|\alpha\|$.

APPLICATIONS

Example 10

Construct an oversimplified model economy in which there are 3 industries: the farming industry, the clothing industry, and the construction industry. There are 6 consumers: the general public, the government, those outside the economy who consume the exports, and the 3 industries themselves. These consumers exert certain demands on each of the industries. For example, suppose that in one unit of time the farming industry uses 3 units of its own production, 1 unit of production from the clothing industry, and 2 units of production from the construction industry. The demand vector then for the farming industry is expressed as $\alpha_f = (3, 1, 2)$. In the same manner the demand vectors for the industries are specified in the form

$$\alpha = (\text{farm products, clothing products, construction products}),$$

and suppose they are

farming industry	$\alpha_f = (3, 1, 2)$,
clothing industry	$\alpha_c = (1, 1, 2)$,
construction industry	$\alpha_b = (2, 2, 1)$,
general public	$\alpha_p = (2, 2, 2)$,
government	$\alpha_g = (1, 1, 1)$,
exports	$\alpha_e = (4, 1, 1)$.

The total demand on all industries then is

$$\alpha_{\text{total}} = \alpha_f + \alpha_c + \alpha_b + \alpha_p + \alpha_g + \alpha_e = (13, 8, 9).$$

If the price of one unit of farm production is \$3, one unit of the clothing industry is \$4, and one unit of the construction industry is \$5, then a price vector β can be written

$$\beta = (3, 4, 5).$$

If the industries produce exactly what is demanded of them, the income of the farming industry is 13 units times a price of \$3 per unit, or \$39. Its operating costs, however, are

$$\alpha_f \cdot \beta = (3, 1, 2) \cdot (3, 4, 5) = \$23.$$

Therefore, the profit of the farming industry is

$$\$39 - \$23 = \$16.$$

A calculation of the profits or losses of the other industries is left for Exercise 19. ▮

Example 11

Suppose that a sum of money is to be distributed by four arbitrators A, B, C, and D into 3 different projects. Suppose that the arbitrators' initial positions on the allocation of the items are represented by the vectors

$$\begin{aligned}\alpha &= (a_1, a_2, a_3),\\ \beta &= (b_1, b_2, b_3),\\ \gamma &= (c_1, c_2, c_3),\\ \delta &= (d_1, d_2, d_3).\end{aligned}$$

Let the actual compromise allocation be represented by the vector

$$\pi = (p_1, p_2, p_3).$$

The four vectors $\pi - \alpha$, $\pi - \beta$, $\pi - \gamma$, and $\pi - \delta$ represent both the magnitude and direction of change in the positions of each arbitrator. The magnitudes $\|\pi - \alpha\|$, $\|\pi - \beta\|$, $\|\pi - \gamma\|$, and $\|\pi - \delta\|$ may be used to rank the degree of position changes of the arbitrators, and, over a period of time, such information may be used to study such things as leadership, influence, and extreme positions. Of course, the basic mathematical model can be interpreted in many ways, such as 3 allocations of property by a committee of four realtors, or appropriations to 3 budget items by a governmental body of four people or four agencies, or the allocation of 3 chemicals for control of a disease by a group of four scientists. ▮

Example 12

The purpose of this example is to point out a few of the many ways that vectors can be applied in the life sciences. All these applications are summarized in greater detail in Batschelet [4] pp. 426–432, where further references may be found. Most of these applications result from the fact that a vector exhibits both magnitude and direction; hence concepts such as force and velocity can be quantified through the use of vectors. Certain information from elementary trigonometry and elementary vector analysis is assumed in the derivation of the equations of this example; an inability to follow these derivations should not preclude the reader from achieving the purpose of the example as stated in the first sentence.

1. *Animal navigation.* Birds and fish are particularly vulnerable to air and water currents, respectively. Consider a bird that attempts to fly in a certain direction with a wind blowing. Let β be the velocity of the bird with respect to the air, and let ω be the velocity of the wind with respect to the ground. If the direction and magnitudes are as indicated in Figure 5.7, then the actual velocity α of the bird with respect to the ground is given by $\alpha = \beta + \omega$. Because of the wind, the bird's body and efforts must be oriented in the direction of β to achieve the direction of α. The ground speed of the bird is $\|\alpha\|$.

2. *Limb traction.* If pulleys are arranged as shown in Figure 5.8, the stretching force (vector) σ is the sum of the forces (vectors) σ_1 and σ_2. If μ is a vector of length 1 parallel to σ, then the magnitude of the stretching force σ can be calculated to be $2(\sigma_1 \cdot \mu) = 2\|\sigma_1\| \cos\theta$, where $\|\sigma_1\|$ equals w. The stretching force can be increased by making the angle θ smaller.

3. *Limb action.* In Figure 5.9 one can see that the force (vector) α exerted by the muscle that helps lift the horizontal forearm must be as

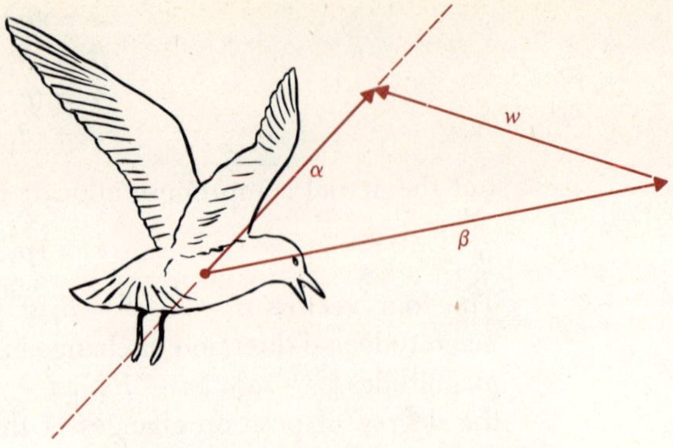

Figure 5.7. *Wind effect on bird flight.*

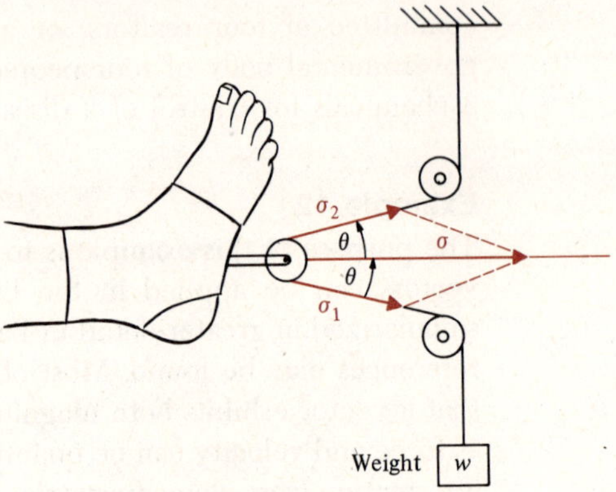

Figure 5.8. *Force of traction.*

large or larger than the lifting force β because $\alpha = \beta + \gamma$; how much larger depends on the angle of the arm at the elbow. The force γ is exerted on the elbow with corresponding stress on the bones and ligaments therein. Similar studies can be made to investigate forces causing broken bones and torn ligaments.

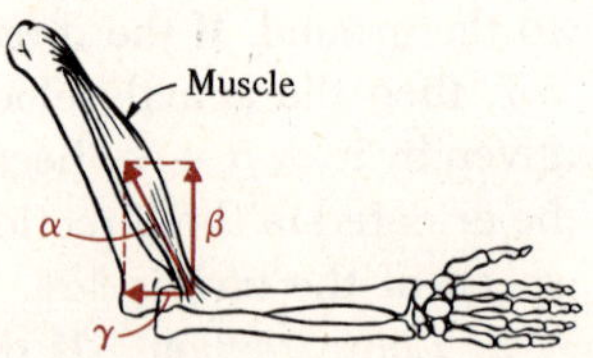

Figure 5.9. *Forces exerted upon flex of arm muscle.*

4. *Animal locomotion.* Consider a grasshopper whose center of gravity moves from point A to point B during the time his feet are on the ground. See Figure 5.10. Represent the directed line segment from A to B by the

vector β. Let the average force exerted by the grasshopper to get from A to B be designated by the vector α. The product of the vertical components of these two vectors is the amount of kinetic energy that is transformed into potential energy. Neglecting air resistance, we have

$$(\|\alpha\| \sin\theta)\ (\|\beta\| \sin\theta) = mgh,$$

where m is the mass of the grasshopper, g is the acceleration due to gravity, and h is the maximum height of the jump above level ground. The height of the jump then is given by

$$h = \frac{1}{mg} \|\alpha\|\ \|\beta\| \sin^2\theta. \ \blacksquare$$

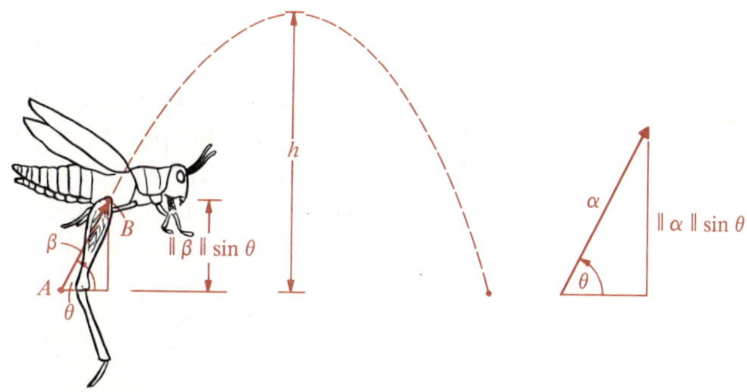

Figure 5.10. *Forces and motion involved in a grasshopper jump.*

EXERCISES

Suggested minimum assignment: Exercises 1, 3, 5, 7, 9, 12, 14, and 18.

In each of Exercises 1–8 let $\alpha = (1, 4, -2)$ and $\beta = (6, 0, 2)$.

1. Find $\alpha + \beta$.
2. Find 3α.
3. Find $3\alpha - \beta$.
4. Find $4\alpha - 2\beta$.
5. Find $\alpha \cdot \beta$.
6. Find $\beta \cdot 2\alpha$.
7. Find $\|\alpha\|$.
8. Find $\|\beta\|$.
9. If $\alpha = (2, 3, 5)$ represents the coordinates of a point in 3-space, graph α and interpret $\|\alpha\|$ geometrically.
10. If $\beta = (4, 3, 7)$ represents the coordinates of a point in 3-space, graph β and interpret $\|\beta\|$ geometrically.
11. Find the sum and the dot product of $\alpha = (6, 3, 4)$ and the zero vector of the same dimension.
12. If we define vectors $\alpha = (2, 1, 4)$ and $\gamma = (3, 2)$, find the sum and the dot product of α and γ if possible. If not possible, give reason.
13. If an arbitrary nonzero vector α is multiplied by the real number $1/\|\alpha\|$, then the resulting vector has a magnitude of 1 unit. Verify that this is the case for $\alpha = [1 \quad 3 \quad 5]$.

14. If

$$A = [2 \quad 3 \quad 4] \qquad \text{and} \qquad B = \begin{bmatrix} 1 \\ 4 \\ 2 \end{bmatrix},$$

find $A \cdot B$.

15. Prove that the addition of vectors of the same dimension is commutative.
16. Prove that the dot product of vectors of the same dimension is commutative.
17. Is subtraction of vectors of the same dimension commutative? Why?
18. In Example 7 of this section suppose that one week later the stockbroker sold all the same shares that he had purchased at prices given by the vector

$$[\$70 \quad \$40 \quad \$10 \quad \$50].$$

Use the dot product to calculate his profit or loss.

19. In Example 10 find the profits or losses of the other two industries.
20. In a situation similar to Example 11, suppose that 4 councilmen must allocate $2000 among two budgetary items. Suppose that the initial positions of the councilmen were

$$\alpha_1 = (\$1000, \$1000), \qquad \alpha_2 = (\$1500, \$500),$$
$$\alpha_3 = (\$1400, \$600), \qquad \alpha_4 = (\$900, \$1100).$$

The final allocation was ($1200, $800). Represent the change in position of each councilman by a vector. If the magnitudes of the vectors just calculated are used as the criteria, which councilman made the greatest change in position and which made the least change in position?

5.3 Matrix Multiplication

We now turn our attention to another matrix operation that has proved to be very useful in a variety of applications. The **product** of two matrices is simply the result of a succession of dot products of vectors.

▶ **Definition 11**

Let A be an m by p matrix, and let B be a p by n matrix. The **product** $C = AB$ is an m by n matrix, each entry of which is determined by the following dot product:

$$c_{ij} = (i\text{th row of } A) \cdot (j\text{th column of } B).$$

Example 1

Let

$$A = \begin{bmatrix} 2 & 1 \\ 3 & 4 \end{bmatrix} \qquad \text{and} \qquad B = \begin{bmatrix} 0 & 1 \\ 5 & 7 \end{bmatrix}.$$

Then the product AB can be found by taking four dot products as shown below:

$$1.\quad \begin{bmatrix} 2 & 1 \\ 3 & 4 \end{bmatrix}\begin{bmatrix} 0 & 1 \\ 5 & 7 \end{bmatrix} = \begin{bmatrix} (2\cdot 0 + 1\cdot 5) & (\qquad\qquad) \\ (\qquad\qquad) & (\qquad\qquad) \end{bmatrix}.$$

$$2.\quad \begin{bmatrix} 2 & 1 \\ 3 & 4 \end{bmatrix}\begin{bmatrix} 0 & 1 \\ 5 & 7 \end{bmatrix} = \begin{bmatrix} (\qquad\qquad) & (2\cdot 1 + 1\cdot 7) \\ (\qquad\qquad) & (\qquad\qquad) \end{bmatrix}.$$

$$3.\quad \begin{bmatrix} 2 & 1 \\ 3 & 4 \end{bmatrix}\begin{bmatrix} 0 & 1 \\ 5 & 7 \end{bmatrix} = \begin{bmatrix} (\qquad\qquad) & (\qquad\qquad) \\ (3\cdot 0 + 4\cdot 5) & (\qquad\qquad) \end{bmatrix}.$$

$$4.\quad \begin{bmatrix} 2 & 1 \\ 3 & 4 \end{bmatrix}\begin{bmatrix} 0 & 1 \\ 5 & 7 \end{bmatrix} = \begin{bmatrix} (\qquad\qquad) & (\qquad\qquad) \\ (\qquad\qquad) & (3\cdot 1 + 4\cdot 7) \end{bmatrix}.$$

Thus

$$C = AB = \begin{bmatrix} 5 & 9 \\ 20 & 31 \end{bmatrix}. \blacksquare$$

Example 2

Consider the matrix equation

$$\begin{bmatrix} 2 & 3 & 1 \\ 1 & 0 & -1 \\ 1 & 1 & 1 \end{bmatrix}\begin{bmatrix} x \\ y \\ z \end{bmatrix} = \begin{bmatrix} 2 \\ 3 \\ 4 \end{bmatrix}.$$

Multiplying the matrices on the left side, we get

$$\begin{bmatrix} (2x + 3y + z) \\ (x \qquad - z) \\ (x + \ y + z) \end{bmatrix} = \begin{bmatrix} 2 \\ 3 \\ 4 \end{bmatrix}.$$

From the definition of equality of matrices,

$$\begin{aligned} 2x + 3y + z &= 2, \\ x \qquad - z &= 3, \\ x + \ y + z &= 4. \end{aligned}$$

Thus we see that matrix multiplication allows the use of a single matrix equation to represent a system of linear equations. $\blacksquare$

Because each entry of the product AB is a dot product of a row of A with a column of B, the product exists only when the number of components of a row of A is the same as the number of components of a column of B. In other words, the number of columns of A must equal the number of rows of B in order for AB to exist.

Example 3
Let

$$A = \begin{bmatrix} 3 & 2 & 1 \\ 4 & 6 & 1 \end{bmatrix} \quad \text{and} \quad B = \begin{bmatrix} 2 & 1 \\ 3 & 2 \end{bmatrix}.$$

The product AB does not exist because we cannot find the dot product of a row of A with a column of B. For instance,

$$[3 \quad 2 \quad 1] \cdot \begin{bmatrix} 2 \\ 3 \end{bmatrix}$$

does not exist. The product BA does exist, however, and is

$$BA = \begin{bmatrix} 2 & 1 \\ 3 & 2 \end{bmatrix}\begin{bmatrix} 3 & 2 & 1 \\ 4 & 6 & 1 \end{bmatrix} = \begin{bmatrix} 10 & 10 & 3 \\ 17 & 18 & 5 \end{bmatrix}.$$

We say that **B is conformable to A for multiplication** when the product BA exists. ▌

Example 3 illustrates the very important fact that in general, $AB \neq BA$. In other words, multiplication is *not* commutative. Exercises 1–8 further illustrate this fact. Multiplication is associative, however, as the following theorem states.

▶ **Theorem 3**
(Associative Property) If A, B, and C are matrices, and if A is conformable to B and B is conformable to C for multiplication, then

$$A(BC) = (AB)C.$$

A proof of Theorem 3, for the case where A, B, and C are 2 by 2 matrices, is left for Exercise 14. A proof for the general case (using abbreviated notation) is similar. ◀

Another important property that can be stated now is the distributive property, which involves both addition and multiplication.

▶ **Theorem 4**
(Distributive Property) If A, B, and C are matrices, and if the necessary conformability for multiplication and addition exists, then

$$(i) \quad A(B + C) = AB + AC.$$
$$(ii) \quad (A + B)C = AC + BC.$$

A proof of Theorem 4(*i*), for the case where A, B, and C are 2 by 2 matrices, is left for Exercise 15. A proof of each part for the general case (using abbreviated notation) is similar.

Example 4
The purpose of this example is to illustrate Theorems 3 and 4.

Let

$$A = \begin{bmatrix} 2 & 1 \\ 0 & 7 \end{bmatrix}, \quad B = \begin{bmatrix} 4 & 3 \\ 2 & 1 \end{bmatrix}, \quad \text{and} \quad C = \begin{bmatrix} 0 & 2 \\ 1 & 4 \end{bmatrix}.$$

$$(AB)C = \left(\begin{bmatrix} 2 & 1 \\ 0 & 7 \end{bmatrix}\begin{bmatrix} 4 & 3 \\ 2 & 1 \end{bmatrix}\right)\begin{bmatrix} 0 & 2 \\ 1 & 4 \end{bmatrix} = \begin{bmatrix} 10 & 7 \\ 14 & 7 \end{bmatrix}\begin{bmatrix} 0 & 2 \\ 1 & 4 \end{bmatrix} = \begin{bmatrix} 7 & 48 \\ 7 & 56 \end{bmatrix};$$

$$A(BC) = \begin{bmatrix} 2 & 1 \\ 0 & 7 \end{bmatrix}\left(\begin{bmatrix} 4 & 3 \\ 2 & 1 \end{bmatrix}\begin{bmatrix} 0 & 2 \\ 1 & 4 \end{bmatrix}\right) = \begin{bmatrix} 2 & 1 \\ 0 & 7 \end{bmatrix}\begin{bmatrix} 3 & 20 \\ 1 & 8 \end{bmatrix} = \begin{bmatrix} 7 & 48 \\ 7 & 56 \end{bmatrix}.$$

Notice that $(AB)C = A(BC)$. Also

$$A(B + C) = \begin{bmatrix} 2 & 1 \\ 0 & 7 \end{bmatrix}\left(\begin{bmatrix} 4 & 3 \\ 2 & 1 \end{bmatrix} + \begin{bmatrix} 0 & 2 \\ 1 & 4 \end{bmatrix}\right) = \begin{bmatrix} 2 & 1 \\ 0 & 7 \end{bmatrix}\begin{bmatrix} 4 & 5 \\ 3 & 5 \end{bmatrix} = \begin{bmatrix} 11 & 15 \\ 21 & 35 \end{bmatrix};$$

$$AB + AC = \begin{bmatrix} 2 & 1 \\ 0 & 7 \end{bmatrix}\begin{bmatrix} 4 & 3 \\ 2 & 1 \end{bmatrix} + \begin{bmatrix} 2 & 1 \\ 0 & 7 \end{bmatrix}\begin{bmatrix} 0 & 2 \\ 1 & 4 \end{bmatrix}$$

$$= \begin{bmatrix} 10 & 7 \\ 14 & 7 \end{bmatrix} + \begin{bmatrix} 1 & 8 \\ 7 & 28 \end{bmatrix} = \begin{bmatrix} 11 & 15 \\ 21 & 35 \end{bmatrix}.$$

Notice that $A(B + C) = AB + AC$. ∎

Now that a type of multiplication of matrices has been defined, we can introduce positive integral powers of square matrices. If A is a square matrix, then

$$A^1 = A, \quad A^2 = AA, \quad A^3 = AA^2, \quad \ldots, \quad A^{k+1} = AA^k,$$

where k is a positive integer.

APPLICATIONS

Example 5

Consider a particular ecological system having vegetation providing food for herbivorous animals. Construct a matrix

$$A = \overset{\text{Species of herbivorous animals}}{\overbrace{\begin{matrix} 1 & 2 & & p \end{matrix}}} \atop \begin{bmatrix} a_{11} & a_{12} & \cdots & a_{1p} \\ a_{21} & a_{22} & \cdots & a_{2p} \\ \vdots & \vdots & & \vdots \\ a_{m1} & a_{m2} & \cdots & a_{mp} \end{bmatrix} \begin{matrix} 1 \\ 2 \\ \\ m \end{matrix} \left.\vphantom{\begin{matrix} 1 \\ 2 \\ \\ m \end{matrix}}\right\} \text{Species of vegetation}$$

where a_{ij} represents the average amount of species i of vegetation eaten by each individual of species j of herbivorous animals during a season. Also construct a matrix

$$
B = \overbrace{\begin{bmatrix} b_{11} & b_{12} & \cdots & b_{1n} \\ b_{21} & b_{22} & \cdots & b_{2n} \\ \vdots & \vdots & & \vdots \\ b_{p1} & b_{p2} & \cdots & b_{pn} \end{bmatrix}}^{\substack{\text{Species of carnivorous} \\ \text{animals} \\ 1 \quad 2 \quad \cdots \quad n}} \begin{matrix} 1 \\ 2 \\ \\ p \end{matrix} \Bigg\} \begin{matrix} \text{Species of} \\ \text{herbivorous animals} \end{matrix}
$$

where b_{ij} represents the average amount of species i of herbivorous animals eaten by each individual of species j of carnivorous animals during a season.

Carnivorous species 1, by feeding on herbivorous species 1, indirectly consumes $a_{11}b_{11}$ amount of plant species 1. Also, the same carnivorous species 1, by feeding on herbivorous species 2, indirectly consumes $a_{12}b_{21}$ amount of plant species 1. In general, an individual of carnivorous species 1, by feeding on all p of the herbivorous species, will indirectly consume

$$a_{11}b_{11} + a_{12}b_{21} + \cdots + a_{1p}b_{p1}$$

units of plant species 1. The latter result is simply the first entry of the product AB. Other entries of AB will similarly produce the amounts of each of various plant species indirectly consumed by an individual of each of the various carnivorous species.

Similar uses of matrix multiplication may be made in representing the interaction of components in an environmental system. ▮

Example 6

A **probability vector** can be defined as a row matrix with nonnegative entries whose sum is 1. A matrix consisting of rows of probability vectors is called a **stochastic matrix.** As the name suggests, these matrices can be used to display the probabilities of a stochastic experiment illustrated by the tree diagram of Figure 5.11. The probabilities of the first trial can be displayed in the stochastic row matrix $P = [p_{11} \quad p_{12} \quad p_{13}]$; the probabilities of the second trial can be displayed in the stochastic matrix

$$Q = \begin{bmatrix} q_{11} & q_{12} \\ q_{21} & q_{22} \\ q_{31} & q_{32} \end{bmatrix},$$

where q_{ij} is the probability that trial outcome Q_j will follow trial outcome P_i. Consider the matrix product

$$PQ = [p_{11} \quad p_{12} \quad p_{13}] \begin{bmatrix} q_{11} & q_{12} \\ q_{21} & q_{22} \\ q_{31} & q_{32} \end{bmatrix}$$

$$= [(p_{11}q_{11} + p_{12}q_{21} + p_{13}q_{31}) \ (p_{11}q_{12} + p_{12}q_{22} + p_{13}q_{32})].$$

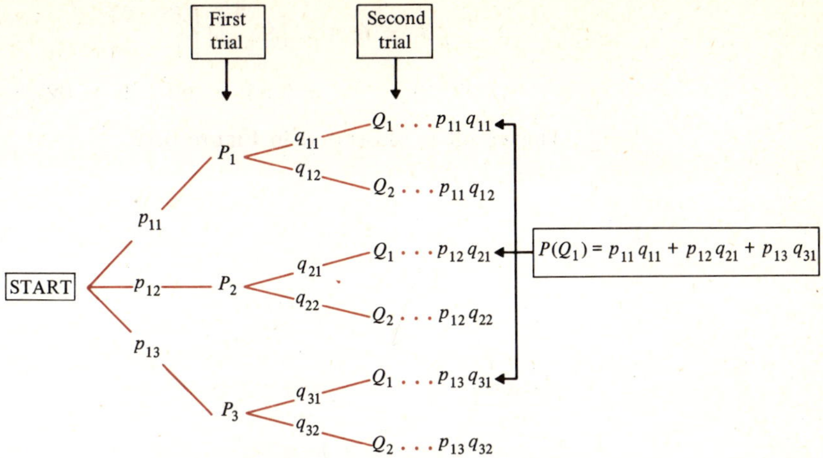

Figure 5.11. *Stochastic experiment.*

Notice that PQ is a 1 by 2 matrix, where the first entry is the probability that after 2 trials, outcome Q_1 will be obtained (see Figure 5.11), and the second entry is the probability that after 2 trials, outcome Q_2 will be obtained (see Figure 5.11). The use of these ideas in the study of anthropology may be found in Kay [25], pp. 185–190. ▮

Example 7

In this example we consider a special type of stochastic experiment, where a finite set of trial outcomes (called **states**) is the same for every trial, and where the probability p_{ij} that trial outcome a_j will follow trial outcome a_i is the same for every trial. Such stochastic experiments are called **Markov chains.** As an illustration reconsider Example 8 of Section 5.1, where the set of trial outcomes or states were

{system performing satisfactorily   system not performing satisfactorily}.

The probabilities of changing or not changing states is displayed in the transition matrix

$$A = \begin{array}{c} \\ \\ \end{array}\overset{\text{To}}{\overbrace{\begin{array}{cc} \text{satis-} & \text{unsatis-} \\ \text{factory} & \text{factory} \end{array}}}$$
$$A = \begin{bmatrix} .6 & .4 \\ .8 & .2 \end{bmatrix} \begin{array}{l} \text{satisfactory} \\ \text{unsatisfactory} \end{array} \Bigg\} \text{From}$$

A row matrix showing the probabilities of the respective elements of the set of trial outcomes is called a **state vector.** If there is a probability of .9 that the system begins in the state of performing satisfactorily, then the initial state vector B_0 is $[.9 \quad .1]$ and the state vector B_1 after one period is given by the product

$$B_1 = B_0 A = [.9 \quad .1]\begin{bmatrix} .6 & .4 \\ .8 & .2 \end{bmatrix}$$
$$= [(.54 + .08) \; (.36 + .02)] = [.62 \quad .38].$$

This result is illustrated in Figure 5.12.

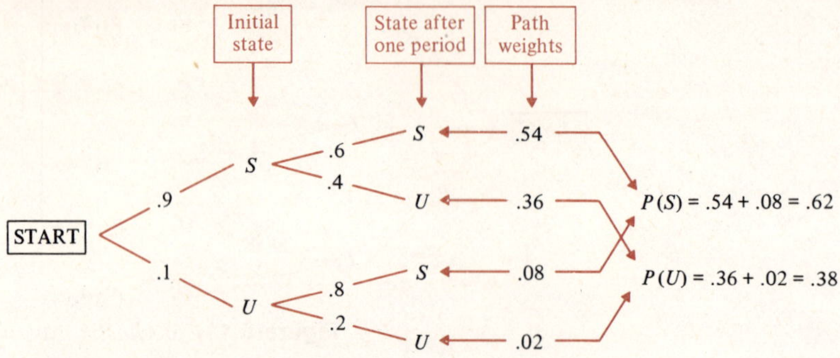

Figure 5.12. *Tree diagram computation of components of state vector.*

Next, to find the state vector B_2 after 2 periods, the process is repeated.

$$B_2 = B_1 A = [.62 \quad .38]\begin{bmatrix} .6 & .4 \\ .8 & .2 \end{bmatrix}$$
$$= [(.372 + .304) \; (.248 + .076)]$$
$$= [.676 \quad .324].$$

An alternative way of calculating B_2 is to recognize that

$$B_2 = B_1 A = (B_0 A)A = B_0(AA) = B_0 A^2.$$

In other words, A^2 represents a transition matrix of a 2-stage change. Using this same approach, one can develop a formula for B_k:

$$B_k = B_0 A^k. \blacksquare$$

Example 8

In Example 9 of Section 5.1 we found that the matrix

$$\begin{array}{c} \text{Genotype of child} \\ \overbrace{\begin{array}{ccc} AA & Aa & aa \end{array}} \end{array}$$
$$A = \begin{bmatrix} p & 1-p & 0 \\ \frac{1}{2}p & \frac{1}{2} & \frac{1}{2}(1-p) \\ 0 & p & 1-p \end{bmatrix} \left.\begin{matrix} AA \\ Aa \\ aa \end{matrix}\right\} \begin{matrix}\text{Genotype} \\ \text{of parent}\end{matrix}$$

can be used to display the various probabilities of offspring genotypes for each parental genotype. Suppose that an investigator wants to determine the probabilities of the various genotypes of the grandchildren of a member of the same population used in Example 9 of Section 5.1; using the information of that example, we can use a tree diagram and the methods of Chapter 4 to calculate the desired probabilities as shown in Figure 5.13.

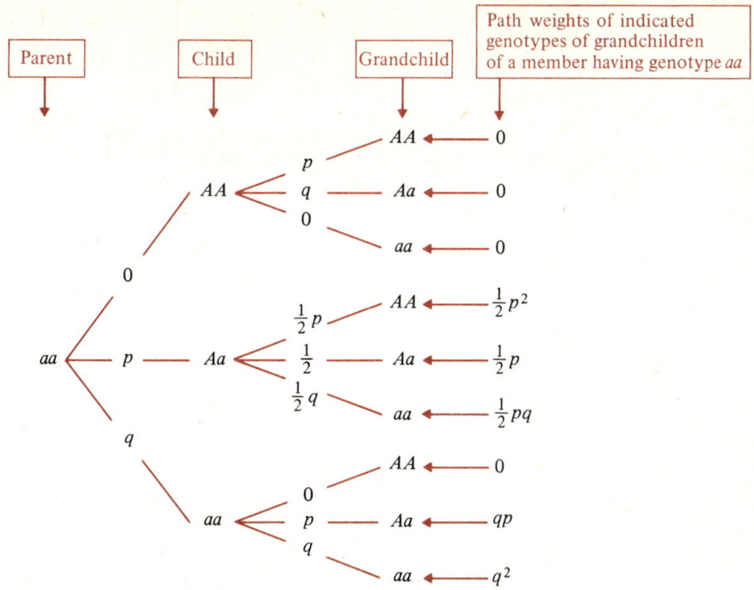

Figure 5.13. *Tree diagram computation of weights of genotypes.*

(In Figure 5.13 let $q = 1 - p$.) An alternative way of obtaining these desired probabilities is to recognize that A is a transition matrix of a Markov chain, and hence the probabilities after 2 transitions are displayed as the entries of A^2. Let $q = 1 - p$; then

$$A^2 = \overbrace{\begin{bmatrix} p^2 + \frac{1}{2}pq & pq + \frac{1}{2}q & \frac{1}{2}q^2 \\ \frac{1}{2}p^2 + \frac{1}{4}p & pq + \frac{1}{4} & \frac{1}{4}q + \frac{1}{2}q^2 \\ \frac{1}{2}p^2 & \frac{1}{2}p + pq & \frac{1}{2}pq + q^2 \end{bmatrix}}^{\substack{\text{Genotype of grandchild} \\ AA \qquad Aa \qquad aa}} \begin{matrix} AA \\ Aa \\ aa \end{matrix} \left.\vphantom{\begin{matrix} AA \\ Aa \\ aa \end{matrix}}\right\} \text{Genotype of parent}$$

Notice that the last row of A^2 provides the same information that can be calculated from the tree diagram of Figure 5.13. The first two rows of A^2 provide information that can be illustrated by tree diagrams similar to Figure 5.13. Probabilities of genotypes of later generations of offspring can be calculated by considering higher powers of A. ∎

EXERCISES

Suggested minimum assignment: Exercises 1, 3, 5, 7, 9, 12, 13, and 16.

In Exercises 1–8 find AB and BA if possible. If AB or BA does not exist, give reasons.

1. $A = \begin{bmatrix} 2 & 1 \\ 3 & 0 \end{bmatrix}; \quad B = \begin{bmatrix} 0 & 1 \\ 3 & -4 \end{bmatrix}.$

2. $A = \begin{bmatrix} 3 & 0 & 0 \\ 0 & 2 & 1 \\ 6 & 9 & 0 \end{bmatrix}$; $B = \begin{bmatrix} 0 & 1 & 4 \\ 0 & 2 & 3 \\ -1 & 0 & 5 \end{bmatrix}$.

3. $A = \begin{bmatrix} 2 & 1 & 3 \\ 0 & 4 & 2 \end{bmatrix}$; $B = \begin{bmatrix} 1 & 2 \\ 1 & 5 \end{bmatrix}$.

4. $A = \begin{bmatrix} 3 & 1 & 4 \\ 0 & 2 & 1 \\ 6 & 5 & 0 \end{bmatrix}$; $B = \begin{bmatrix} 3 & 2 \\ 0 & 1 \\ -3 & 5 \end{bmatrix}$.

5. $A = [1 \quad 3]$; $B = \begin{bmatrix} 8 \\ 3 \end{bmatrix}$.

6. $A = \begin{bmatrix} 1 \\ 4 \\ 3 \end{bmatrix}$; $B = [2 \quad 1 \quad 5]$.

7. $A = \begin{bmatrix} 2 & 3 \\ 4 & 5 \end{bmatrix}$; $B = \begin{bmatrix} x \\ y \end{bmatrix}$.

8. $A = \begin{bmatrix} 3 & 4 & 5 \\ 6 & 9 & 2 \\ 3 & 5 & 2 \end{bmatrix}$; $B = \begin{bmatrix} x \\ y \\ z \end{bmatrix}$.

9. Find A^2 if $A = \begin{bmatrix} 2 & 3 \\ 4 & 6 \end{bmatrix}$.

10. Find A^3 if $A = \begin{bmatrix} 3 & 2 \\ 1 & 0 \end{bmatrix}$.

11. Illustrate the associative property of multiplication with the matrices

$$A = \begin{bmatrix} 2 & 3 \\ 4 & 5 \end{bmatrix}, \quad B = \begin{bmatrix} 0 & 1 \\ 2 & 3 \end{bmatrix}, \quad C = \begin{bmatrix} 4 & 2 \\ 0 & 1 \end{bmatrix}.$$

12. Illustrate Theorem 4(i) with the matrices of Exercise 11.

13. Given: $A = [a_{ij}]_{(m,n)}$ and $B = [b_{ij}]_{(p,q)}$.
   (a) Under what conditions does AB exist?
   (b) If AB exists, what is its order?
   (c) Under what conditions does BA exist?
   (d) If BA exists, what is its order?
   (e) Under what conditions does A^2 exist?
   (f) If A^2 exists, what is its order?

14. Prove Theorem 3 for the case where A, B, and C are 2 by 2 matrices.

15. Prove Theorem 4(i) for the case where A, B, and C are 2 by 2 matrices.

16. The production per day of 2 products by 3 factories is given by the following matrix:

$$\begin{array}{c} \text{Factories} \\ \overbrace{\begin{matrix} A & B & C \end{matrix}} \end{array}$$

$$M = \begin{bmatrix} 1 & 3 & 4 \\ 6 & 9 & 2 \end{bmatrix} \begin{matrix} P \\ Q \end{matrix} \Big\} \text{Products}$$

If the entries of

$$X = \begin{bmatrix} 3 \\ 2 \\ 1 \end{bmatrix}$$

represent the number of days that the respective factories operate, find MX and state what the entries represent.

17. If 5 judges on an appellate court influence each other according to Figure 5.14, write a matrix that shows the number of ways 1 judge can influence another judge through 1 intermediary. *Hint*: Reconsider Example 7 of Section 5.1, and then investigate the significance of the product AA, where A is an influence matrix.

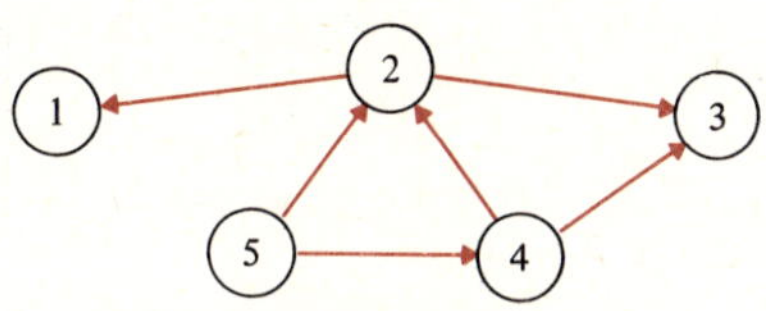

Figure 5.14

18. Use the data of Exercise 17 to write a matrix that shows the number of ways that 1 judge can influence another through *at most* 1 intermediary. Rank the judges according to their influence through *at most* 1 intermediary.

5.4 Special Matrices

The purpose of this section is to introduce some special matrices that are important in the theory and applications of matrices.

▶ **Definition 12**

An m by n matrix with entries that are all 0's is called a **zero matrix** and is designated by **0**.

Example 1

Each of the following is a zero matrix:

$$\begin{bmatrix} 0 & 0 & 0 \\ 0 & 0 & 0 \end{bmatrix}, \quad \begin{bmatrix} 0 \\ 0 \end{bmatrix}, \quad [0 \quad 0 \quad 0], \quad \begin{bmatrix} 0 & 0 \\ 0 & 0 \end{bmatrix}. \ \blacksquare$$

A zero matrix has the property that when it is added to any other matrix A of the same order, the sum is the same matrix A; that is,

$$A + \mathbf{0} = \mathbf{0} + A = A.$$

Because of this property the zero matrix is called the **identity element for matrix addition.** The special matrix I_n that is defined next has the property that when it is *multiplied* on either side by a matrix A of the same order, the product is A; that is,

$$AI_n = I_nA = A.$$

Because of this property the **identity matrix** I_n is called the **identity element for matrix multiplication.**

▶ **Definition 13**

An n by n matrix with 1's on the main diagonal and 0's elsewhere is called an **identity matrix** and is designated by I_n. The columns of I_n are designated $I_{(1)}, I_{(2)}, \ldots, I_{(n)}$.

Example 2

$$I_2 = \begin{bmatrix} 1 & 0 \\ 0 & 1 \end{bmatrix}, \quad I_3 = \begin{bmatrix} 1 & 0 & 0 \\ 0 & 1 & 0 \\ 0 & 0 & 1 \end{bmatrix}, \quad I_4 = \begin{bmatrix} 1 & 0 & 0 & 0 \\ 0 & 1 & 0 & 0 \\ 0 & 0 & 1 & 0 \\ 0 & 0 & 0 & 1 \end{bmatrix}.$$

Moreover

$$AI_2 = \begin{bmatrix} 2 & 1 \\ 3 & 4 \end{bmatrix}\begin{bmatrix} 1 & 0 \\ 0 & 1 \end{bmatrix} = \begin{bmatrix} 2 & 1 \\ 3 & 4 \end{bmatrix} = A$$

and

$$I_2A = \begin{bmatrix} 1 & 0 \\ 0 & 1 \end{bmatrix}\begin{bmatrix} 2 & 1 \\ 3 & 4 \end{bmatrix} = \begin{bmatrix} 2 & 1 \\ 3 & 4 \end{bmatrix} = A.$$

Also, for I_3 we have

$$I_{(1)} = \begin{bmatrix} 1 \\ 0 \\ 0 \end{bmatrix}, \quad I_{(2)} = \begin{bmatrix} 0 \\ 1 \\ 0 \end{bmatrix}, \quad I_{(3)} = \begin{bmatrix} 0 \\ 0 \\ 1 \end{bmatrix}. \blacksquare$$

Notice that the identity matrix I_n with respect to multiplication corresponds to the real number 1 with respect to real number multiplication, and the zero matrix $\mathbf{0}$ with respect to matrix addition corresponds to the real number 0 with respect to real number addition.

▶ **Definition 14**

If A is a square matrix with $a_{ij} = 0$ for all $i \neq j$, then A is a **diagonal matrix.**

Example 3
The following matrices are diagonal.

$$\begin{bmatrix} 2 & 0 & 0 \\ 0 & 1 & 0 \\ 0 & 0 & 3 \end{bmatrix}, \quad \begin{bmatrix} 2 & 0 \\ 0 & 1 \end{bmatrix}, \quad \begin{bmatrix} 0 & 0 & 0 \\ 0 & 3 & 0 \\ 0 & 0 & 2 \end{bmatrix}, \quad I_n, \quad \begin{bmatrix} 0 & 0 & 0 \\ 0 & 0 & 0 \\ 0 & 0 & 0 \end{bmatrix}. \ \blacksquare$$

Those diagonal matrices that have the same entries on the main diagonal are called **scalar matrices;** an identity matrix multiplied by any real number is an example of a scalar matrix.

▶ **Definition 15**
Let A be a square matrix. If $a_{ij} = 0$ for all $i > j$ or for all $i < j$, then A is a **triangular matrix.**

Example 4
The following matrices are triangular. For emphasis, the 0's required to make each matrix triangular are placed in parentheses.

$$\begin{bmatrix} 8 & 2 & 3 \\ (0) & 4 & 2 \\ (0) & (0) & 6 \end{bmatrix}, \quad \begin{bmatrix} 2 & (0) \\ 3 & 4 \end{bmatrix}, \quad \begin{bmatrix} 0 & 2 & 1 \\ (0) & 4 & 3 \\ (0) & (0) & 5 \end{bmatrix}, \quad \begin{bmatrix} 0 & (0) & (0) \\ 0 & 0 & (0) \\ 0 & 2 & 0 \end{bmatrix}. \ \blacksquare$$

▶ **Definition 16**
If some rows or columns of a matrix A are deleted, the remaining array is called a **submatrix** of A. Also, it is customary to treat A as a submatrix of itself.

Example 5
For the matrix

$$A = \begin{bmatrix} 2 & 1 & 3 \\ 5 & 0 & 4 \end{bmatrix},$$

the following matrices are submatrices of A:

$$\begin{bmatrix} 2 & 1 & 3 \\ 5 & 0 & 4 \end{bmatrix}, \quad \begin{bmatrix} 2 & 1 \\ 5 & 0 \end{bmatrix}, \quad \begin{bmatrix} 2 & 3 \\ 5 & 4 \end{bmatrix}, \quad \begin{bmatrix} 1 & 3 \\ 0 & 4 \end{bmatrix}, \quad \begin{bmatrix} 2 \\ 5 \end{bmatrix}, \quad \begin{bmatrix} 1 \\ 0 \end{bmatrix}, \quad \begin{bmatrix} 3 \\ 4 \end{bmatrix},$$

$$[2 \ \ 1 \ \ 3], \quad [5 \ \ 0 \ \ 4], \quad [2 \ \ 1], \quad [2 \ \ 3], \quad [1 \ \ 3],$$

$$[5 \ \ 0], \quad [5 \ \ 4], \quad [0 \ \ 4], \quad [2], \quad [1], \quad [3], \quad [5], \quad [0], \quad [4]. \ \blacksquare$$

In some problems, it is helpful to separate a matrix into submatrices by means of dashed lines. For example, we can write

$$A = \left[\begin{array}{c|c} A_{11} & A_{12} \\ \hline A_{21} & A_{22} \end{array}\right] = \left[\begin{array}{c|cc} a_{11} & a_{12} & a_{13} \\ \hline a_{21} & a_{22} & a_{23} \\ a_{31} & a_{32} & a_{33} \end{array}\right].$$

When this has been done, we say that A has been **partitioned,** and the partitioned submatrices, designated by A_{ij}, are called **blocks of A;** A is then called a **block matrix.** In Exercises 12 and 13 diagonal and triangular block matrices are defined.

A very simple but useful operation that can be performed on a single matrix is that of interchanging the rows and columns of the matrix. This operation will lead to more special matrices.

▶ **Definition 17**

The **transpose** of an m by n matrix A is the n by m matrix A^T that is formed when the ith row of A becomes the ith column of A^T ($i = 1, 2, 3, \ldots, m$).

Example 6

If $A = \begin{bmatrix} 3 & 2 & 1 \\ 6 & 4 & 5 \end{bmatrix}$, then $A^T = \begin{bmatrix} 3 & 6 \\ 2 & 4 \\ 1 & 5 \end{bmatrix}$. ∎

Once a new operation has been defined, it is natural to wonder about its properties. A few of the properties of this operation are stated in Exercise 7.

▶ **Definition 18**

A square matrix A is said to be **symmetric** if $A = A^T$, and **skew-symmetric** if $A = -A^T$.

The names "symmetric" and "skew-symmetric" arise because of the physical symmetry that exists with respect to the main diagonal as shown in Figures 5.15 and 5.16.

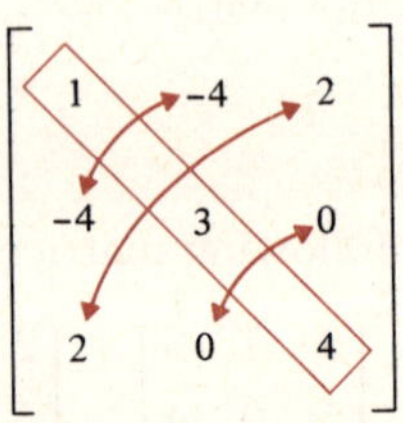

Figure 5.15. *Symmetric matrix.*

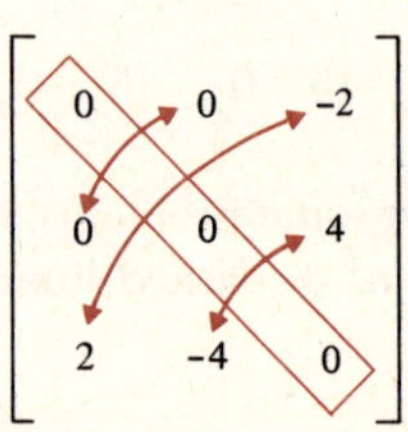

Figure 5.16. *Skew-symmetric matrix.*

Example 7

The matrix

$$A = \begin{bmatrix} 4 & 2 & 1 \\ 2 & 7 & 3 \\ 1 & 3 & 8 \end{bmatrix}$$

is symmetric because $A = A^T$. The matrix

$$B = \begin{bmatrix} 0 & 1 & 6 \\ -1 & 0 & -2 \\ -6 & 2 & 0 \end{bmatrix}$$

is not symmetric because $B \neq B^T$, but B is skew-symmetric because $B = -B^T$. The matrix

$$\begin{bmatrix} 2 & 1 \\ 3 & 4 \end{bmatrix}$$

is neither symmetric nor skew-symmetric. ▮

A few of the many properties of symmetric and skew-symmetric matrices will be given in Exercises 14 and 16.

At this point it may be helpful to review the special matrices presented in this section:

1. The *zero* and *identity matrices* which correspond to the numbers 0 and 1 in the real number system.
2. A *diagonal matrix* whose off-diagonal entries must all be 0, and the special diagonal matrix called a *scalar matrix,* where all the diagonal entries are the same.
3. A *triangular matrix,* all of whose entries either above or below the main diagonal are 0's.
4. A *submatrix* of a matrix, which is obtained by deleting some rows or columns of the given matrix. Also a matrix is a submatrix of itself.
5. *Symmetric* and *skew-symmetric matrices,* which have the respective properties that $A = A^T$ and $A = -A^T$.

APPLICATIONS

Example 8

Many applications of matrix algebra in accounting may be found in Ijiri [19]. Background for this example may be found in Chapter 5 of [19]. Consider the problem of computing the closing balance for a firm that has an initial balance sheet given in Table 5.1. The firm, which is in the merchandising business, buys on credit and makes only cash sales; we assume that the only accounts in use are those listed above. Arrange the

Table 5.1. Initial balance sheet.

Debit balances		Credit balances	
Cash	\$30	Accounts payable	\$15
Inventories	20	Equity	35
	\$50		\$50

initial balances as a column matrix

$$U = \begin{bmatrix} u_1 \\ u_2 \\ u_3 \\ u_4 \end{bmatrix} = \begin{bmatrix} 30 \\ 20 \\ -15 \\ -35 \end{bmatrix}.$$

Over a certain time period suppose that the following transactions took place:

\$40	purchases of goods on credit
\$35	cash payments to suppliers
\$15	fixed costs paid in cash
\$25	contribution margin, cash sales
\$30	cost of goods sold
\$ 5	goods returned to suppliers

The dollar representations of the transactions over a period of time are arranged into a "spread sheet matrix" W with a row for each account and a column for each account. The rows represent the accounts to be debited, and the columns represent accounts to be credited with the transactions of the time period.

The first transaction listed above is $w_{23} = 40$ because, by accounting convention, item 2 (inventory) is debited and item 3 (accounts payable) is credited. The second transaction is $w_{31} = 35$ because item 3 (accounts payable) is debited and item 1 (cash) is credited. In the same manner the remaining transactions are represented by $w_{41} = 15$, $w_{14} = 25$, $w_{12} = 30$, and $w_{32} = 5$. The spread matrix W is therefore

$$W = \begin{bmatrix} 0 & 30 & 0 & 25 \\ 0 & 0 & 40 & 0 \\ 35 & 5 & 0 & 0 \\ 15 & 0 & 0 & 0 \end{bmatrix}.$$

The main diagonal of W consists of all 0's because it is not customary to recognize a debit and a credit to the same account as a genuine transaction. By transposing matrix W, we form matrix W^T. Then the change in U, called ΔU, is equal to the expression $(W - W^T)E$, where E is a column vector that has all its elements equal to 1 and has as many elements as there are accounts in the system. (Notice that $W - W^T$ is a skew-symmetric matrix.)

$$\Delta U = (W - W^T)E = \left(\begin{bmatrix} 0 & 30 & 0 & 25 \\ 0 & 0 & 40 & 0 \\ 35 & 5 & 0 & 0 \\ 15 & 0 & 0 & 0 \end{bmatrix} - \begin{bmatrix} 0 & 0 & 35 & 15 \\ 30 & 0 & 5 & 0 \\ 0 & 40 & 0 & 0 \\ 25 & 0 & 0 & 0 \end{bmatrix}\right)\begin{bmatrix} 1 \\ 1 \\ 1 \\ 1 \end{bmatrix}$$

$$= \begin{bmatrix} 0 & 30 & -35 & 10 \\ -30 & 0 & 35 & 0 \\ 35 & -35 & 0 & 0 \\ -10 & 0 & 0 & 0 \end{bmatrix}\begin{bmatrix} 1 \\ 1 \\ 1 \\ 1 \end{bmatrix} = \begin{bmatrix} 5 \\ 5 \\ 0 \\ -10 \end{bmatrix}.$$

The closing balance U' can now be found by

$$U' = U + \Delta U = \begin{bmatrix} 30 \\ 20 \\ -15 \\ -35 \end{bmatrix} + \begin{bmatrix} 5 \\ 5 \\ 0 \\ -10 \end{bmatrix} = \begin{bmatrix} 35 \\ 25 \\ -15 \\ -45 \end{bmatrix},$$

from which the closing balance sheet can be written; see Table 5.2. ▮

Table 5.2. CLOSING BALANCE SHEET.

Debit balances		Credit balances	
Cash	\$35	Accounts payable	\$15
Inventories	25	Equity	45
	\$60		\$60

Example 9

Suppose that each of 8 patients suffering from a certain psychological disorder is given 3 different treatments. Furthermore, suppose that the response of these patients to each of these treatments is classified as improvement, no change, or deterioration. The response of each patient to each treatment can be characterized by three matrices. Suppose that the "improvement" response matrix is

$$R_1 = \overbrace{\begin{matrix} 1 & 2 & 3 & 4 & 5 & 6 & 7 & 8 \end{matrix}}^{\text{Patients}}$$
$$R_1 = \begin{bmatrix} 1 & 0 & 1 & 0 & 1 & 0 & 0 & 1 \\ 1 & 1 & 1 & 0 & 0 & 1 & 0 & 0 \\ 0 & 0 & 1 & 1 & 0 & 1 & 0 & 0 \end{bmatrix} \begin{matrix} 1 \\ 2 \\ 3 \end{matrix}\Bigg\} \text{Treatment}$$

where a 1 in the ij position means that improvement was shown when treatment i was given to patient j. Otherwise, a 0 is given. Next suppose that the "no change" response matrix is

$$
R_2 = \overbrace{\begin{bmatrix} 0 & 0 & 0 & 1 & 0 & 0 & 1 & 0 \\ 0 & 0 & 0 & 1 & 1 & 0 & 1 & 1 \\ 1 & 0 & 0 & 0 & 0 & 0 & 0 & 0 \end{bmatrix}}^{\text{Patients: } 1\ 2\ 3\ 4\ 5\ 6\ 7\ 8} \left.\begin{matrix} 1 \\ 2 \\ 3 \end{matrix}\right\} \text{Treatment}
$$

where a 1 in the *ij* position means that no change was shown when treatment *i* was given to patient *j*. Otherwise a 0 is given. Finally, suppose that the "deterioration" response matrix is

$$
R_3 = \overbrace{\begin{bmatrix} 0 & 1 & 0 & 0 & 0 & 1 & 0 & 0 \\ 0 & 0 & 0 & 0 & 0 & 0 & 0 & 0 \\ 0 & 1 & 0 & 0 & 1 & 0 & 1 & 1 \end{bmatrix}}^{\text{Patients: } 1\ 2\ 3\ 4\ 5\ 6\ 7\ 8} \left.\begin{matrix} 1 \\ 2 \\ 3 \end{matrix}\right\} \text{Treatment}
$$

where the *ij* entry is 1 if deterioration occurs when treatment *i* is given to patient *j*, and is 0 otherwise. Now form the partitioned matrix

$$
R = \left[\begin{array}{c} R_1 \\ \hline R_2 \\ \hline R_3 \end{array}\right],
$$

and then the product

$$
A = RR^T = \left[\begin{array}{c} R_1 \\ \hline R_2 \\ \hline R_3 \end{array}\right] \left[\begin{array}{c|c|c} R_1^T & R_2^T & R_3^T \end{array}\right] = \left[\begin{array}{c|c|c} R_1R_1^T & R_1R_2^T & R_1R_3^T \\ \hline R_2R_1^T & R_2R_2^T & R_2R_3^T \\ \hline R_3R_1^T & R_3R_2^T & R_3R_3^T \end{array}\right]
$$

$$
= \overbrace{\left[\begin{array}{ccc|ccc|ccc}
\multicolumn{3}{c|}{\text{improve}} & \multicolumn{3}{c|}{\text{no change}} & \multicolumn{3}{c}{\text{deteriorate}} \\
\multicolumn{3}{c|}{\text{Treatment}} & \multicolumn{3}{c|}{\text{Treatment}} & \multicolumn{3}{c}{\text{Treatment}} \\
1 & 2 & 3 & 1 & 2 & 3 & 1 & 2 & 3 \\ \hline
4 & 2 & 1 & 0 & 2 & 1 & 0 & 0 & 2 \\
2 & 4 & 2 & 0 & 0 & 1 & 2 & 0 & 1 \\
1 & 2 & 3 & 1 & 1 & 0 & 1 & 0 & 0 \\ \hline
0 & 0 & 1 & 2 & 2 & 0 & 0 & 0 & 1 \\
2 & 0 & 1 & 2 & 4 & 0 & 0 & 0 & \underline{3} \\
1 & 1 & 0 & 0 & 0 & 1 & 0 & 0 & 0 \\ \hline
0 & 2 & 1 & 0 & 0 & 0 & 2 & 0 & 1 \\
0 & 0 & 0 & 0 & 0 & 0 & 0 & 0 & 0 \\
2 & 1 & 0 & 1 & \underline{3} & 0 & 1 & 0 & 4
\end{array}\right]}^{\text{Responses}}
\begin{array}{l}
\\ \\ \\
\left.\begin{matrix} 1 \\ 2 \\ 3 \end{matrix}\right\} \text{Treatment} \quad \text{improve} \\
\left.\begin{matrix} 1 \\ 2 \\ 3 \end{matrix}\right\} \text{Treatment} \quad \text{no change} \\
\left.\begin{matrix} 1 \\ 2 \\ 3 \end{matrix}\right\} \text{Treatment} \quad \text{deteriorate}
\end{array}
\Bigg\} \text{Responses}
$$

It can be shown that the ijth entry of the submatrix $R_h R_k^T$ of A represents the number of patients in which treatment i produced response h and treatment j produced response k. For instance, from matrix A we can determine at a glance that in exactly 3 patients, treatment 2 produced no change and treatment 3 produced deterioration. (The entries that convey this fact are underlined; there are two such entries because A is a symmetric matrix.) Also the ijth entry of the sum $R_1R_1^T + R_2R_2^T + R_3R_3^T$, which in this problem is

$$\begin{bmatrix} 8 & 4 & 2 \\ 4 & 8 & 2 \\ 2 & 2 & 8 \end{bmatrix},$$

represents the number of times that treatments i and j produced the *same* response; for instance, treatments 1 and 2 produced the same response in 4 of the 8 patients. Although the matrix method illustrated in this example derives results that can be obtained by other methods, this matrix method has the advantage of displaying and manipulating vast amounts of data in a systematic way. Of course, an added advantage is that computer programs needed to perform the necessary matrix operations are readily available. Another application of this method to a study of responses to various appeals by the judges of an appellate court may be found in Campbell [8], pp. 83–84, 207–209. ▮

Example 10

The matrix method of the last example applies to any problem in which m objects yield p responses when subjected to n stimuli; moreover, the responses need not be mutually exclusive as they were in the last example. Also, it can be observed that objects, stimuli, and responses need only to be listed, numbered, and counted. No other quantification is necessary. This matrix method might apply to such diverse problems as (1) studies of responses of certain business or economic indicators when subjected to various national or international events; (2) studies of responses of legislatures or legislators, or nations of the UN, when subjected to votes on certain issues; (3) studies of responses of animals or plants when subjected to certain diseases, medication, weather, or food, and so on; and (4) psychological studies of responses of individuals or groups of individuals when subjected to certain emotional stresses or physical stresses or mental testing. ▮

Example 11

In this example we further extend Example 7 of Section 5.3 to illustrate the use of the transpose of a matrix to achieve a desired result. Recall that the transition matrix from one stage to another stage was

$$A = \overbrace{\begin{bmatrix} .6 & .4 \\ .8 & .2 \end{bmatrix}}^{\text{To: satisfactory, unsatisfactory}} \left.\begin{matrix} \text{satisfactory} \\ \text{unsatisfactory} \end{matrix}\right\} \text{From}$$

Suppose that there is some value, say profit, associated with each transition; these profits are given by the matrix

$$V = \begin{bmatrix} \$100 & \$20 \\ \$30 & -\$50 \end{bmatrix},$$

where v_{ij} is the value or profit to the owner of the system for a transition from the ith state to the jth state. The two main diagonal entries of the product

$$AV^T = \begin{bmatrix} .6 & .4 \\ .8 & .2 \end{bmatrix}\begin{bmatrix} 100 & 30 \\ 20 & -50 \end{bmatrix} = \begin{bmatrix} (60+8) & (\quad) \\ (\quad) & (24-10) \end{bmatrix} = \begin{bmatrix} 68 & (\ ) \\ (\ ) & 14 \end{bmatrix}$$

represent the expected profit to the owner for each of the two respective trial outcomes after the first transition.

In other words, if the system is working at the outset, then the expected profit in one transition is \$68, but if the system is not working at the outset, the expected profit is \$14. A more complete discussion, including that of expected values after repeated transitions and optimal decisions, may be found in Searle and Hausman [39], pp. 201–204. ▮

EXERCISES

Suggested minimum assignment: Exercises 1, 2, 3, 5, 6 and 7.

1. Given:

$$A = \begin{bmatrix} 3 & 0 & 0 \\ 0 & 2 & 0 \\ 0 & 0 & 0 \end{bmatrix}, \quad B = \begin{bmatrix} 1 & 0 & 0 \\ 0 & 1 & 0 \\ 0 & 0 & 1 \end{bmatrix}, \quad C = \begin{bmatrix} 3 & 2 & 0 \\ 2 & 1 & 0 \\ 0 & 0 & 0 \end{bmatrix},$$

$$D = \begin{bmatrix} 1 & 1 & 0 \\ 0 & 0 & 0 \\ 0 & 0 & 1 \end{bmatrix}, \quad E = \begin{bmatrix} 0 & 0 & 0 \end{bmatrix}, \quad F = \begin{bmatrix} 1 & 0 & 0 \\ 0 & 1 & 0 \end{bmatrix}.$$

(*a*) Which matrices, if any, are zero matrices?
(*b*) Which matrices, if any, are identity matrices for multiplication?
(*c*) Which matrices, if any, are diagonal matrices?
(*d*) Which matrices, if any, are scalar matrices?
(*e*) Which matrices, if any, are triangular matrices?
(*f*) I_2 is a submatrix of which matrices?
(*g*) Which matrices, if any, are symmetric?
(*h*) Which matrices, if any, are skew-symmetric?

2. Answer the same questions raised in Exercise 1 for the matrices

$$A = \begin{bmatrix} 0 & 0 & 6 \\ 0 & 3 & 0 \\ 6 & 0 & 0 \end{bmatrix}, \quad B = \begin{bmatrix} 1 & 0 \\ 0 & 0 \end{bmatrix}, \quad C = \begin{bmatrix} 0 & 0 \\ 0 & 0 \\ 0 & 0 \end{bmatrix},$$

$$D = \begin{bmatrix} 1 \\ 0 \\ 0 \end{bmatrix}, \qquad E = \begin{bmatrix} 2 & 0 \\ 0 & 2 \end{bmatrix}, \qquad F = \begin{bmatrix} 3 & 0 & 0 \\ 1 & 4 & 0 \\ 3 & 2 & 0 \end{bmatrix}.$$

3. Verify that $I_2A = AI_2 = A$ if

$$A = \begin{bmatrix} a_{11} & a_{12} \\ a_{21} & a_{22} \end{bmatrix}.$$

4. Verify that $A + \mathbf{0} = \mathbf{0} + A = A$ if $\mathbf{0}$ is a 2 by 2 zero matrix and

$$A = \begin{bmatrix} a_{11} & a_{12} \\ a_{21} & a_{22} \end{bmatrix}.$$

5. Partition the matrix

$$\begin{bmatrix} 3 & 5 & 1 \\ 4 & 0 & 6 \\ 9 & 2 & 1 \end{bmatrix}$$

so that the $[5 \quad 1]$ is one block and $\begin{bmatrix} 4 \\ 9 \end{bmatrix}$ is another block.

6. Tell whether each of the following is true or false. If false, give an example illustrating your conclusion.
   (*a*) The transpose of a triangular matrix is a triangular matrix.
   (*b*) Every diagonal matrix is a triangular matrix.
   (*c*) Every scalar matrix is square.
   (*d*) Every scalar matrix is also a diagonal matrix.
   (*e*) The product of two diagonal matrices of the same order is a diagonal matrix.
7. A few of the properties of the transpose of a matrix follow: Let A and B be matrices, let k be a real number, and let the orders of A and B be such that conformability requirements are met. Then

$$\begin{array}{ll} (i) & (A^T)^T = A, \\ (ii) & (A + B)^T = A^T + B^T, \\ (iii) & (AB)^T = B^TA^T, \\ (iv) & (kA)^T = kA^T. \end{array}$$

Illustrate these properties with

$$A = \begin{bmatrix} 1 & 4 \\ 3 & 5 \end{bmatrix}, \quad B = \begin{bmatrix} 6 & 0 \\ 2 & 3 \end{bmatrix}, \quad \text{and} \quad k = 3.$$

8. Prove property (*i*) of Exercise 7.
9. Prove property (*ii*) of Exercise 7. *Hint*: Let $C = A + B$. Show that the (ij)th entry of C^T is the same as the (ij)th entry of $A^T + B^T$.
10. Prove property (*iii*) of Exercise 7. *Hint*: Let $C = AB$. Show that the (ij)th entry of C^T is the same as the (ij)th entry of B^TA^T. The first two

steps are:

$$((ij)\text{th entry of } C^T) = ((ji)\text{th entry of } C)$$
$$= (j\text{th row of } A) \cdot (i\text{th column of } B).$$

11. Prove property (*iv*) of Exercise 7.
12. Let A_{ij} represent blocks of A. If a square matrix A can be partitioned in such a way

$$A = \begin{bmatrix} A_{11} & A_{12} & \cdots & A_{1m} \\ \vdots & \vdots & & \vdots \\ A_{m1} & A_{m2} & \cdots & A_{mm} \end{bmatrix}$$

that all A_{ij} are square when $i = j$, and all A_{ij} are zero matrices when $i \neq j$, then A is a **diagonal block matrix.** Which, if any, of the matrices in Exercise 1 are diagonal block matrices?
13. Let A_{ij} represent blocks of A. If a square matrix A can be partitioned in such a way

$$A = \begin{bmatrix} A_{11} & A_{12} & \cdots & A_{1m} \\ \vdots & \vdots & & \vdots \\ A_{m1} & A_{m2} & \cdots & A_{mm} \end{bmatrix}$$

that all A_{ii} are square, and if, for $i > j$ (or $i < j$), all A_{ij} are zero matrices, then A is a **triangular block matrix.** Which, if any, of the matrices in Exercise 1 are triangular block matrices?
14. Construct a pair of 2 by 2 matrices and illustrate the following properties of symmetric matrices. If A and B are symmetric matrices of the same order and if k is a real number, then
    (*a*) $A + B$ is symmetric,
    (*b*) kA is symmetric,
    (*c*) A^2 is symmetric,
    (*d*) $AA^T = A^TA$ and is symmetric.
15. Prove that each of the properties listed in Exercise 14 is valid. Use properties given in Exercise 7.
16. Construct a pair of 2 by 2 matrices and illustrate the following properties of skew-symmetric matrices. If A and B are skew-symmetric matrices of the same order and if k is a real number, then
    (*a*) $A + B$ is skew-symmetric,
    (*b*) kA is skew-symmetric,
    (*c*) A^2 is symmetric,
    (*d*) $AA^T = A^TA$ and is symmetric.
17. Prove that each of the properties listed in Exercise 16 is valid. Use properties given in Exercise 7.
18. Prove that if A is an m by n matrix, then AA^T is symmetric. Use properties given in Exercise 7.

5.5 Systems of Linear Equations Having a Unique Solution

A **system of linear equations over the real numbers** is a set of m equations in n unknowns of the form

$$\begin{aligned} a_{11}x_1 + a_{12}x_2 + \cdots + a_{1n}x_n &= b_1, \\ a_{21}x_1 + a_{22}x_2 + \cdots + a_{2n}x_n &= b_2, \\ \vdots \qquad \vdots \qquad\qquad \vdots \quad &\quad \vdots \\ a_{m1}x_1 + a_{m2}x_2 + \cdots + a_{mn}x_n &= b_m, \end{aligned} \tag{5.1}$$

where the coefficients $a_{11}, a_{12}, \cdots, a_{mn}$ and the right-hand numbers $b_1, b_2, \cdots, b_m$ are real numbers. A **solution** of the linear system (5.1) is a vector $(c_1, c_2, \cdots, c_n)$ such that m identities result when its n components are substituted for the respective unknowns.

For the linear system (5.1) the matrix of coefficients

$$A = \begin{bmatrix} a_{11} & a_{12} & \cdots & a_{1n} \\ a_{21} & a_{22} & \cdots & a_{2n} \\ \vdots & \vdots & & \vdots \\ a_{m1} & a_{m2} & \cdots & a_{mn} \end{bmatrix}$$

is known as the **coefficient matrix** of the system, and this matrix augmented by the members of the right side, namely,

$$[A \mid B] = \left[\begin{array}{cccc|c} a_{11} & a_{12} & \cdots & a_{1n} & b_1 \\ a_{21} & a_{22} & \cdots & a_{2n} & b_2 \\ \vdots & \vdots & & \vdots & \vdots \\ a_{m1} & a_{m2} & \cdots & a_{mn} & b_m \end{array}\right],$$

is known as the **augmented matrix** of the system.

Example 1

The vector $(3, 4)$ is a solution of the linear system

$$\begin{aligned} x + y &= 7, \\ 2x - y &= 2, \end{aligned}$$

because if $x = 3$ and $y = 4$ are substituted in the system, we get the identities

$$\begin{aligned} (3) + (4) &= 7, \\ 2(3) - (4) &= 2. \end{aligned}$$

A graph of the system with the solution is shown in Figure 5.17; the graph of each equation is a line, and the intersection of the two lines represents the solution. The coefficient matrix of the given system is

$$\begin{bmatrix} 1 & 1 \\ 2 & -1 \end{bmatrix},$$

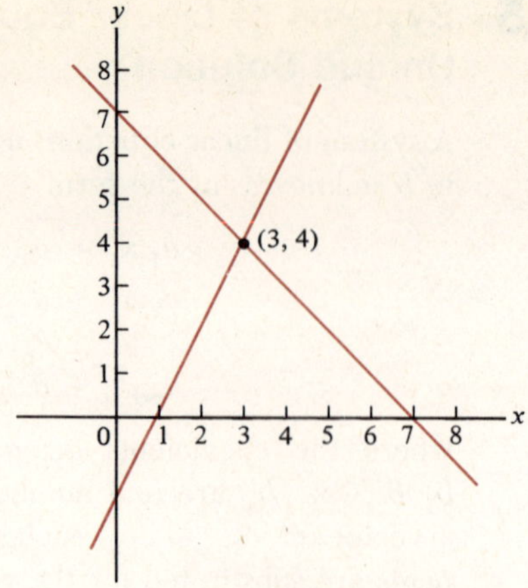

Figure 5.17

and the augmented matrix of the given system is

$$\left[\begin{array}{cc|c} 1 & 1 & 7 \\ 2 & -1 & 2 \end{array}\right]. \blacksquare$$

The primary purpose of this section is to introduce a method for finding a solution of a linear system if a solution exists.

From the old addition–subtraction method of elementary algebra the reader probably remembers that if, on a linear system, we perform the operations

(1) interchange any two equations,
(2) multiply the terms of any equation by a nonzero real number, (5.2)
(3) add to the terms of any equation k times the corresponding terms of any other equation (k is a real number),

then the resulting linear system will have the same solution as the original linear system. In other words, the solution of a system remains unchanged when the system is modified by the three operations listed above. (A proof of this assertion can be found in Campbell [8], pp. 18–20.)

Example 2

Find a solution, if one exists, of the system

$$\begin{aligned} x_1 + 2x_2 - 6x_3 &= 4, \\ -3x_1 - 6x_2 + 9x_3 &= -3, \\ x_2 - 3x_3 &= 1. \end{aligned}$$

We shall solve this system by applying the operations (5.2) to the given system; we shall also apply corresponding operations to the augmented matrix of the system in order to introduce the method that we are developing in this section. We start with

$$\begin{aligned} x_1 + 2x_2 - 6x_3 &= 4, \\ -3x_1 - 6x_2 + 9x_3 &= -3, \\ x_2 - 3x_3 &= 1, \end{aligned} \qquad \left[\begin{array}{rrr|r} 1 & 2 & -6 & 4 \\ -3 & -6 & 9 & -3 \\ 0 & 1 & -3 & 1 \end{array}\right].$$

Add 3 times the terms of the first equation to the corresponding terms of the second equation (denoted by $3R_1 + R_2$, where R_i represents the ith equation or the ith row of the augmented matrix).

$$\begin{aligned} x_1 + 2x_2 - 6x_3 &= 4, \\ - 9x_3 &= 9, \\ x_2 - 3x_3 &= 1, \end{aligned} \qquad \left[\begin{array}{rrr|r} (1) & 2 & -6 & 4 \\ (0) & 0 & -9 & 9 \\ (0) & 1 & -3 & 1 \end{array}\right].$$

Parentheses are used in the matrix to indicate when an entry is in the desired form. Next interchange the second and third equations (denoted by $R_2 \leftrightarrow R_3$).

$$\begin{aligned} x_1 + 2x_2 - 6x_3 &= 4, \\ x_2 - 3x_3 &= 1, \\ - 9x_3 &= 9, \end{aligned} \qquad \left[\begin{array}{rrr|r} (1) & 2 & -6 & 4 \\ (0) & (1) & -3 & 1 \\ (0) & (0) & -9 & 9 \end{array}\right].$$

Add -2 times the terms of the second equation to the corresponding terms of the first equation (denoted $-2R_2 + R_1$).

$$\begin{aligned} x_1 \qquad\qquad &= 2, \\ x_2 - 3x_3 &= 1, \\ - 9x_3 &= 9, \end{aligned} \qquad \left[\begin{array}{rrr|r} (1) & (0) & 0 & 2 \\ (0) & (1) & -3 & 1 \\ (0) & (0) & -9 & 9 \end{array}\right].$$

Multiply the terms of the third equation by $-\frac{1}{9}$ (denoted $-\frac{1}{9}R_3$).

$$\begin{aligned} x_1 \qquad\qquad &= 2, \\ x_2 - 3x_3 &= 1, \\ x_3 &= -1, \end{aligned} \qquad \left[\begin{array}{rrr|r} (1) & (0) & 0 & 2 \\ (0) & (1) & -3 & 1 \\ (0) & (0) & (1) & -1 \end{array}\right].$$

Add 3 times the terms of the third equation to the corresponding terms of the second equation (denoted $3R_3 + R_2$).

$$\begin{aligned} x_1 \qquad\qquad &= 2, \\ x_2 \qquad &= -2, \\ x_3 &= -1, \end{aligned} \qquad \left[\begin{array}{rrr|r} (1) & (0) & (0) & 2 \\ (0) & (1) & (0) & -2 \\ (0) & (0) & (1) & -1 \end{array}\right].$$

Because only the three allowed operations were used, we can say that the solution of the original system is the same as the obvious solution of the last system. Notice that this solution is prominently displayed in the last column of the augmented matrix. Geometrically, this solution represents the intersection of three planes in 3-space at a single point—namely, $(2, -2, -1)$. ∎

By performing the operations on the augmented matrix in each step of Example 2, we are simply illustrating the fact that the operations of (5.2) are in reality performed on the coefficients of the system.

The operations performed on the augmented matrices of Example 2 are called **elementary row operations.** They obviously correspond to the operations (5.2) that are used to solve linear systems of equations.

▶ Definition 19

The following operations performed on matrices are called **elementary row operations:**

1. the interchange of any two rows,
2. the multiplication of the entries of any row by a nonzero real number,
3. the addition to the entries of any row of k times the corresponding entries of any other row, where k is a real number.

▶ Definition 20

If a matrix A can be transformed into a matrix B by means of one or more elementary row operations, then we say that A is **row equivalent** to B; this relation will be denoted by

$$A \overset{\text{row}}{\sim} B.$$

Example 3

If the first and second rows of the matrix

$$A = \left[\begin{array}{cc|c} 2 & 9 & 6 \\ 1 & 4 & 2 \end{array}\right]$$

are interchanged, denoted $(R_1 \leftrightarrow R_2)$, we get

$$B = \left[\begin{array}{cc|c} 1 & 4 & 2 \\ 2 & 9 & 6 \end{array}\right].$$

We can say that $A \overset{\text{row}}{\sim} B$. Furthermore, if we add -2 times the entries of the first row to the corresponding entries of the second row, denoted $(-2R_1 + R_2)$, we get

$$C = \left[\begin{array}{cc|c} 1 & 4 & 2 \\ 0 & 1 & 2 \end{array}\right].$$

Thus we say $B \overset{\text{row}}{\sim} C$. We can also say $A \overset{\text{row}}{\sim} C$. Finally, if we add -4 times the entries of the second row to the corresponding entries of the first row (denoted $-4R_2 + R_1$), we get

$$D = \left[\begin{array}{cc|c} 1 & 0 & -6 \\ 0 & 1 & 2 \end{array}\right].$$

We can say $C \overset{\text{row}}{\sim} D$ and we can also say $A \overset{\text{row}}{\sim} D$. Without the explanations the work above can be expressed very simply as

$$\left[\begin{array}{cc|c} 2 & 9 & 6 \\ 1 & 4 & 2 \end{array}\right] \underset{R_1 \leftrightarrow R_2}{\overset{\text{row}}{\sim}} \left[\begin{array}{cc|c} (1) & 4 & 2 \\ 2 & 9 & 6 \end{array}\right] \underset{-2R_1 + R_2}{\overset{\text{row}}{\sim}} \left[\begin{array}{cc|c} (1) & 4 & 2 \\ (0) & 1 & 2 \end{array}\right]$$

$$\underset{-4R_2 + R_1}{\overset{\text{row}}{\sim}} \left[\begin{array}{cc|c} (1) & (0) & -6 \\ (0) & (1) & 2 \end{array}\right].$$

Therefore, if we had been asked to solve the system

$$\begin{aligned} 2x_1 + 9x_2 &= 6, \\ x_1 + 4x_2 &= 2, \end{aligned}$$

with the corresponding augmented matrix

$$\left[\begin{array}{cc|c} 2 & 9 & 6 \\ 1 & 4 & 2 \end{array}\right],$$

we could read the answer in the last column of the row equivalent augmented matrix

$$\left[\begin{array}{cc|c} (1) & (0) & -6 \\ (0) & (1) & 2 \end{array}\right]. \blacksquare$$

In summary, we have illustrated a method of solution of systems of linear equations that have a single solution. The reader should be aware, however, that some systems of linear equations have more than one solution and some have no solution; a method for dealing with such systems is similar to that used in this section and will be discussed later.

APPLICATIONS

Example 4

In this example we introduce an economic model that has been used extensively since its invention in 1936 by Leontief. The model is frequently referred to as the Leontief input–output model. Consider an economy (system) with n industries (components) that are dependent upon each other. Each industry enjoys a demand for its products—an *internal* demand from the other industries and an *external* demand from outside the economy. Once an agreement has been made concerning a time interval and the units of measure (such as dollar value) of each of the products, then the following notation can be introduced. Let

x_j = output of industry j,
c_{ij} = number of units of output of industry i used in the production of one unit of output of industry j,
d_i = external demand on industry i.

If c_{51} represents the number of units of output of industry 5 used in the production of *one* unit of output of industry 1, and if x_1 represents the number of units of output of industry 1, then the product $c_{51}x_1$ represents the demand on industry 5 by industry 1, the product $c_{52}x_2$ represents the demand on industry 5 by industry 2, and so on. The sum of these products

$$c_{51}x_1 + c_{52}x_2 + \cdots + c_{5n}x_n$$

represents the total internal demand on industry 5. This internal demand plus the external demand d_5 should equal the production, x_5. Thus we have the equation

$$c_{51}x_1 + c_{52}x_2 + \cdots + c_{5n}x_n + d_5 = x_5$$

for the fifth industry. Analogous equations for the other industries produce the system

$$\begin{aligned} c_{11}x_1 + c_{12}x_2 + \cdots + c_{1n}x_n + d_1 &= x_1, \\ c_{21}x_1 + c_{22}x_2 + \cdots + c_{2n}x_n + d_2 &= x_2, \\ &\vdots \\ c_{n1}x_1 + c_{n2}x_2 + \cdots + c_{nn}x_n + d_n &= x_n. \end{aligned}$$

If rearranged in standard form, the system is

$$\begin{aligned} (1 - c_{11})x_1 - c_{12}x_2 - \cdots - c_{1n}x_n &= d_1, \\ -c_{21}x_1 + (1 - c_{22})x_2 - \cdots - c_{2n}x_n &= d_2, \\ &\vdots \\ -c_{n1}x_1 - c_{n2}x_2 - \cdots + (1 - c_{nn})x_n &= d_n. \end{aligned}$$ ∎

Example 5

In a wide variety of applications it is desirable to generate polynomial equations that will approximate a possible relationship between two variables when pairs of corresponding values of these variables are known. One method of obtaining such an equation is the method of **least squares.** We shall illustrate the method by means of an example in which we generate a first-degree polynomial equation $y = ax + b$ subject to the observed data of Table 5.3. The least-squares method requires that we find numbers a and b in the equations

$$\begin{aligned} y_1 &= ax_1 + b + r_1, \\ y_2 &= ax_2 + b + r_2, \\ y_3 &= ax_3 + b + r_3, \\ y_4 &= ax_4 + b + r_4, \end{aligned} \tag{5.3}$$

in such a way that $S = r_1^2 + r_2^2 + r_3^2 + r_4^2$ is a minimum (whence the name "least squares"). The quantities r_1, r_2, r_3, and r_4 are called **residuals** and are pictured in Figure 5.18. The four equations of (5.3) can be expressed in matrix notation as

$$Y = C\alpha + \rho,$$

where

$$Y = \begin{bmatrix} y_1 \\ y_2 \\ y_3 \\ y_4 \end{bmatrix}, \quad C = \begin{bmatrix} x_1 & 1 \\ x_2 & 1 \\ x_3 & 1 \\ x_4 & 1 \end{bmatrix}, \quad \alpha = \begin{bmatrix} a \\ b \end{bmatrix}, \quad \rho = \begin{bmatrix} r_1 \\ r_2 \\ r_3 \\ r_4 \end{bmatrix}.$$

Table 5.3

x	1	3	4	6
y	3	7	8	8

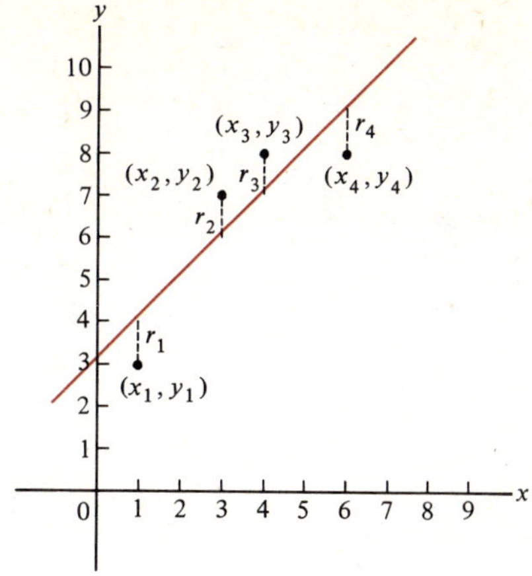

Figure 5.18

Using calculus, one can prove (Campbell [8], p. 145) that S assumes its minimum value only if

$$(C^T C)\alpha = C^T Y.$$

Using the data of Table 5.3, we get

$$\begin{bmatrix} 1 & 3 & 4 & 6 \\ 1 & 1 & 1 & 1 \end{bmatrix} \begin{bmatrix} 1 & 1 \\ 3 & 1 \\ 4 & 1 \\ 6 & 1 \end{bmatrix} \begin{bmatrix} a \\ b \end{bmatrix} = \begin{bmatrix} 1 & 3 & 4 & 6 \\ 1 & 1 & 1 & 1 \end{bmatrix} \begin{bmatrix} 3 \\ 7 \\ 8 \\ 8 \end{bmatrix},$$

or

$$\begin{bmatrix} 62 & 14 \\ 14 & 4 \end{bmatrix} \begin{bmatrix} a \\ b \end{bmatrix} = \begin{bmatrix} 104 \\ 26 \end{bmatrix}.$$

By the methods of this section the solution of this system is

$$\begin{bmatrix} a \\ b \end{bmatrix} = \begin{bmatrix} 1 \\ 3 \end{bmatrix}.$$

Hence if S is a minimum, then the desired equation of the line of Figure 5.18 is

$$y = x + 3.$$

In summary, data such as Table 5.3 generally produces a system $C\alpha = Y$ that has no solutions, but by multiplying both sides of this matrix equation by C^T and by solving for α, we obtain the least-squares solution of the first-degree polynomial equation of best fit. Generalizations of the method, including weighted residuals and multiple variables, are outlined in Campbell [8], pp. 144–145. ▮

EXERCISES

Suggested minimum assignment: Exercises 1, 5, 7, 9, 13, and 15.

In Exercises 1–4 write the coefficient and augmented matrices of the given system.

1. $x_1 - x_2 = 2,$
   $2x_1 + 4x_2 = 1.$

2. $x_1 + 3x_2 = 6,$
   $2x_1 - x_2 = 1.$

3. $x_1 + 3x_2 + x_3 - 2 = 0,$
   $- x_2 - 2x_3 + 3 = 0.$

4. $x_1 - 3x_2 - x_3 + 5 = 0,$
   $x_1 + x_3 - 7 = 0.$

In Exercises 5–8 find the solution of the given system by following the format of Example 2. That is, solve the given system by using the operations of (5.2), and by performing the corresponding elementary row operations on the associated augmented matrices. Check each answer by substituting it in the original system to see if m identities are obtained.

5. $x_1 + 3x_2 = 3,$
   $2x_1 + x_2 = 6.$

6. $x_1 - x_2 = 4,$
   $-3x_1 + x_2 = 6.$

7. $x_1 - x_3 = 2,$
   $x_1 - 2x_2 = 7,$
   $x_2 + 3x_3 = 1.$

8. $x_1 + 2x_2 - x_3 = 6,$
   $- 5x_2 + 5x_3 = -25,$
   $3x_1 - 2x_2 + 3x_3 = -16.$

In Exercises 9–12 find the solution of the given system by using elementary row operations on augmented matrices. Check your answers.

9. $x_1 + 2x_2 = 5,$
   $-2x_1 + 4x_2 = 6.$

10. $2x_1 + 3x_2 = 8,$
    $x_1 + x_2 = 2.$

11. $x_2 + x_3 = 1,$
    $x_1 + x_2 + x_3 = 2,$
    $x_2 - x_3 = 5.$

12. $x_1 + 4x_3 = 12,$
    $2x_1 + x_2 + 4x_3 = 13,$
    $-2x_1 + 2x_3 = 6.$

13. Graph the equations of the system in Exercise 9 and give a geometric interpretation of the solution.
14. Graph the equations of the system in Exercise 10 and give a geometric interpretation of the solution.
15. In finding the solution of Exercise 9, one must write a sequence of row equivalent matrices. Each of these matrices represents an augmented matrix of a corresponding system; graph the equations of each of these systems. Give a geometric interpretation of the algebraic procedure used to find the solution.
16. In finding the solution of Exercise 10, one must write a sequence of row equivalent matrices. Each of these matrices represents an augmented matrix of a corresponding system; graph the equations of each of these systems. Give a geometric interpretation of the algebraic procedure used to find the solution.

5.6 The Inverse Matrix

In this section we shall present an alternative method to that given in the last section for solving a system of m linear equations in n unknowns when that system has a single solution, and when $m = n$.

As we illustrated in Example 2 of Section 5.3, multiplication can be used to express a system of linear equations

$$\begin{aligned} a_{11}x_1 + a_{12}x_2 + \cdots + a_{1n}x_n &= b_1, \\ a_{21}x_1 + a_{22}x_2 + \cdots + a_{2n}x_n &= b_2, \\ \vdots \qquad \vdots \qquad\quad\;\; \vdots \quad\; &\quad \vdots \\ a_{m1}x_1 + a_{m2}x_2 + \cdots + a_{mn}x_n &= b_m, \end{aligned}$$

by means of a single matrix equation

$$AX = B,$$

where

$$A = \begin{bmatrix} a_{11} & a_{12} & \cdots & a_{1n} \\ a_{21} & a_{22} & \cdots & a_{2n} \\ \vdots & \vdots & & \vdots \\ a_{m1} & a_{m2} & \cdots & a_{mn} \end{bmatrix}, \quad X = \begin{bmatrix} x_1 \\ x_2 \\ \vdots \\ x_n \end{bmatrix}, \quad \text{and} \quad B = \begin{bmatrix} b_1 \\ b_2 \\ \vdots \\ b_m \end{bmatrix}.$$

In the algebra of real numbers, when one was faced with the question of solving a single equation

$$ax = b \qquad \text{where } a \neq 0,$$

the standard approach was to premultiply both sides by $1/a$ or a^{-1} to obtain

$$a^{-1}(ax) = a^{-1}b,$$

which by the associative law for multiplication of real numbers is

$$(a^{-1}a)x = a^{-1}b.$$

Since $(a^{-1}a) = 1$, we have

$$1 \cdot x = a^{-1}b,$$

and because 1 is the identity element for multiplication we have

$$x = a^{-1}b.$$

It is therefore natural to wonder if we can use an analogous procedure to solve for X in the matrix equation $AX = B$. In order to provide an answer to this speculation, we must be sure that the term $1/a$ or a^{-1} is clearly understood and then extend this concept to matrix algebra. If for a given real number a there is a real number $1/a$ or a^{-1} such that

$$a \cdot a^{-1} = a^{-1} \cdot a = 1,$$

we say that a^{-1} is the **multiplicative inverse** of a. The reader should remember that all real numbers except 0 have a multiplicative inverse. For example, the multiplicative inverse of 3 is $\frac{1}{3}$ because $3 \cdot \frac{1}{3} = \frac{1}{3} \cdot 3 = 1$. We shall define a multiplicative inverse of a matrix in a similar way.

▶ Definition 21

If for a given n by n matrix A, there is a matrix designated A^{-1} such that

$$AA^{-1} = A^{-1}A = I_n,$$

then A^{-1} is an **inverse** of A with respect to multiplication.

Example 1

For the matrix $A = \begin{bmatrix} 3 & 1 \\ 5 & 2 \end{bmatrix}$ the matrix $\begin{bmatrix} 2 & -1 \\ -5 & 3 \end{bmatrix}$ can be designated as A^{-1} because

$$\begin{bmatrix} 3 & 1 \\ 5 & 2 \end{bmatrix}\begin{bmatrix} 2 & -1 \\ -5 & 3 \end{bmatrix} = \begin{bmatrix} 2 & -1 \\ -5 & 3 \end{bmatrix}\begin{bmatrix} 3 & 1 \\ 5 & 2 \end{bmatrix} = \begin{bmatrix} 1 & 0 \\ 0 & 1 \end{bmatrix}. \blacksquare$$

Example 2

Consider the system

$$\begin{aligned} 3x_1 + x_2 &= 3, \\ 5x_1 + 2x_2 &= 2, \end{aligned}$$

which can be expressed as the matrix equation

$$AX = B,$$

where

$$A = \begin{bmatrix} 3 & 1 \\ 5 & 2 \end{bmatrix}, \quad X = \begin{bmatrix} x_1 \\ x_2 \end{bmatrix}, \quad \text{and} \quad B = \begin{bmatrix} 3 \\ 2 \end{bmatrix}.$$

From Example 1 we know that

$$A^{-1} = \begin{bmatrix} 2 & -1 \\ -5 & 3 \end{bmatrix};$$

hence we can premultiply both sides of $AX = B$ by A^{-1} to get

$$\begin{aligned} A^{-1}(AX) &= A^{-1}B, \\ (A^{-1}A)X &= A^{-1}B, \\ I_2X &= A^{-1}B, \\ X &= A^{-1}B. \end{aligned}$$

Thus a solution of the given system is

$$X = A^{-1}B = \begin{bmatrix} 2 & -1 \\ -5 & 3 \end{bmatrix}\begin{bmatrix} 3 \\ 2 \end{bmatrix} = \begin{bmatrix} 4 \\ -9 \end{bmatrix}. \blacksquare$$

From the last two examples, three rather obvious questions emerge. (1) Do all matrices have multiplicative inverses? (2) How does one calculate a multiplicative inverse if one exists? (3) Is there more than one multiplicative inverse of a matrix? Answers to these questions are provided by the next two theorems.

▶ Theorem 5

An n by n matrix A has a multiplicative inverse if and only if $A \overset{\text{row}}{\sim} I_n$. Moreover, if $[A \mid I_n] \overset{\text{row}}{\sim} [I_n \mid P]$, then P is a multiplicative inverse of A.

A proof of Theorem 5 is beyond the scope of this text and may be found in more advanced texts. (One source is Section 3.4 of Campbell [8].) ◀

Example 3

Calculate an inverse of $A = \begin{bmatrix} 1 & 4 \\ 2 & 9 \end{bmatrix}$ if possible. According to Theorem 5 we construct the matrix $[A \mid I_2]$ and then determine if we can find a matrix $[I_2 \mid P]$ to which $[A \mid I_2]$ is row equivalent. If we can find such a matrix, then $A^{-1} = P$.

$$[A \mid I_2] = \left[\begin{array}{cc|cc} 1 & 4 & 1 & 0 \\ 2 & 9 & 0 & 1 \end{array}\right] \underset{-2R_1+R_2}{\overset{\text{row}}{\sim}} \left[\begin{array}{cc|cc} 1 & 4 & 1 & 0 \\ 0 & 1 & -2 & 1 \end{array}\right]$$

$$\underset{-4R_2+R_1}{\overset{\text{row}}{\sim}} \left[\begin{array}{cc|cc} 1 & 0 & 9 & -4 \\ 0 & 1 & -2 & 1 \end{array}\right] = [I_2 \mid P].$$

Therefore,

$$A^{-1} = P = \begin{bmatrix} 9 & -4 \\ -2 & 1 \end{bmatrix}. \blacksquare$$

Example 4

Calculate an inverse of $A = \begin{bmatrix} 1 & 2 \\ 3 & 6 \end{bmatrix}$ if possible.

$$[A \mid I_2] = \left[\begin{array}{cc|cc} 1 & 2 & 1 & 0 \\ 3 & 6 & 0 & 1 \end{array}\right] \underset{-3R_1+R_2}{\overset{\text{row}}{\sim}} \left[\begin{array}{cc|cc} 1 & 2 & 1 & 0 \\ 0 & 0 & -3 & 1 \end{array}\right].$$

We quickly see that A is *not* row equivalent to I_2; hence, by Theorem 5, A does *not* have a multiplicative inverse. ▮

A square matrix that has a multiplicative inverse is called an **invertible** matrix, and a square matrix that does not have a multiplicative inverse is called a **noninvertible** matrix.†

▶ Theorem 6

If a square matrix is invertible, then the inverse is unique.

† Invertible matrices are often called *nonsingular* matrices and noninvertible matrices are often called *singular* matrices.

Proof: Assume that there are two multiplicative inverses of an n by n matrix A and show that they must be equal. Assume that B and C are inverses of A; hence

$$AB = BA = I_n \quad \text{and} \quad AC = CA = I_n.$$

Statements: $C = CI_n = C(AB) = (CA)B = I_nB = B$. The reasons for the statements of the proof are left as Exercise 20. ◀

Other properties of the inverse matrix are stated in Exercise 15.

Because of Theorem 6, we can be sure that in Example 2 the given multiplicative inverse was the only multiplicative inverse, and hence the solution of the system given in Example 2 was unique. We now conclude the material of this section with the following important theorem.

▶ **Theorem 7**

If A is invertible, then the linear system $AX = B$ has the unique solution

$$X = A^{-1}B.$$

The proof of Theorem 7 is left as Exercise 21. ◀

APPLICATIONS

Example 5

Theorem 7 is particularly useful in those applications where the coefficient matrix remains fixed and the constants on the right-hand side vary. For example, suppose that a certain company has two machines A and B that manufacture two products C and D. The time needed by each machine to produce one unit of each product is given in the following chart.

	Products	
	C	D
Machine A	5 hours used	3 hours used
Machine B	1 hour used	7 hours used

The management wishes to decide how many units (or fractions of units) of each product to make if each machine is to be used exactly 16 hours per day. (The remaining hours are reserved for the maintenance of the machines.) They also want to know the effect on production if additional machines are purchased.

Solution: Let x_1 be the number of units of product C produced each day, and let x_2 be the number of units of product D produced each day. The time spent by machine A is

(number of hours per unit on product C) (number of units of C)
+ (number of hours per unit on product D) (number of units of D)

or

$$5x_1 + 3x_2,$$

and the amount must be equal to 16 hours. Hence we have the equation

$$5x_1 + 3x_2 = 16.$$

For machine B we have the equation

$$1x_1 + 7x_2 = 16.$$

These two equations can be expressed as the single matrix equation

$$AX = B,$$

where

$$A = \begin{bmatrix} 5 & 3 \\ 1 & 7 \end{bmatrix}, \quad X = \begin{bmatrix} x_1 \\ x_2 \end{bmatrix}, \quad \text{and} \quad B = \begin{bmatrix} 16 \\ 16 \end{bmatrix}.$$

By Theorem 7, if A^{-1} exists,

$$X = A^{-1}B,$$

and by use of Theorem 5 we find

$$A^{-1} = \begin{bmatrix} \frac{7}{32} & \frac{-3}{32} \\ \frac{-1}{32} & \frac{5}{32} \end{bmatrix}.$$

Hence

$$X = \begin{bmatrix} \frac{7}{32} & \frac{-3}{32} \\ \frac{-1}{32} & \frac{5}{32} \end{bmatrix} \begin{bmatrix} 16 \\ 16 \end{bmatrix} = \begin{bmatrix} 2 \\ 2 \end{bmatrix}.$$

Therefore two units of product C and two units of product D should be produced each day to satisfy the conditions stated initially. Now suppose that the management wants to know the effect on production if more machines are purchased. For example, if one additional machine of type A is purchased, and three additional machines of type B are purchased, then 32 hours are available on a type A machine and 64 hours are available on a type B machine.

We now recalculate $A^{-1}B$, noticing that A^{-1} does not change.

$$X = \begin{bmatrix} \frac{7}{32} & \frac{-3}{32} \\ \frac{-1}{32} & \frac{5}{32} \end{bmatrix} \begin{bmatrix} 32 \\ 64 \end{bmatrix} = \begin{bmatrix} 1 \\ 9 \end{bmatrix}. \blacksquare$$

Example 6

In this example we illustrate an application of the Leontief input–output model presented in Example 4 of Section 5.5. Suppose that a certain utility corporation has three components of operation: it mines coal, produces gasoline, and generates electricity. Each component makes use of the products of the other components as shown in Table 5.4. The entries of

Table 5.4. CONSUMPTION TABLE.

To produce one unit of		
coal requires	gasoline requires	electricity requires
0 units of coal	0 units of coal	$\frac{1}{5}$ unit of coal
1 unit of gasoline	$\frac{1}{5}$ unit of gasoline	$\frac{2}{5}$ unit of gasoline
1 unit of electricity	$\frac{2}{5}$ unit of electricity	$\frac{1}{5}$ unit of electricity

Table 5.4 form what is called a consumption matrix, C. If for a given time interval, we let x_1 be the output of coal, x_2 be the output of gasoline, and x_3 be the output of electricity, then the product

$$CX = \begin{bmatrix} 0 & 0 & \frac{1}{5} \\ 1 & \frac{1}{5} & \frac{2}{5} \\ 1 & \frac{2}{5} & \frac{1}{5} \end{bmatrix} \begin{bmatrix} x_1 \\ x_2 \\ x_3 \end{bmatrix}$$

represents the internal consumption of each of the products of the 3 components. Finally, we let $d_1 = 100$, $d_2 = 100$, and $d_3 = 100$ be the external consumption of the products of each component; then we have

$$(\text{production}) = (\text{internal consumption}) + (\text{external consumption})$$

or

$$X = CX + D \qquad \text{or} \qquad (I_3 - C)X = D.$$

The latter matrix equation can be solved for X provided $(I_3 - C)^{-1}$ exists. We get

$$X = (I_3 - C)^{-1}D,$$

which for our data becomes

$$X = \left(\begin{bmatrix} 1 & 0 & 0 \\ 0 & 1 & 0 \\ 0 & 0 & 1 \end{bmatrix} - \begin{bmatrix} 0 & 0 & \frac{1}{5} \\ 1 & \frac{1}{5} & \frac{2}{5} \\ 1 & \frac{2}{5} & \frac{1}{5} \end{bmatrix} \right)^{-1} \begin{bmatrix} 100 \\ 100 \\ 100 \end{bmatrix} = \begin{bmatrix} 300 \\ 1000 \\ 1000 \end{bmatrix}.$$

Once the internal consumption is established and $(I_3 - C)^{-1}$ is calculated, then variations in X will be a function of D only. ∎

Example 7

In Chapter 6 of Johnston et al. [23] the reader may find several variations of the model presented in Example 6; these variations are applied to problems in business and economics. Particular attention is called to the model for allocation of service charges in accounting that is begun on page 262 of [23]. Students of business and economics are encouraged to at least glance at Chapter 6 of [23] to assure them of the relevance of the material they are studying in this chapter. The bibliography and bibliographical notes on pages 272–273 of [23] may also be helpful. ▮

EXERCISES

Suggested minimum assignment: Exercises 1, 4, 6, 8, 10, 12, and 14.

1. Write the multiplicative inverses of the real numbers 7, $\frac{2}{5}$, and $\sqrt{2}$. Why does the real number 0 not have an inverse?
2. Write the multiplicative inverses of the 1 by 1 matrices [7], [$\frac{2}{5}$], and [$\sqrt{2}$]. Why does the matrix [0] not have an inverse?

In Exercises 3–10 calculate the inverses of the given matrices if possible. If not possible, give the reason. Check your answer by using Definition 21.

3. $\begin{bmatrix} 2 & 1 \\ 6 & 2 \end{bmatrix}$.

4. $\begin{bmatrix} 4 & 5 \\ 6 & 7 \end{bmatrix}$.

5. $\begin{bmatrix} 1 & -2 \\ -3 & 6 \end{bmatrix}$.

6. $\begin{bmatrix} -1 & 3 \\ -3 & 9 \end{bmatrix}$.

7. $\begin{bmatrix} 2 & 1 \\ 3 & 2 \\ 1 & 4 \end{bmatrix}$.

8. $\begin{bmatrix} 1 & 2 & 6 \\ 0 & 1 & 5 \end{bmatrix}$.

9. $\begin{bmatrix} 1 & 1 & 7 \\ 0 & 1 & 2 \\ 2 & 4 & 17 \end{bmatrix}$.

10. $\begin{bmatrix} 1 & 2 & 4 \\ 2 & 4 & 7 \\ 0 & 1 & 3 \end{bmatrix}$.

In Exercises 11 and 12 find the solution of the system by finding and using the multiplicative inverse of the coefficient matrix. Check your solution by substituting it into the original system.

11. $x_1 + x_2 = 5,$
    $2x_1 - x_2 = 1.$

12. $x_1 - 2x_2 = -5,$
    $3x_1 + 5x_2 = 18.$

In Exercises 13 and 14 find the solution of the systems $AX = B$, where A^{-1} is given. Check your solution by substituting it into the original system.

13. $$\begin{aligned} 2x_1 + 4x_2 \qquad &= 60, \\ 2x_2 + x_3 &= 20, \\ 3x_1 \qquad + 2x_3 &= 40. \end{aligned}$$

$$A^{-1} = \begin{bmatrix} \frac{4}{20} & \frac{-8}{20} & \frac{4}{20} \\ \frac{3}{20} & \frac{4}{20} & \frac{-2}{20} \\ \frac{-6}{20} & \frac{12}{20} & \frac{4}{20} \end{bmatrix}.$$

14. $\begin{aligned} x_1 - \quad x_2 - \quad x_3 &= 0, \\ 5x_1 \qquad + 20x_3 &= 50, \\ 10x_2 - 20x_3 &= 30. \end{aligned}$ $\quad A^{-1} = \begin{bmatrix} \frac{4}{7} & \frac{3}{35} & \frac{4}{70} \\ \frac{-2}{7} & \frac{2}{35} & \frac{5}{70} \\ \frac{-1}{7} & \frac{1}{35} & \frac{-1}{70} \end{bmatrix}.$

15. A few of the properties involving invertible and noninvertible matrices follow. (In all four cases A and B are square matrices.)
    (*i*) If A is an invertible matrix, then $(A^{-1})^{-1} = A$;
    (*ii*) If A and B are invertible matrices, then $(AB)^{-1} = B^{-1}A^{-1}$;
    (*iii*) If $AB = \mathbf{0}$, then $A = \mathbf{0}$, or $B = \mathbf{0}$, or both A and B are noninvertible;
    (*iv*) If $AB = I_n$, then $B = A^{-1}$, $A = B^{-1}$, and $AB = BA$.

    (*a*) Illustrate (*i*) and (*ii*) with $A = \begin{bmatrix} 1 & 3 \\ 2 & 7 \end{bmatrix}$, $\quad B = \begin{bmatrix} 1 & 3 \\ 0 & 1 \end{bmatrix}$.

    (b) Does $(AB)^{-1} = A^{-1}B^{-1}$ and why?

    (*c*) Illustrate (*iii*) with $A = \begin{bmatrix} 1 & 1 \\ 2 & 2 \end{bmatrix}$, $\quad B = \begin{bmatrix} 1 & -3 \\ -1 & 3 \end{bmatrix}$.

    (*d*) Illustrate (*iv*) with $A = \begin{bmatrix} 1 & 3 \\ 0 & 1 \end{bmatrix}$, $\quad B = \begin{bmatrix} 1 & -3 \\ 0 & 1 \end{bmatrix}$.

    (*e*) If A and B are nonzero matrices and $AB = \mathbf{0}$, is $B \overset{\text{row}}{\sim} I_n$ and why?

16. Prove property (*i*) of Exercise 15. *Hint:* Let $B = A^{-1}$ and hence $AB = BA = I_n$; then $A = B^{-1}$.
17. Prove property (*ii*) of Exercise 15. *Hint:* Show that $(AB)(B^{-1}A^{-1}) = I_n$ and that $(B^{-1}A^{-1})(AB) = I_n$; then use Definition 21 and Theorem 6.
18. Prove property (*iii*) of Exercise 15. *Hint:* Show that the existence of A^{-1} implies $B = \mathbf{0}$; show that the existence of B^{-1} implies $A = \mathbf{0}$, and then consider the only other possibility.
19. Prove property (*iv*) of Exercise 15. *Hint:* Start with $B = I_nB = (A^{-1}A)B$ and continue on to show $B = A^{-1}$. Use a similar procedure to show $A = B^{-1}$, and then use Definition 21 to show $AB = BA$.
20. Give reasons for the statements of the proof of Theorem 6.
21. Prove Theorem 7. *Hint:* See Example 2.
22. Rework Example 5 if there are 4 type A machines and 2 type B machines.

5.7 Systems of Linear Equations Not Having a Unique Solution

In general, a system of linear equations over the real numbers may have (1) a unique solution, (2) an infinite number of solutions, or (3) no solution.

In the last two sections we have concentrated on the first of these three possibilities; in Section 5.5 we used elementary row operations on the augmented matrix to obtain the unique solution, and in Section 5.6 we used the inverse matrix and multiplication to determine the unique solution. In this and the next section we shall extend and formalize the use of elementary operations on the augmented matrix of a linear system in such a way that the same procedure may be used on any linear system, regardless of the nature of the solution. We shall begin by illustrating the method in the following two examples.

Example 1

Consider the following linear system over the real numbers:

$$\begin{aligned}
(1)\quad x_1 + x_2 - 2x_3 &= -1,\\
(2)\quad x_1 + x_2 + 2x_3 &= 3,\\
(3)\quad 2x_1 + 2x_2 + x_3 &= 3.
\end{aligned}$$

We will find that this system has an infinite number of solutions; this case is illustrated in Figure 5.19, where the graphs of the equations of the system are planes which are seen to intersect along a line.

The points of the line of intersection correspond to the solutions of the system; algebraically, we can find those solutions in the following way. The augmented matrix of the system is

$$\left[\begin{array}{ccc|c} 1 & 1 & -2 & -1 \\ 1 & 1 & 2 & 3 \\ 2 & 2 & 1 & 3 \end{array}\right].$$

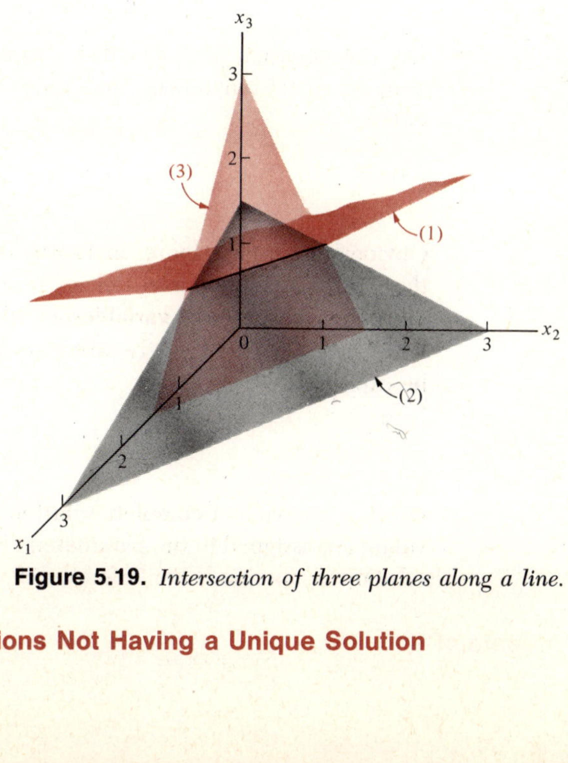

Figure 5.19. *Intersection of three planes along a line.*

First we perform the elementary row operations necessary to make the first entry in the first row a 1 and to make the other entries in the first column a 0. The operations $(-R_1 + R_2)$ and $(-2R_1 + R_3)$ will accomplish this task, and we get the row equivalent matrix

$$\left[\begin{array}{ccc|c} (1) & 1 & -2 & -1 \\ (0) & 0 & 4 & 4 \\ (0) & 0 & 5 & 5 \end{array}\right].$$

We now move to the second row and perform the elementary row operations necessary to make the first nonzero entry of the second row a 1. The operation $(\frac{1}{4}R_2)$ will do the job, and we obtain the row equivalent matrix

$$\left[\begin{array}{ccc|c} (1) & 1 & -2 & -1 \\ (0) & 0 & (1) & 1 \\ (0) & 0 & 5 & 5 \end{array}\right].$$

We then make the other entries in the third column be 0. The operations $(2R_2 + R_1)$ and $(-5R_2 + R_3)$ will produce the row equivalent matrix

$$\left[\begin{array}{ccc|c} (1) & 1 & (0) & 1 \\ (0) & 0 & (1) & 1 \\ (0) & 0 & (0) & 0 \end{array}\right].$$

Because there are no other nonzero rows, the process is complete, and we write the system for which the latter matrix is the augmented matrix;

$$\begin{aligned} x_1 + x_2 \qquad &= 1, \\ x_3 &= 1, \\ 0x_1 + 0x_2 + 0x_3 &= 0. \end{aligned} \tag{5.4}$$

The last equation of (5.4) will be an identity no matter what values are assigned to the unknowns, and hence (5.4) has the same solution as the system

$$\begin{aligned} x_1 + x_2 \qquad &= 1, \\ x_3 &= 1. \end{aligned} \tag{5.5}$$

Obviously there are more unknowns than there are equations in (5.5). If there are r equations in n unknowns at this stage, then we can *express r unknowns* (called **basic variables**) *in terms of the remaining n–r unknowns* which are considered to be arbitrary parameters. System (5.5) can then be expressed as

$$\begin{aligned} x_1 &= 1 - x_2, \\ x_3 &= 1 - 0x_2, \end{aligned}$$

which is known as a **complete solution** of the original system. If particular values are assigned to the parameters (x_2 in this example), then a **particular solution** is obtained. For example, if we let $x_2 = 1$, then we obtain the particular solution (0, 1, 1), which can be interpreted as the list of coordinates of a single point on the line of intersection shown in Figure 5.19. ∎

The next example will illustrate how we can use elementary row operations to reveal the case in which no solution exists for a linear system.

Example 2

Consider the system

$$\begin{aligned}
(1)\quad & x_1 \phantom{{}+x_2} + 2x_3 = 4,\\
(2)\quad & + x_2 - 2x_3 = 2,\\
(3)\quad & x_1 + x_2 \phantom{{}-2x_3} = 7,
\end{aligned}$$

whose graph (see Figure 5.20) consists of 3 planes oriented in such a way that no point is common to all 3 planes. Construct the augmented matrix of the system and perform the indicated elementary row operations

$$\left[\begin{array}{ccc|c} 1 & 0 & 2 & 4 \\ 0 & 1 & -2 & 2 \\ 1 & 1 & 0 & 7 \end{array}\right] \underset{-R_1+R_3}{\overset{\text{row}}{\sim}} \left[\begin{array}{ccc|c} (1) & 0 & 2 & 4 \\ (0) & 1 & -2 & 2 \\ (0) & 1 & -2 & 3 \end{array}\right]$$

$$\underset{-R_2+R_3}{\overset{\text{row}}{\sim}} \left[\begin{array}{ccc|c} (1) & (0) & 2 & 4 \\ (0) & (1) & -2 & 2 \\ (0) & (0) & 0 & 1 \end{array}\right] \underset{\substack{-2R_3+R_2\\ -4R_3+R_1}}{\overset{\text{row}}{\sim}} \left[\begin{array}{ccc|c} (1) & (0) & 2 & (0) \\ (0) & (1) & -2 & (0) \\ (0) & (0) & 0 & (1) \end{array}\right].$$

The last matrix is the augmented matrix of the system

$$\begin{aligned}
x_1 \phantom{{}+0x_2} + 2x_3 &= 0,\\
x_2 - 2x_3 &= 0,\\
0x_1 + 0x_2 + 0x_3 &= 1,
\end{aligned}$$

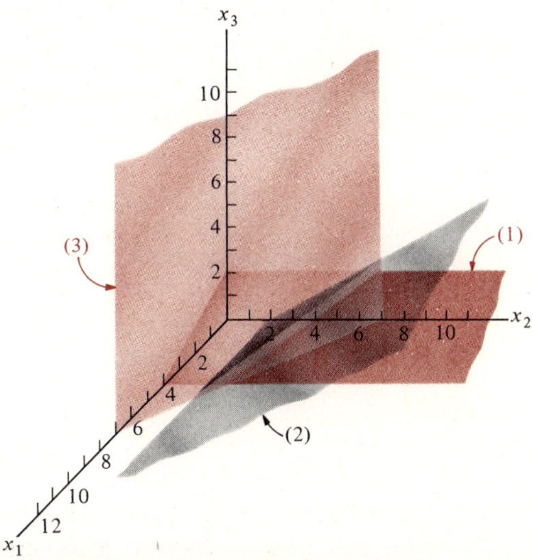

Figure 5.20. *Three planes having no point in common.*

which we know must have the same solution as the original system. But the last system has no solution because there are no real numbers for x_1, x_2, and x_3 such that $0x_1 + 0x_2 + 0x_3 = 1$. Therefore, the original system has no solution. ▮

APPLICATIONS

Example 3

Consider the system of linear equations

$$\begin{aligned} p_{11}x_1 + p_{12}x_2 + p_{13}x_3 &= \lambda x_1, \\ p_{21}x_1 + p_{22}x_2 + p_{23}x_3 &= \lambda x_2, \\ p_{31}x_1 + p_{32}x_2 + p_{33}x_3 &= \lambda x_3, \end{aligned}$$

where λ is a constant, p_{ij} is the probability that a purchaser who bought brand j of a certain product last time will buy brand i next time, and where x_j represents the numbers of buyers that bought brand j the first time. If the probabilities p_{ij} are known, then we have a linear system

$$PX = \lambda X,$$

whose solution would be the number of initial buyers of each brand that would result in the same proportion of buyers of each brand at the next purchase time. A stability in the proportions of buyers of each brand would be in effect as long as the probabilities did not change. To illustrate, suppose that the probabilities are as given in Figure 5.21, where an arrow extending from brand j to brand i represents p_{ij}. For some values of λ the system will have meaningful solutions; the means for determining such values of λ will be presented in Section 5.10. For the given probabilities, $\lambda = 1$ will produce meaningful solutions. The resulting system is

$$\begin{aligned} .3x_1 + .6x_2 + .3x_3 &= x_1, \\ .6x_1 + .2x_2 + .2x_3 &= x_2, \\ .1x_1 + .2x_2 + .5x_3 &= x_3, \end{aligned}$$

or

$$\begin{aligned} -.7x_1 + .6x_2 + .3x_3 &= 0, \\ .6x_1 - .8x_2 + .2x_3 &= 0, \\ .1x_1 + .2x_2 - .5x_3 &= 0. \end{aligned}$$

By the methods of this section we find that the system has an infinite number of solutions and that a complete solution is

$$\begin{aligned} x_1 &= \tfrac{9}{5}x_3, \\ x_2 &= \tfrac{8}{5}x_3. \end{aligned}$$

Therefore, $X = [\tfrac{9}{5} \quad \tfrac{8}{5} \quad 1]^T x_3$ is a complete solution of the original system; for this solution there is a stability in the proportion of buyers of each brand. ▮

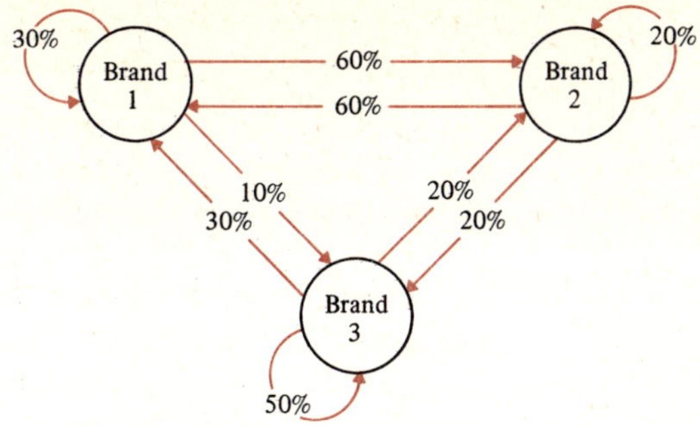

Figure 5.21. *Transition diagram.*

Example 4
The matrix model developed in Example 3 can also be used to find stable proportions of n genetic characteristics of a population with respect to certain breeding probabilities, or to find stable proportions of n political affiliations of a population of voters with respect to certain probabilities of voting transition. ▮

EXERCISES

Suggested minimum assignment: Exercises 1, 2, 5, 9, and 11.

1. Find a complete solution and a particular solution of the system

$$\begin{aligned} x_1 + 2x_2 + 2x_3 &= 4, \\ 2x_1 \qquad\quad + \ x_3 &= 4. \end{aligned}$$

Interpret your complete and particular solutions geometrically.

2. Show algebraically that the system

$$\begin{aligned} x_1 + 2x_2 &= 6, \\ 2x_1 + 4x_2 &= 8, \end{aligned}$$

has no solution. Interpret your algebraic results geometrically.

In each of Exercises 3–10, by the methods of this section, solve the given system or determine that there is no solution. If the system has an infinite number of solutions, find a complete solution and one particular solution.

3. $$\begin{aligned} x_1 + 2x_2 + x_3 &= 4, \\ 2x_1 + x_2 + 6x_3 &= 5. \end{aligned}$$

4. $$\begin{aligned} x_1 + 3x_2 + x_3 &= 4, \\ 2x_1 + 6x_2 + 2x_3 &= 6. \end{aligned}$$

5. $$\begin{aligned} x_1 - x_2 + x_3 &= 4, \\ x_1 + x_2 + 2x_3 &= 2, \\ -2x_2 - x_3 &= 2. \end{aligned}$$

6. $$\begin{aligned} x_1 + x_2 + x_3 &= 4, \\ x_2 - 2x_3 &= 1, \\ 2x_1 + 3x_2 \qquad &= 9. \end{aligned}$$

7. $$\begin{aligned} x_1 + x_2 + x_3 &= 4, \\ x_1 - x_2 \qquad &= 3, \\ 2x_1 \qquad + x_3 &= 8. \end{aligned}$$

8. $$\begin{aligned} x_1 + 3x_2 + x_3 &= 4, \\ x_1 + 4x_2 + 5x_3 &= 2. \end{aligned}$$

9. $x_1 + 2x_2 + x_3 + x_4 = 0,$
$x_1 - x_2 + x_3 = 7.$

10. $x_1 + x_2 - x_3 + x_4 = 0,$
$x_1 + 2x_2 + x_3 + x_4 = 0.$

11. In order to control a certain crop disease, someone determines that it is necessary to use 12 units of chemical A and 16 units of chemical B. One barrel of commercial spray P contains 1 unit of A and 2 units of B. One barrel of spray Q contains 1 unit of A and 1 unit of B, and one barrel of spray R contains 3 units of A and 1 unit of B. How many barrels of each type of spray could be used to spread the exact amount of chemicals needed to control the disease?

12. Three types of trucks are equipped to haul 2 different types of machines per load according to the following chart:

Machine	Truck 1	Truck 2	Truck 3
A	3	1	2
B	2	1	1

How many trucks of each type should be sent to haul exactly 16 of the type A machines and 12 of the type B machines? (Assume that each truck is fully loaded.)

5.8 The Gauss–Jordan Method

As we mentioned before, in general, a linear system over the real numbers has (1) a unique solution, (2) an infinite number of solutions, or (3) no solution. In previous sections we have illustrated by means of examples how elementary row operations can be applied to the augmented matrix of any linear system; we now formalize the method of solution used in those examples. In the first two examples of Section 5.7, which had three linear equations in three unknowns, elementary row operations were used to obtain the columns

$$I_{(1)} = \begin{bmatrix} 1 \\ 0 \\ 0 \end{bmatrix}, \quad I_{(2)} = \begin{bmatrix} 0 \\ 1 \\ 0 \end{bmatrix}, \quad \text{and if possible } I_{(3)} = \begin{bmatrix} 0 \\ 0 \\ 1 \end{bmatrix}$$

in the augmented matrix; these columns $I_{(1)}$, $I_{(2)}$, and $I_{(3)}$ need not be adjacent to each other, nor need they appear in that order, although in practice they frequently do. We also observe in Example 1 of Section 5.7 that when $I_{(3)}$ did not appear, the last row consisted entirely of 0's. In general we formalize these observations with the following definition.

▶ **Definition 22**

Let A be an m by n matrix, and let r be an integer such that $1 \leq r \leq m$. If A has $I_{(1)}, I_{(2)}, \ldots, I_{(r)}$ among its columns and if A has $(m - r)$ rows consisting entirely of 0 entries, then A is said to be a **reduced matrix.**

Example 1
The following matrices *are* reduced matrices:

$$\begin{bmatrix}1&0&0&2\\0&1&0&3\\0&0&1&4\end{bmatrix},\quad \begin{bmatrix}1&0&4&6\\0&1&3&2\\0&0&0&0\end{bmatrix},\quad \begin{bmatrix}1&0&3&0\\0&1&2&0\\0&0&0&1\end{bmatrix},$$

$$\begin{bmatrix}1&0&2&0&2\\0&1&3&0&3\\0&0&4&1&2\end{bmatrix},\quad \begin{bmatrix}1&0&3\\0&1&2\\0&0&0\end{bmatrix},\quad \begin{bmatrix}2&0&5&1&2\\3&1&6&0&3\\0&0&0&0&0\end{bmatrix}.$$

The matrix

$$\begin{bmatrix}1&0&2&2\\0&1&3&1\\0&0&1&5\end{bmatrix}$$

is *not* a reduced matrix because $I_{(3)}$ is not a column and the third row does not have all 0 entries; one or the other must be true. ▮

Because it can be proved that every nonzero matrix is row equivalent to a reduced matrix and that the resulting reduced matrix is unique for specific positions of $I_{(1)}, I_{(2)}, \ldots, I_{(r)}$, we can define the general method of solution as *a sequence of elementary row operations necessary to transform the augmented matrix of a linear system to a row equivalent reduced matrix;* this method of solving a linear system is usually called the **Gauss–Jordan method.**

Linear systems $AX = B$ in which $B = \mathbf{0}$ are called **homogeneous** systems; such systems always have a solution because the zero vector always satisfies the system. The question when solving a homogeneous system is whether the system has only a unique solution, namely, the zero vector, or an infinite number of solutions.

Example 2
Solve the homogeneous system

$$\begin{aligned} x_1 + \ x_2 - \ x_3 &= 0,\\ 2x_2 - \ x_3 &= 0,\\ 4x_1 + 6x_2 - 5x_3 &= 0,\end{aligned}$$

by the Gauss–Jordan method. Elementary row operations necessary to change the augmented matrix of the given system to a reduced matrix are as follows:

$$\left[\begin{array}{ccc|c}1&1&-1&0\\0&2&-1&0\\4&6&-5&0\end{array}\right] \underset{-4R_1+R_3}{\overset{\text{row}}{\sim}} \left[\begin{array}{ccc|c}(1)&1&-1&0\\(0)&2&-1&0\\(0)&2&-1&0\end{array}\right] \underset{\frac{1}{2}R_2}{\overset{\text{row}}{\sim}} \left[\begin{array}{ccc|c}(1)&1&-1&0\\(0)&(1)&-\frac{1}{2}&0\\(0)&2&-1&0\end{array}\right]$$

$$\underset{\substack{-2R_2+R_3\\-R_2+R_1}}{\overset{\text{row}}{\sim}} \left[\begin{array}{ccc|c}(1)&(0)&-\frac{1}{2}&0\\(0)&(1)&-\frac{1}{2}&0\\(0)&(0)&0&0\end{array}\right].$$

The reduced matrix represents the augmented matrix of a system that has the same solution as that of the original system. The system is

$$\begin{aligned} x_1 \qquad - \tfrac{1}{2}x_3 &= 0, \\ x_2 - \tfrac{1}{2}x_3 &= 0, \\ 0x_1 + 0x_2 + 0x_3 &= 0, \end{aligned}$$

and the complete solution of this homogeneous system is

$$\begin{aligned} x_1 &= \tfrac{1}{2}x_3, \\ x_2 &= \tfrac{1}{2}x_3. \end{aligned}$$

Thus we have a system with an infinite number of solutions, depending upon the value that is assigned to the parameter x_3. If we let the parameter x_3 equal 2, then a particular solution is $(1, 1, 2)$. ▮

APPLICATIONS

Example 3

A primary and significant use of the Gauss–Jordan method occurs in solving linear programming problems. This use, and applications of linear programming problems, will be demonstrated in Chapters 6 and 7. ▮

Example 4

In Example 7 of Section 5.3 we learned that the state vector B_k after k transitions of a Markov chain is found by

$$B_n = B_{n-1}A \qquad \text{or by} \qquad B_n = B_0A^n,$$

where A is the transition matrix. In some applications, for large values of n, B_n will differ very little from B_{n-1}. In such applications it is natural to wonder, to what vector the sequence of vectors $B_1, B_2, B_3, \ldots, B_n, \ldots$ will approach for increasing values of n; it can be proved that if there is such a row vector X, it is the solution of the equation

$$X = XA$$

and is called a **steady state vector** of the matrix A.

In this example we shall illustrate the calculation of a steady state vector by again using the example in which a system performs satisfactorily or unsatisfactorily at given time intervals (Example 7 of Section 5.3). Thus

$$X = XA$$

becomes

$$[x_1 \quad x_2] = [x_1 \quad x_2]\begin{bmatrix} .6 & .4 \\ .8 & .2 \end{bmatrix},$$

which is equivalent to the system

$$\begin{aligned} x_1 &= .6x_1 + .8x_2, \\ x_2 &= .4x_1 + .2x_2, \end{aligned}$$

or

$$\begin{aligned} .4x_1 - .8x_2 &= 0, \\ -.4x_1 + .8x_2 &= 0. \end{aligned}$$

The latter system has a complete solution $x_1 = 2x_2$; since x_1 and x_2 must add up to 1, we choose $x_2 = \frac{1}{3}$, and hence $x_1 = \frac{2}{3}$. Thus in the long run after a sufficient number of transitions, and independent of the initial state, the state vector will eventually approach

$$[\tfrac{2}{3} \;\; \tfrac{1}{3}].$$

This means that in the long run the system will operate satisfactorily $\frac{2}{3}$ of the time and unsatisfactorily $\frac{1}{3}$ of the time. It can be proved that a steady state vector exists for any Markov chain having a transition matrix with all positive entries.

Because the sum of the components of X must be 1, some efficiency can be gained by applying the Gauss–Jordan method initially to the nonhomogeneous system

$$\begin{aligned} (I_n - A)X^T &= \mathbf{0}, \\ [1\,1 \cdots 1]X^T &= 1. \end{aligned}$$ ▮

Example 5

The use of matrices in the study of Markov chains has been illustrated by several examples in this chapter. For a more comprehensive study of Markov chains and their applications, the reader is referred to Kemeny et al. [27], pp. 215–253. Markov decision processes with applications are discussed in Chapter 8 of [27]. ▮

EXERCISES

Suggested minimum assignment: Exercises 1, 4, 5, and 7.

1. Which of the following matrices are reduced matrices? If they are not reduced matrices, state the reason.

$$(a) \begin{bmatrix} 1 & 0 & 0 & 3 \\ 2 & 1 & 0 & 2 \\ 0 & 0 & 1 & 4 \end{bmatrix}; \quad (b) \begin{bmatrix} 1 & 4 & 0 & 4 & 1 \\ 0 & 0 & 1 & 3 & 6 \\ 0 & 0 & 0 & 0 & 0 \end{bmatrix}; \quad (c) \begin{bmatrix} 1 & 3 & 0 & 4 \\ 0 & 0 & 1 & 5 \\ 0 & 0 & 0 & 1 \end{bmatrix}.$$

2. Which of the following are reduced matrices? If they are not reduced matrices, state the reason.

$$(a) \begin{bmatrix} 2 & 0 & 0 & 3 \\ 0 & 1 & 0 & 2 \\ 0 & 0 & 1 & 4 \end{bmatrix}; \quad (b) \begin{bmatrix} 0 & 1 & 0 & 3 \\ 1 & 0 & 0 & 2 \\ 0 & 0 & 1 & 1 \end{bmatrix}; \quad (c) \begin{bmatrix} 1 & 3 & 2 & 6 \\ 0 & 0 & 0 & 4 \\ 0 & 0 & 0 & 0 \end{bmatrix}.$$

In each of Exercises 3–8 use the Gauss–Jordan method to solve the given system or to determine that there is no solution. If the system has an infinite number of solutions, find a complete solution and one particular solution.

3. $$\begin{aligned} x_1 + x_2 \qquad &= 4, \\ x_1 - x_2 + 2x_3 &= 2, \\ x_2 + x_3 &= 5. \end{aligned}$$

4. $$\begin{aligned} x_1 \qquad + x_3 &= 5, \\ 2x_2 + 4x_3 &= 22, \\ x_1 - 2x_2 \qquad &= -2. \end{aligned}$$

5. $$\begin{aligned} x_1 + 2x_2 + x_3 &= 0, \\ x_2 + x_3 &= 0, \\ 3x_1 + 5x_2 + 2x_3 &= 0. \end{aligned}$$

6. $$\begin{aligned} x_1 + x_2 - x_3 - 1 &= 0, \\ 2x_2 + 4x_3 - 3 &= 0, \\ x_1 + x_2 + 5x_3 - 2 &= 0. \end{aligned}$$

7. $$\begin{aligned} x_1 - x_2 \qquad + x_4 &= 2, \\ x_2 + x_3 + x_4 &= 5, \\ -x_1 + 2x_2 + x_3 \qquad &= 3, \\ -x_1 + 3x_2 + 2x_3 + x_4 &= 8. \end{aligned}$$

8. $$\begin{aligned} 2x_2 + 2x_3 \qquad &= 2, \\ x_1 \qquad + x_4 &= 2, \\ 2x_1 \qquad + 2x_4 &= 6, \\ x_3 + x_4 &= 7. \end{aligned}$$

9. Which of Exercises 3–8 is a homogeneous system?

10. If A is a square matrix, prove that the homogeneous system $AX = \mathbf{0}$ has *only* the solution $X = \mathbf{0}$ if and only if A is invertible.

5.9 The Determinant of a Matrix (Optional†)

The reader should have noticed that the definition of a matrix does *not* assign a numerical value to a matrix. The purpose of this section is to define an operation on a matrix in such a way that to every square matrix A there corresponds a single real number. The result of this operation is known as the **determinant of A** and is designated by $|A|$ or $\det A$.

We shall begin by defining the **determinant of a 1 by 1 matrix** to be equal to the single entry of that matrix.

Example 1

$$\det[6] = 6, \qquad \det[-3] = -3, \qquad \det[0] = 0. \blacksquare$$

The **determinant of a 2 by 2 matrix** is defined by

$$\det\begin{bmatrix} a_{11} & a_{12} \\ a_{21} & a_{22} \end{bmatrix} = a_{11}a_{22} - a_{12}a_{21}.$$

Example 2

$$\det\begin{bmatrix} 3 & 2 \\ 4 & 6 \end{bmatrix} = 3\cdot 6 - 2\cdot 4 = 10;$$

$$\det\begin{bmatrix} 0 & -2 \\ 6 & 3 \end{bmatrix} = 0\cdot 3 - (-2)\cdot 6 = 12. \blacksquare$$

Before defining the determinant of a 3 by 3 matrix, we first define a **minor** and a **cofactor** of an entry.

Definition 23

If the ith row and the jth column of a 3 by 3 matrix A are deleted, then the determinant of the resulting 2 by 2 submatrix is called the **minor** of

† This section is required for the optional section 5.10.

the entry a_{ij} and is denoted by M_{ij}. The quantity $A_{ij} = (-1)^{i+j}M_{ij}$ is known as the **cofactor** of a_{ij}.

Example 3

Let

$$A = \begin{bmatrix} 2 & 3 & 0 \\ 4 & 2 & 1 \\ 6 & 5 & 2 \end{bmatrix}.$$

$$M_{11} = \det \begin{bmatrix} 2 & 3 & 0 \\ 4 & 2 & 1 \\ 6 & 5 & 2 \end{bmatrix} = \det \begin{bmatrix} 2 & 1 \\ 5 & 2 \end{bmatrix} = 2 \cdot 2 - 1 \cdot 5 = -1;$$

$$M_{12} = \det \begin{bmatrix} 2 & 3 & 0 \\ 4 & 2 & 1 \\ 6 & 5 & 2 \end{bmatrix} = \det \begin{bmatrix} 4 & 1 \\ 6 & 2 \end{bmatrix} = 4 \cdot 2 - 1 \cdot 6 = 2;$$

$$M_{13} = \det \begin{bmatrix} 2 & 3 & 0 \\ 4 & 2 & 1 \\ 6 & 5 & 2 \end{bmatrix} = \det \begin{bmatrix} 4 & 2 \\ 6 & 5 \end{bmatrix} = 4 \cdot 5 - 2 \cdot 6 = 8;$$

$$A_{11} = (-1)^{1+1}M_{11} = (+1)(-1) = -1;$$
$$A_{12} = (-1)^{1+2}M_{12} = (-1)(2) = -2;$$
$$A_{13} = (-1)^{1+3}M_{13} = (+1)(8) = 8.$$

The **vector of cofactors** corresponding to the first row of A is

$$(-1, \; -2, \; 8). \; \blacksquare$$

▶ **Definition 24**

The **determinant of a 3 by 3 matrix A** is the dot product of the ith row (or jth column) of A with the vector of cofactors corresponding to the ith row (or jth column) of A.†

The validity of Definition 24 depends upon the fact that $\det A$ is the same no matter which row i (or column j) is chosen. The latter assertion is illustrated in Exercise 3.

Example 4

If we use the matrix

$$A = \begin{bmatrix} 2 & 3 & 0 \\ 4 & 2 & 1 \\ 6 & 5 & 2 \end{bmatrix}$$

† Notice that the definition of the determinant of a 2 by 2 matrix can be expressed in a manner that is analogous to Definition 24.

of Example 3, and if we let $i = 1$ in Definition 24, then, using the cofactors found in Example 3, we get

$$\begin{aligned}\det A &= (a_{11}, a_{12}, a_{13}) \cdot (A_{11}, A_{12}, A_{13})\\ &= a_{11}A_{11} + a_{12}A_{12} + a_{13}A_{13}\\ &= 2(-1) + 3(-2) + 0(8)\\ &= -8.\end{aligned}$$

The dot product of any other row or column with its vector of cofactors will also produce the number -8. ▮

We have defined the determinant of matrix A, where A was 1 by 1, 2 by 2, or 3 by 3. In general, if n is any positive integer such that the determinant of an $(n-1)$ by $(n-1)$ matrix is defined, then the **determinant of an *n* by *n* matrix A** is defined as the dot product of the ith row (or jth column) of A with the vector of cofactors corresponding to the ith row (or jth column); the cofactor of an entry of an n by n matrix is defined in the manner of Definition 23. Thus if A is an n by n matrix,

$$\begin{aligned}\det A &= (a_{i1}, a_{i2}, \ldots, a_{in}) \cdot (A_{i1}, A_{i2}, \ldots, A_{in})\\ &= a_{i1}A_{i1} + a_{i2}A_{i2} + \cdots + a_{in}A_{in},\end{aligned} \tag{5.6}$$

or

$$\begin{aligned}\det A &= (a_{1j}, a_{2j}, \ldots, a_{nj}) \cdot (A_{1j}, A_{2j}, \ldots, A_{nj})\\ &= a_{1j}A_{1j} + a_{2j}A_{2j} + \cdots + a_{nj}A_{nj}.\end{aligned} \tag{5.7}$$

Expression (5.6) is said to be the **expansion** of $\det A$ about the ith row, whereas expression (5.7) is said to be the **expansion** of $\det A$ about the jth column. In Example 4, $\det A$ was calculated by means of the expansion about the first row; any other row or column could be chosen to expand about, and the value of $\det A$ would be the same. The first row was chosen because the presence of a zero in the upper right-hand corner shortened the work.

Elementary row operations have been used frequently in this chapter; it seems reasonable, therefore, to ascertain what effect, if any, these operations have on the value of the determinant of a matrix.

▶ **Theorem 8**

If a matrix B is formed from a square matrix A by the interchange of two rows, then $\det B = -\det A$. ◀

▶ **Theorem 9**

If a matrix B is formed from a square matrix A by multiplying each entry of any row of A by a real number k, then $\det B = k \det A$. ◀

Notice the difference in $\det kA$ and $k \det A$.

▶ **Theorem 10**

If a matrix B is formed from a square matrix A by adding to the entries of any row of A a constant real number multiple of the corresponding entries of another row, then $\det B = \det A$.

In Exercises 20–22 the reader will be asked to prove Theorems 8–10 for the case where $n = 3$. ◀

In practice, the calculation of the determinant of a matrix often can be handled more efficiently if the following theorem is used.

▶ **Theorem 11**

The determinant of a triangular matrix is the product of the entries on the main diagonal.

The proof of Theorem 11 is left as Exercise 23. ◀

We now illustrate the use of Theorem 11 in calculating the determinant of a matrix.

Example 5

If a matrix is triangular, then by Theorem 11

$$\det\begin{bmatrix} 4 & 2 & 3 & 4 \\ 0 & 6 & 2 & 4 \\ 0 & 0 & 1 & 7 \\ 0 & 0 & 0 & 3 \end{bmatrix} = 4 \cdot 6 \cdot 1 \cdot 3 = 72. \blacksquare$$

Example 6

If a matrix is not triangular, we can first repeatedly use Theorem 10 to triangularize it. Using the elementary row operations $(-2R_1 + R_3)$, $(2R_2 + R_3)$, and $(-R_3 + R_4)$ in that order, we evaluate the following determinant;

$$\det\begin{bmatrix} 1 & 3 & 0 & 2 \\ 0 & 2 & 1 & 3 \\ 2 & 2 & 1 & 3 \\ 0 & 0 & 3 & -2 \end{bmatrix} \underset{-2R_1+R_3}{=} \det\begin{bmatrix} 1 & 3 & 0 & 2 \\ 0 & 2 & 1 & 3 \\ 0 & -4 & 1 & -1 \\ 0 & 0 & 3 & -2 \end{bmatrix}$$

$$\underset{2R_2+R_3}{=} \det\begin{bmatrix} 1 & 3 & 0 & 2 \\ 0 & 2 & 1 & 3 \\ 0 & 0 & 3 & 5 \\ 0 & 0 & 3 & -2 \end{bmatrix} \underset{-R_3+R_4}{=} \det\begin{bmatrix} 1 & 3 & 0 & 2 \\ 0 & 2 & 1 & 3 \\ 0 & 0 & 3 & 5 \\ 0 & 0 & 0 & -7 \end{bmatrix}$$

$$= 1 \cdot 2 \cdot 3 \cdot -7$$
$$= -42. \blacksquare$$

APPLICATIONS

Example 7

The determinant of a matrix can be used in an alternative method for calculating the multiplicative inverse of a square matrix. If

$$A = \begin{bmatrix} a_{11} & a_{12} & \cdots & a_{1n} \\ a_{21} & a_{22} & \cdots & a_{2n} \\ \vdots & \vdots & & \vdots \\ a_{n1} & a_{n2} & \cdots & a_{nn} \end{bmatrix},$$

then the **cofactor matrix** of A, designated by cof A, is defined by

$$\text{cof } A = \begin{bmatrix} A_{11} & A_{12} & \cdots & A_{1n} \\ A_{21} & A_{22} & \cdots & A_{2n} \\ \vdots & \vdots & & \vdots \\ A_{n1} & A_{n2} & \cdots & A_{nn} \end{bmatrix}.$$

It can be proved (Campbell [7], p. 99) that a square matrix A is invertible if and only if $\det A \neq 0$, and moreover

$$A^{-1} = \left(\frac{1}{\det A}\right)(\text{cof } A)^T.$$

Let

$$A = \begin{bmatrix} 1 & 0 & 4 \\ 0 & 0 & 1 \\ 0 & 2 & 3 \end{bmatrix}.$$

By Theorems 8 and 11,

$$\det A \underset{R_2 \leftrightarrow R_3}{=} -\det \begin{bmatrix} 1 & 0 & 4 \\ 0 & 2 & 3 \\ 0 & 0 & 1 \end{bmatrix} = -(1 \cdot 2 \cdot 1) = -2.$$

Then

$$\text{cof } A = \begin{bmatrix} \det \begin{bmatrix} 0 & 1 \\ 2 & 3 \end{bmatrix} & -\det \begin{bmatrix} 0 & 1 \\ 0 & 3 \end{bmatrix} & \det \begin{bmatrix} 0 & 0 \\ 0 & 2 \end{bmatrix} \\ -\det \begin{bmatrix} 0 & 4 \\ 2 & 3 \end{bmatrix} & \det \begin{bmatrix} 1 & 4 \\ 0 & 3 \end{bmatrix} & -\det \begin{bmatrix} 1 & 0 \\ 0 & 2 \end{bmatrix} \\ \det \begin{bmatrix} 0 & 4 \\ 0 & 1 \end{bmatrix} & -\det \begin{bmatrix} 1 & 4 \\ 0 & 1 \end{bmatrix} & \det \begin{bmatrix} 1 & 0 \\ 0 & 0 \end{bmatrix} \end{bmatrix} = \begin{bmatrix} -2 & 0 & 0 \\ 8 & 3 & -2 \\ 0 & -1 & 0 \end{bmatrix}.$$

Therefore, $A^{-1} = \left(\frac{1}{-2}\right) \begin{bmatrix} -2 & 0 & 0 \\ 8 & 3 & -2 \\ 0 & -1 & 0 \end{bmatrix}^T = \begin{bmatrix} 1 & -4 & 0 \\ 0 & -\frac{3}{2} & \frac{1}{2} \\ 0 & 1 & 0 \end{bmatrix}$. ∎

EXERCISES

Suggested minimum assignment: Exercises 1, 3, 5, and 9.

1. Find: (*a*) $\det \begin{bmatrix} 2 & 1 \\ 3 & 6 \end{bmatrix}$; (*b*) $\det \begin{bmatrix} -3 & 2 \\ -9 & 0 \end{bmatrix}$; (*c*) $\det \begin{bmatrix} -1 & -2 \\ 6 & 12 \end{bmatrix}$.

2. Find: (*a*) $\det\begin{bmatrix}0 & 2\\1 & 4\end{bmatrix}$; (*b*) $\det\begin{bmatrix}2 & -1\\3 & 4\end{bmatrix}$; (*c*) $\det\begin{bmatrix}2 & 1\\8 & 4\end{bmatrix}$.

3. Given:

$$A = \begin{bmatrix}1 & 3 & 4\\0 & 1 & 1\\2 & 4 & 3\end{bmatrix}.$$

Without the aid of the theorems of this section:

(*a*) Write the expansion of $\det A$ about the first row and evaluate $\det A$.

(*b*) Write the expansion of $\det A$ about the second column and evaluate $\det A$.

(*c*) Write the expansion of $\det A$ about the second row and evaluate $\det A$.

(*d*) Write the expansion of $\det A$ about the first column and evaluate $\det A$.

4. Given:

$$A = \begin{bmatrix}3 & 2 & 1\\4 & 2 & 1\\6 & 7 & 2\end{bmatrix}.$$

(*a*) Write the minor of the entry in the first row and second column.

(*b*) Write the cofactor of the entry in the first row and second column.

(*c*) Write the minor of the entry in the second row and third column.

(*d*) Write the cofactor of the entry in the second row and third column.

In Exercises 5–10 use the theorems of this section to evaluate the determinants of the following matrices.

5. $\begin{bmatrix}2 & 6 & 5\\0 & 3 & 2\\4 & 12 & 1\end{bmatrix}$.

6. $\begin{bmatrix}2 & 1 & 2\\4 & 3 & 1\\6 & 7 & 4\end{bmatrix}$.

7. $\begin{bmatrix}1 & 3 & 2\\2 & 6 & 7\\1 & 2 & 4\end{bmatrix}$.

8. $\begin{bmatrix}2 & 4 & 3\\4 & 8 & 4\\4 & 2 & 1\end{bmatrix}$.

9. $\begin{bmatrix}2 & 1 & 0 & 2\\0 & 1 & 3 & 4\\2 & 4 & 13 & 3\\0 & 2 & 2 & 4\end{bmatrix}$.

10. $\begin{bmatrix}3 & 3 & 2 & 1\\6 & 7 & 3 & 2\\0 & 0 & 1 & 4\\9 & 2 & 1 & 0\end{bmatrix}$.

11. An important property is that the determinant of the product of two square matrices of the same order is the product of the determinants of the two matrices; that is,

$$\det(AB) = (\det A)(\det B).$$

Illustrate this property for

$$A = \begin{bmatrix} 2 & 1 \\ 3 & 2 \end{bmatrix} \quad \text{and} \quad B = \begin{bmatrix} 3 & 1 \\ 6 & 0 \end{bmatrix}.$$

12. Prove, using the definition of a determinant, that if all the entries of a given row or column of a square matrix A are 0's, then $\det A = 0$.
13. (*a*) Use Theorem 8 to prove that if any two rows of a square matrix A are identical, then $\det A = 0$.
    (*b*) Prove that if any two rows of a square matrix A are proportional, then $\det A = 0$. *Hint:* Use Theorem 9 and part (*a*).
14. A property of determinants is that if A is square, then $\det A^T = \det A$. Prove the validity of this property for $n = 1, 2, 3$.
15. Prove that in Theorems 8, 9, and 10, the word "row" may be replaced by the word "column." *Hint:* Use Exercise 14.

In Exercises 16–19, calculate A^{-1} by the formula given in Example 7.

16. $A = \begin{bmatrix} 2 & 1 \\ 6 & 2 \end{bmatrix}$.
17. $A = \begin{bmatrix} 1 & 4 \\ 2 & 10 \end{bmatrix}$.
18. $A = \begin{bmatrix} 1 & 2 & 0 \\ 0 & 1 & 0 \\ 1 & 0 & 2 \end{bmatrix}$.
19. $A = \begin{bmatrix} 0 & 0 & 1 \\ 1 & 3 & 0 \\ 0 & 2 & 1 \end{bmatrix}$.
20. Prove Theorem 8 for the case $n = 3$. *Hint:* Expand about the row not involved in the switch.
21. Prove Theorem 9 for the case $n = 3$. *Hint:* Expand about the row that is multiplied by the number k.
22. Prove Theorem 10 for the case $n = 3$. *Hint:* Expand about the row that is changed by the row operation, and use Exercise 13(a).
23. Prove Theorem 11.

5.10 The Characteristic Value Problem (Optional)

A problem that is common in many applications of matrix algebra is that of finding a real number λ and a nonzero column vector X such that

$$AX = \lambda X,$$

where A is a square matrix that is known.

Example 1

If $A = \begin{bmatrix} 2 & 1 \\ 3 & 4 \end{bmatrix}$, find a real number λ and a nonzero column vector X such that

$$AX = \lambda X.$$

In other words, because

$$\begin{bmatrix} 2 & 1 \\ 3 & 4 \end{bmatrix}\begin{bmatrix} x_1 \\ x_2 \end{bmatrix} = \lambda\begin{bmatrix} x_1 \\ x_2 \end{bmatrix}$$

is the matrix representation of the linear system

$$\begin{aligned} 2x_1 + x_2 &= \lambda x_1, \\ 3x_1 + 4x_2 &= \lambda x_2, \end{aligned}$$

we must first find λ such that nonzero solutions of the system then can be found.

Solution: After transposing the terms of the latter system from the right side to the left side and combining terms, we obtain

$$\begin{aligned} (2 - \lambda)x_1 + \qquad\quad x_2 &= 0, \\ 3x_1 + (4 - \lambda)x_2 &= 0. \end{aligned} \tag{5.8}$$

The last system is a linear homogeneous system

$$CX = \mathbf{0},$$

where

$$C = \begin{bmatrix} 2 - \lambda & 1 \\ 3 & 4 - \lambda \end{bmatrix} \quad \text{and} \quad X = \begin{bmatrix} x_1 \\ x_2 \end{bmatrix}.$$

In Section 5.8 we learned that a homogeneous system such as

$$CX = \mathbf{0}$$

has either a unique solution, which is the zero vector, or an infinite number of solutions. Because we are looking for nonzero solutions, we must find those values of λ in matrix C that will make the system $CX = \mathbf{0}$ have an infinite number of solutions. It can be proved that the homogeneous system $CX = \mathbf{0}$ has solutions other than the zero vector if and only if $\det C = 0$. Thus we write

$$\det\begin{bmatrix} 2 - \lambda & 1 \\ 3 & 4 - \lambda \end{bmatrix} = 0,$$

and solve for λ. Expansion of the determinant about the first row gives us

$$\begin{aligned} (2 - \lambda)(4 - \lambda) - 1 \cdot 3 &= 0, \\ 8 - 6\lambda + \lambda^2 - 3 &= 0, \\ \lambda^2 - 6\lambda + 5 &= 0, \\ (\lambda - 1)(\lambda - 5) &= 0, \\ \lambda = 1, \qquad \lambda &= 5. \end{aligned}$$

These values of λ that we have just found are called **characteristic values** (or **eigenvalues**) of matrix A and are substituted into the system (5.8) so that we can solve that system for x_1 and x_2. With $\lambda = 1$ the system (5.8) becomes

$$\begin{aligned} (2 - 1)x_1 + \qquad\quad x_2 &= 0, \\ 3x_1 + (4 - 1)x_2 &= 0, \end{aligned}$$

or

$$\begin{aligned} x_1 + x_2 &= 0, \\ 3x_1 + 3x_2 &= 0. \end{aligned}$$

By the Gauss–Jordan method a complete solution of the latter system is

$$x_1 = -x_2.$$

Hence

$$X = \begin{bmatrix} x_1 \\ x_2 \end{bmatrix} = \begin{bmatrix} -x_2 \\ x_2 \end{bmatrix} = \begin{bmatrix} -1 \\ 1 \end{bmatrix} x_2,$$

where x_2 is a parameter that can be assigned an arbitrary nonzero real number value. If we let $x_2 = 1$, then

$$X = \begin{bmatrix} -1 \\ 1 \end{bmatrix}.$$

The column vector X is called a **characteristic vector** of matrix A.

Next, we substitute the other characteristic value $\lambda = 5$ into the system (5.8) to get

$$\begin{aligned} (2-5)x_1 + x_2 &= 0, \\ 3x_1 + (4-5)x_2 &= 0, \end{aligned}$$

or

$$\begin{aligned} -3x_1 + x_2 &= 0, \\ 3x_1 - x_2 &= 0. \end{aligned}$$

By the Gauss–Jordan method a complete solution of the latter system is

$$x_1 = \tfrac{1}{3}x_2.$$

Hence

$$X = \begin{bmatrix} x_1 \\ x_2 \end{bmatrix} = \begin{bmatrix} \frac{1}{3}x_2 \\ x_2 \end{bmatrix} = \begin{bmatrix} \frac{1}{3} \\ 1 \end{bmatrix} x_2,$$

where x_2 is a parameter which can be assigned an arbitrary nonzero real number value. If we let $x_2 = 3$, then

$$X = \begin{bmatrix} 1 \\ 3 \end{bmatrix},$$

and we have a second characteristic vector of matrix A. To distinguish between the two characteristic vectors that we have found for A, we label them X_1 and X_2, respectively, where the subscript identifies the corresponding characteristic value. Thus

$$X_1 = \begin{bmatrix} -1 \\ 1 \end{bmatrix} \quad \text{and} \quad X_2 = \begin{bmatrix} 1 \\ 3 \end{bmatrix},$$

together with the appropriate λ, satisfy the system

$$AX = \lambda X.$$

That is,

$$\begin{bmatrix} 2 & 1 \\ 3 & 4 \end{bmatrix} \begin{bmatrix} -1 \\ 1 \end{bmatrix} = (1) \begin{bmatrix} -1 \\ 1 \end{bmatrix},$$

and also

$$\begin{bmatrix} 2 & 1 \\ 3 & 4 \end{bmatrix} \begin{bmatrix} 1 \\ 3 \end{bmatrix} = (5) \begin{bmatrix} 1 \\ 3 \end{bmatrix}. \blacksquare$$

In general, we may summarize the characteristic value problem as follows. First, for a given n by n matrix A and for an unknown column vector

$$X = \begin{bmatrix} x_1 \\ x_2 \\ \vdots \\ x_n \end{bmatrix},$$

find values of the real number λ for which the matrix equation

$$AX = \lambda X$$

has nonzero solutions; that is, $X \neq \mathbf{0}$. The matrix equation $AX = \lambda X$, which can be rewritten as

$$(A - \lambda I_n)X = \mathbf{0},$$

is a homogeneous linear system having nonzero solutions if and only if

$$\det (A - \lambda I_n) = 0.$$

The equation $\det (A - \lambda I_n) = 0$ is called the **characteristic equation,** and its roots λ_i are called **characteristic values** of the matrix A. Once the characteristic values have been found, then each homogeneous system

$$(A - \lambda_i I_n)X_i = \mathbf{0}, \qquad i = 1, 2, \ldots, n,$$

can be solved for nonzero X_i. The resulting nonzero column vectors X_i are called **characteristic vectors.**

At this point the reader should be reminded that although the characteristic equation will have n *complex* roots, it may have fewer *real* roots (possibly none). Thus when *real* characteristic values and vectors are sought, there can be from 0 to n of them. (See Exercise 9 for an illustration of a characteristic value problem with less than n real characteristic values.) When a matrix is symmetric, however, the roots are always real. (See Exercise 11.)

APPLICATIONS

Example 2

Consider a population of objects living in a certain environment in which

$x_{i,t}$ = number of living objects of age i at time t,
p_i = probability that an object of age i at time t will survive to time $t + 1$,

f_i = number of new objects created per old object of age i,
n = maximum age.

A column matrix showing the population of each age group at time $t + 1$ can be determined by the following matrix equation:

$$\begin{bmatrix} x_{0,t+1} \\ x_{1,t+1} \\ x_{2,t+1} \\ \vdots \\ x_{n,t+1} \end{bmatrix} = \begin{bmatrix} f_0 & f_1 & f_2 & \cdots & f_{n-1} & f_n \\ p_0 & 0 & 0 & \cdots & 0 & 0 \\ 0 & p_1 & 0 & \cdots & 0 & 0 \\ \vdots & \vdots & \vdots & & \vdots & \vdots \\ 0 & 0 & 0 & \cdots & p_{n-1} & 0 \end{bmatrix} \begin{bmatrix} x_{0,t} \\ x_{1,t} \\ x_{2,t} \\ \vdots \\ x_{n,t} \end{bmatrix}.$$

Let the respective matrices in the latter equation be designated X_{t+1}, A, and X_t so that the equation can be expressed as

$$X_{t+1} = AX_t.$$

Suppose that we want to find a population matrix X_t that will produce a proportional population matrix X_{t+1} one unit of time later in the given environment. In other words, we want to find an X_t such that

$$X_{t+1} = \lambda X_t.$$

In this case we can say that there is a stability of the relative size of age groups. The equation $X_{t+1} = AX_t$ becomes

$$AX_t = \lambda X_t,$$

which is the characteristic value problem. In [6], Bernadelli describes a hypothetical beetle living in a certain environment in which the A matrix is

$$\begin{bmatrix} 0 & 0 & 6 \\ \frac{1}{2} & 0 & 0 \\ 0 & \frac{1}{3} & 0 \end{bmatrix}.$$

The only real characteristic value is 1 and a corresponding characteristic vector of A is

$$X_t = \begin{bmatrix} 6 \\ 3 \\ 1 \end{bmatrix};$$

therefore any population of these beetles with an age distribution proportional to 6:3:1 should keep a stable age distribution in the given environment. ∎

EXERCISES

Suggested minimum assignment: Exercises 1, 5, and 7.

In each of Exercises 1–4 find the characteristic equation, the real characteristic values, and corresponding characteristic vectors for the given matrix.

1. $\begin{bmatrix} 1 & 2 \\ -1 & 4 \end{bmatrix}$.

2. $\begin{bmatrix} 5 & -2 \\ -2 & 5 \end{bmatrix}$.

3. $\begin{bmatrix} 0 & 3 \\ 3 & 0 \end{bmatrix}$.

4. $\begin{bmatrix} 2 & 3 \\ 1 & 4 \end{bmatrix}$.

In each of Exercises 5–8 find a nonzero real solution, if possible, of the equation $AX = \lambda X$, where A is given and X is a corresponding column vector of unknowns.

5. $\begin{bmatrix} 1 & -1 \\ 3 & 5 \end{bmatrix}$.

6. $\begin{bmatrix} 0 & 1 \\ 3 & 2 \end{bmatrix}$.

7. $\begin{bmatrix} 0 & 0 & 2 \\ 0 & 1 & 0 \\ 2 & 0 & 0 \end{bmatrix}$.

8. $\begin{bmatrix} 1 & 1 & 1 \\ 0 & 3 & 1 \\ 0 & 0 & 4 \end{bmatrix}$.

9. In Example 2 verify that the real characteristic value and characteristic vector are as stated for the hypothetical beetle matrix

$$\begin{bmatrix} 0 & 0 & 6 \\ \frac{1}{2} & 0 & 0 \\ 0 & \frac{1}{3} & 0 \end{bmatrix}.$$

10. Prove: If A is square and if a system of linear homogeneous equations $AX = \mathbf{0}$ over the real numbers has a nonzero solution, then $\det A = 0$. *Hint:* Either $\det A = 0$ or $\det A \neq 0$; assume that $\det A \neq 0$, and show that this leads to a contradiction of the hypothesis. Also use Example 7 of Section 5.9.
11. Many applications involving the characteristic value problem require the characteristic values of a symmetric matrix. It is important in such problems to realize that the characteristic values of a symmetric matrix are always real numbers. Prove this assertion for the 2 by 2 symmetric matrix

$$A = \begin{bmatrix} a & b \\ b & c \end{bmatrix}.$$

Appendix: Matrix Operations Using APL

Because many significant applications involve large matrices and because of the tedious nature of matrix operations, computer help is essential in many applied problems. The computer programming language APL, introduced in the appendix of Chapter 1, is particularly adaptable to matrix operations. The purpose of this section is to illustrate the computer calculation of the results of various matrix operations; this will be done by means of examples in the same order that the operations themselves were presented in Sections 5.1–5.4. The appendix of Chapter 1, and in particular the APL keyboard shown in Figure 1.37, is a prerequisite for this section.

Example 1
To *insert* the 2 by 3 matrix

$$A = \begin{bmatrix} 1 & 3 & 5 \\ 4 & 0 & 7 \end{bmatrix}$$

in the computer, type: $A \leftarrow 2\ 3\rho 1\ 3\ 5\ 4\ 0\ 7$, which means "Assign to A ($A \leftarrow$) the entries of a 2 by 3 matrix ($2\ 3\rho$), where those entries are 1, 3, 5, 4, 0, and 7." Remember that the symbol $\leftarrow$ always assigns what follows the symbol to the variable that precedes the symbol. The symbol ρ shapes the entries following the symbol according to the desired order that is stated immediately preceding the symbol. Having typed: $A \leftarrow 2\ 3\rho 1\ 3\ 5\ 4\ 0\ 7$, we push the RETURN key on the console. This will automatically return the carriage. If we want to ask the computer to print A in the correct format, type A and push RETURN. The computer then prints the array

```
1 3 5
4 0 7
```

Remember that the RETURN key on an APL typewriter console works like the return key on a regular electric typewriter, that is, it returns the carriage and spaces down one line. The APL RETURN key has the additional function, however, of requesting computer printout when pushed after a "nonequation." ∎

Example 2
(*Matrix Addition*) To find the sum

$$\begin{bmatrix} 2 & 3 \\ 4 & 5 \end{bmatrix} + \begin{bmatrix} 4 & 5 \\ 1 & 1 \end{bmatrix},$$

the following steps are performed. The APL symbol for matrix addition is +.

Type: $A \leftarrow 2\ 2\rho 2\ 3\ 4\ 5$ (push RETURN)
Type: $B \leftarrow 2\ 2\rho 4\ 5\ 1\ 1$ (push RETURN)
Type: $A + B$ (push RETURN)
Computer types: 6 8
5 6 ■

Example 3

(*Multiplication of a Matrix by a Real Number*) Suppose that we wish to multiply the result of Example 2 by some real number, say 3. The APL symbol × effects this operation.

Type: $C \leftarrow A + B$ (push RETURN)
Type $3 \times C$ (push RETURN)
Computer types: 18 24
15 18 ■

Example 4

To *insert* a vector $V = (4, 0, 5, 6)$ with APL, simply type: $V \leftarrow 4\ 0\ 5\ 6$, and press the RETURN. To get the computer to print the vector, type V and then push RETURN. ■

Example 5

(*Addition of Vectors*) To add the vector $A = (3, 0, 4, 1)$ to the vector $B = (5, 3, 1, 6)$, the following sequence of steps is executed. The symbol + effects this operation.

Type: $A \leftarrow 3\ 0\ 4\ 1$ (push RETURN)
Type: $B \leftarrow 5\ 3\ 1\ 6$ (push RETURN)
Type: $A + B$ (push RETURN)
Computer types: 8 3 5 7 ■

Example 6

(*Multiplication of a Vector by a Real Number*) Suppose that we wish to multiply the vector $A + B$ obtained in Example 5 by some real number, say 4. The symbol × effects this operation.

Type: $C \leftarrow A + B$ (push RETURN)
Type: $4 \times C$ (push RETURN)
Computer types: 32 12 20 28 ■

Example 7

The symbol ×, when used between two vectors, produces a *new* type of vector product in which corresponding components of each vector are multiplied together.

Type: $A \leftarrow 3\ 0\ 4\ 2$ (push RETURN)
Type: $B \leftarrow 5\ 2\ 1\ 6$ (push RETURN)
Type: $A \times B$ (push RETURN)
Computer types: 15 0 4 12 ■

Example 8

(*Dot Product*) The result of the operation of Example 7 is not the dot product of A and B, but the dot product can be obtained if we sum the entries of the result of Example 7. The APL symbol used to sum the entries of a vector is $+/$. To find the dot product of A and B of Example 7,
Type: $C \leftarrow +/(A \times B)$ (push RETURN)
Type: C (push RETURN)
Computer types: 31 ▮

Example 9

(*Matrix Multiplication*) The symbol $+.\times$ is used to accomplish matrix multiplication. To find the product

$$\begin{bmatrix} 2 & 3 & 4 \\ 1 & 0 & 2 \end{bmatrix} \begin{bmatrix} 1 & 5 \\ 7 & 2 \\ 1 & 4 \end{bmatrix},$$

the following steps are performed.
Type: $A \leftarrow 2\ 3\rho 2\ 3\ 4\ 1\ 0\ 2$ (push RETURN)
Type: $B \leftarrow 3\ 2\rho 1\ 5\ 7\ 2\ 1\ 4$ (push RETURN)
Type: $A+.\times B$ (push RETURN)
Computer types: 27 32
3 13 ▮

Example 10

(*Transpose of a Matrix*) The symbol ⍉, is formed by typing the shift of the letter "O" and overstriking that with "\," is used to accomplish the transpose of a matrix. Because this operation acts on only one matrix, it is called a **monadic** operation in APL. Suppose that we want to find the transpose of the matrix

$$\begin{bmatrix} 3 & 4 & 1 \\ 6 & 9 & 5 \end{bmatrix}.$$

Type: $A \leftarrow 2\ 3\rho 3\ 4\ 1\ 6\ 9\ 5$ (push RETURN)
Type: ⍉A (push RETURN)
Computer types: 3 6
4 9
1 5 ▮

It is customary for APL installations to have a "library" of *program packages* that will perform other useful matrix operations such as matrix inversion. Of course, the contents of each "library" may vary, and the user should determine exactly what is available to him.

NEW VOCABULARY

matrix 5.1
order of a matrix 5.1
entry of a matrix 5.1
square matrix 5.1
main diagonal 5.1
equal matrices 5.1
sum of two matrices 5.1
conformable for addition 5.1
multiplication of a matrix by a real number 5.1
subtraction of two matrices 5.1
vector 5.2
dimension of a vector 5.2
components 5.2
row vector 5.2
column vector 5.2
zero vector 5.2
nonzero vector 5.2
equal vectors 5.2
sum of two vectors 5.2
multiplication of a vector by a real number 5.2
subtraction of two vectors 5.2
dot product of two vectors 5.2
magnitude or length of a vector 5.2
matrix product 5.3
conformable for multiplication 5.3
zero or null matrix 5.4
identity element for addition 5.4
identity element for multiplication 5.4
identity matrix 5.4
diagonal matrix 5.4
scalar matrix 5.4
triangular matrix 5.4
submatrix 5.4
partitioned matrix 5.4
blocks of a matrix 5.4
block matrix 5.4
transpose of a matrix 5.4
symmetric matrix 5.4
skew-symmetric matrix 5.4
system of linear equations over the real numbers 5.5
solution of a linear system 5.5
coefficient matrix 5.5
augmented matrix 5.5
elementary row operations 5.5
row equivalent matrices 5.5
multiplicative inverse of a real number 5.6
inverse of a matrix 5.6
invertible matrix 5.6
noninvertible matrix 5.6
basic variables 5.7
complete solution 5.7
particular solution 5.7
reduced matrix 5.8
Gauss–Jordan method 5.8
homogeneous system 5.8
determinant of a matrix 5.9
minor of an entry 5.9
cofactor of an entry 5.9
vector of cofactors 5.9
expansion of a determinant 5.9
characteristic values (or eigenvalues) 5.10
characteristic vector 5.10
characteristic equation 5.10

6 Linear Programming

The purpose of this chapter is to study a special type of optimization problem known as a linear programming problem. Such problems, which are frequently concerned with decision making, have attracted considerable attention in recent years. As an illustration of a linear programming problem, consider the example: Minimize the linear function defined by

$$f(x_1, x_2) = 3x_1 + 5x_2$$

subject to the linear conditions or constraints

$$\begin{aligned} 3x_1 + 4x_2 &\geq 340, \\ x_1 + x_2 &\geq 100, \\ x_1 + 5x_2 &\geq 150, \end{aligned} \qquad \text{and} \qquad \begin{aligned} x_1 &\geq 0, \\ x_2 &\geq 0. \end{aligned}$$

The word "programming" does not refer to the programming of a computer; it refers to the programming or allocation of items or entities. The unknowns x_1 and x_2 represent a measure of those allocations. The word "linear" is used to describe the fact that we wish to optimize a linear function subject to linear constraints.

Prerequisites: 5.1–5.3, 5.5 if only 6.1 and 6.2 are to be covered; 5.1–5.8 if 6.3–6.6 are to be covered.
Suggested Sections for Moderate Emphasis: 6.1–6.4
Suggested Sections for Minimum Emphasis: 6.1, 6.2

6.1 The Feasible Set of a Linear Programming Problem

The basic ideas of this section will be introduced by means of the following example.

Example 1

Consider a hypothetical situation in which a dietician wishes to serve food that provides the necessary calories, vitamins, and minerals. The following chart indicates the amounts of calories, vitamins, and minerals to be provided, and the amounts found in foods F_1 and F_2.

	Food F_1	Food F_2	Units needed
Calories	3 units/pound	4 units/pound	340
Vitamins	1 unit/pound	1 unit/pound	100
Minerals	1 unit/pound	5 units/pound	150

Suppose that foods F_1 and F_2 cost \$3 and \$5, respectively. Further suppose that the dietician wants to minimize the total cost of the foods purchased, subject to the dietary requirements shown in the chart. In other words, how many pounds of each food should be purchased in order to meet the dietary needs at least cost? The solution of this problem requires a determination of the possible combinations of foods that can be used to meet the dietary requirements, and from among these, a choice of that combination of foods that costs the least.

The remainder of this example will treat only the problem of meeting the dietary requirements; the minimization of cost will be the subject of Section 6.2. We now construct a mathematical model of the dietary needs.

Let x_1 be the number of pounds of food F_1 that is ordered, and let x_2 be the number of pounds of food F_2 that is ordered. The number of units of *calories* obtained will be

$$\overbrace{(3 \text{ units per pound}) \cdot (x_1 \text{ pounds})}^{\text{From } F_1} + \overbrace{(4 \text{ units per pound}) \cdot (x_2 \text{ pounds})}^{\text{From } F_2}.$$

Thus from both foods we obtain $(3x_1 + 4x_2)$ units of calories; the fact that we should have at least 340 units of calories produces the first constraint,

$$(1) \quad 3x_1 + 4x_2 \geq 340.$$

Similarly, we obtain a constraint corresponding to the requirement for *vitamins*,

$$(2) \quad x_1 + x_2 \geq 100,$$

and a constraint for the requirement for *minerals*,

$$(3) \quad x_1 + 5x_2 \geq 150.$$

The fact that we must purchase a nonnegative number of pounds of each food produces the constraints

$$(4) \quad x_1 \geq 0 \qquad \text{and} \qquad (5) \quad x_2 \geq 0.$$

Hence the mathematical model of the dietary requirements is the linear system of constraints

$$\begin{array}{ll} (1) & 3x_1 + 4x_2 \geq 340, \\ (2) & x_1 + x_2 \geq 100, \\ (3) & x_1 + 5x_2 \geq 150, \end{array} \qquad \text{and} \qquad \begin{array}{ll} (4) & x_1 \geq 0, \\ (5) & x_2 \geq 0. \end{array}$$

We shall determine values of x_1 and x_2 that satisfy this system. Geometrically the constraint $3x_1 + 4x_2 \geq 340$ represents the set of all points on or above the line

$$3x_1 + 4x_2 = 340,$$

as shown by the colored region in Figure 6.1. Similarly, the other constraints can be graphed, and the set of points whose coordinates satisfy all the constraints simultaneously is shown as the colored region in Figure 6.2. ▮

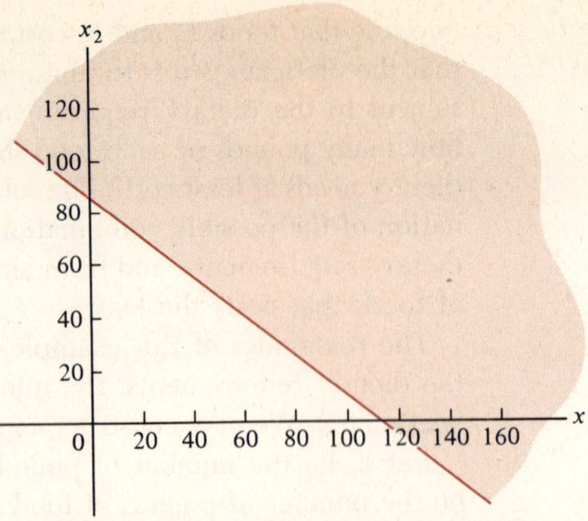

Figure 6.1. *Geometric representation of the inequality* $3x_1 + 4x_2 \geq 340$.

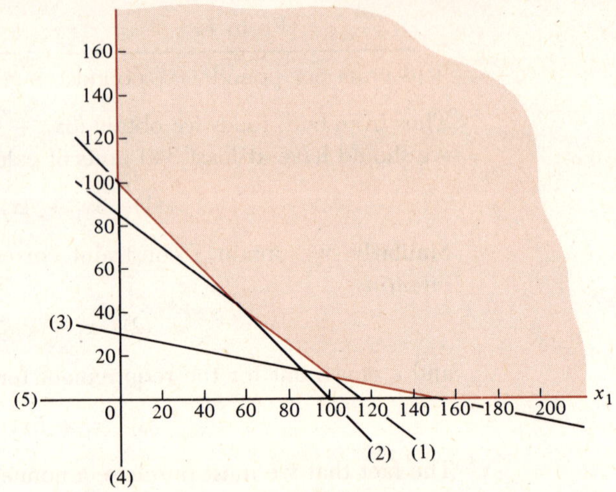

Figure 6.2. *Geometric representation of a feasible set.*

Any one point in the colored region of Figure 6.2 represents a solution of the system of constraints given in Example 1. The set of solutions is known as the **feasible set of the system.** In Figure 6.2 the colored region is the geometric representation of the feasible set of the given system.

▶ **Definition 1**

The feasible set of a system of linear constraints (which may include equations as well as inequalities) is the set of all vectors that satisfy all the constraints simultaneously. Elements of the feasible set are called **feasible solutions.**

It is customary in linear programming problems to require that all the unknowns used in the constraints be nonnegative. If problems are encountered where there may be negative values for an unknown x_i, then x_i can be expressed as $x_i = u_i - v_i$, where u_i and v_i are nonnegative. The requirements that the unknowns be nonnegative are appropriately called the **nonnegativity constraints.** All the other constraints are called **structural constraints.** Structural constraints may be in the form of equations or inequalities. In Example 1, constraints (1), (2), and (3) were structural constraints, and constraints (4) and (5) were nonnegativity constraints.

EXERCISES

Suggested minimum assignment: Exercises 1, 3, 5, and 7.

In Exercises 1–4 graph the feasible sets for each of the given systems of linear constraints.

1. $x_1 + x_2 \leq 6,\quad 3x_1 - x_2 \leq 10,$ and $x_1 \geq 0,\quad x_2 \geq 0.$
2. $x_1 - x_2 \leq 4,\quad x_2 \leq 3,$ and $x_1 \geq 0,\quad x_2 \geq 0.$
3. $2x_1 + x_2 \geq 8,\quad x_1 + x_2 \geq 6,\quad x_1 + 2x_2 \geq 9,$ and $x_1 \geq 0,\quad x_2 \geq 0.$
4. $x_1 + x_2 \geq 1,\quad x_1 - x_2 \geq -1,\quad x_1 + 2x_2 \geq 4,$ and $x_1 \geq 0,\quad x_2 \geq 0.$
5. Identify the structural and nonnegativity constraints of Exercises 1 and 3.
6. Identify the structural and nonnegativity constraints of Exercises 2 and 4.

In Exercises 7–8 graph the feasible set for each of the given systems of linear constraints.

7. $4x_1 + 3x_2 \geq 18,\quad 2x_1 + 5x_2 \geq 16,$ and $x_1 \geq 0,\quad x_2 \geq 0.$
8. $-x_1 + x_2 \leq 1,\quad x_1 + x_2 \leq 3,$ and $x_1 \geq 0,\quad x_2 \geq 0.$

6.2 A Geometric Method of Solution

So far we have discussed only the *constraints* of a linear programming problem. In various applied problems one often wishes to optimize some linear function subject to such a system of linear constraints.

Example 1

In the dietary problem (Example 1) of the last section, remember that foods F_1 and F_2 cost \$3 and \$5 per pound, respectively. The dietician wants to minimize the total cost of the foods purchased, subject to the constraints imposed by the dietary requirements discussed in the last section. We let x_1 and x_2 represent the number of pounds of the respective foods that are to be purchased. Hence the total cost will be

$$\overbrace{(\$3 \text{ per pound})(x_1 \text{ pounds})}^{\text{Cost of } F_1} + \overbrace{(\$5 \text{ per pound})(x_2 \text{ pounds})}^{\text{Cost of } F_2}.$$

Therefore, the dietician seeks to minimize $(3x_1 + 5x_2)$ subject to

$$\begin{array}{ll} (1) & 3x_1 + 4x_2 \geq 340, \\ (2) & x_1 + x_2 \geq 100, \\ (3) & x_1 + 5x_2 \geq 150, \end{array} \quad \text{and} \quad \begin{array}{ll} (4) & x_1 \geq 0, \\ (5) & x_2 \geq 0. \end{array}$$

Values of the cost $(3x_1 + 5x_2)$ will be represented by the letter f. We now illustrate a geometric method of solution of this linear programming problem.

If arbitrary values are assigned to f in the equation $f = 3x_1 + 5x_2$, we obtain a family of straight lines imposed on the feasible set as shown by the dashed lines in Figure 6.3. Notice that for increasing values of f, the corresponding lines appear farther upward and away from the origin. From Figure 6.3 it appears that the line corresponding to $f = 350$ is the "lowest" line that has a point in common with the shaded region; any lower line does not intersect the shaded region, and higher lines represent larger values of f than the minimum value. Hence the point (100, 10), which is the intersection of

$$(1) \quad 3x_1 + 4x_2 = 340 \qquad \text{and} \qquad (3) \quad x_1 + 5x_2 = 150,$$

gives the minimum f, namely,

$$f = 3(100) + 5(10) = 350.$$

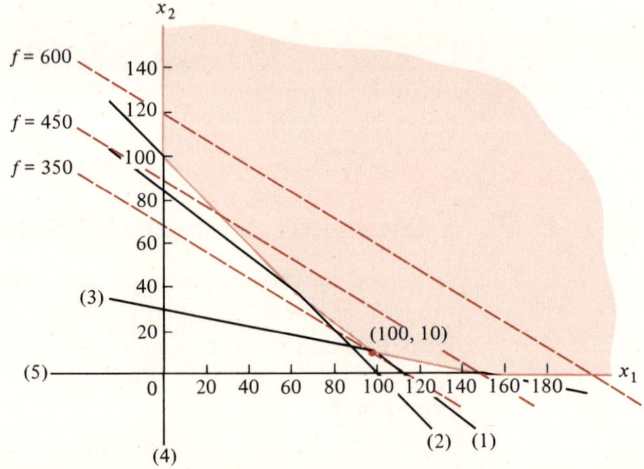

Figure 6.3. *Geometric representation of objective equation showing minimum f.*

Thus if 100 pounds of F_1 and 10 pounds of F_2 are purchased, the dietary requirements will be met at less cost than with any other combination.

Notice that if $x_1 = 100$ and $x_2 = 10$, then $x_1 + x_2 = 110$, or 110 units of vitamins are acquired. This is more than the minimum requirement of vitamins, and at first it seems strange that a surplus of some item can be obtained at minimum cost. ∎

In general the linear function, whose values are determined by

$$c_1x_1 + c_2x_2 + \cdots + c_nx_n$$

and which is to be optimized, is called the **objective function;** the corresponding equation

$$f = c_1x_1 + c_2x_2 + \cdots + c_nx_n$$

is called the **objective equation.**

A further illustration of a linear programming problem and the geometric method of solution follows.

Example 2

A local television network has found that program A, with 20 minutes of music and 1 minute of advertisement, draws 30,000 viewers; and program B, with 10 minutes of music and 1 minute of advertisement, draws 10,000 viewers. Within one week the advertiser insists that at least 6 minutes be devoted to his advertisement, and the network can afford no more than 80 minutes of music. How many times per week should each program be given to obtain the maximum number of viewers while satisfying the requirements of the advertiser and the network?

Solution: Arrange the given information in chart form.

	A	B	Requirements
Music	20	10	at most 80
Advertisement	1	1	at least 6
Viewers	30,000	10,000	

Let x_1 represent the number of times program A is put on the air, and let x_2 represent the number of times program B is put on the air. From the information in the chart we form the constraints

$$\begin{aligned} 20x_1 + 10x_2 &\le 80, \\ x_1 + x_2 &\ge 6, \end{aligned} \quad \text{and} \quad \begin{aligned} x_1 &\ge 0, \\ x_2 &\ge 0. \end{aligned}$$

The objective function to be maximized, subject to the constraints, has values given by $(30{,}000x_1 + 10{,}000x_2)$. The feasible set and the family of lines representing the objective function are graphed in Figure 6.4. From the graph we see that f increases as the lines representing the objective function move away from the origin. The member of the family of lines

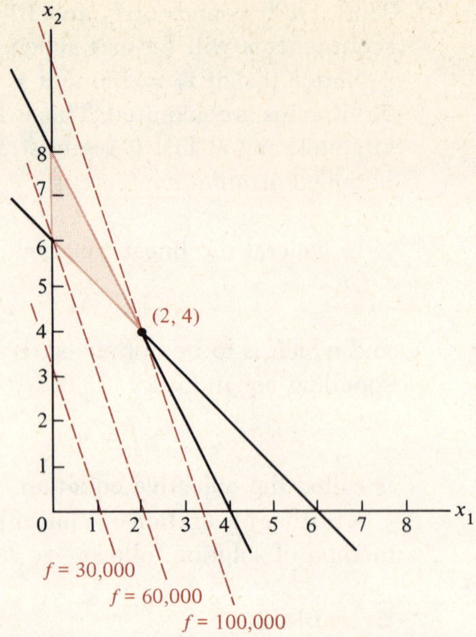

Figure 6.4. *Geometric representation of Example 2.*

representing the largest f that intersects the feasible set is the line that passes through the point (2, 4). Therefore, we have

$$\text{maximum } f = 100{,}000 \text{ at } (2, 4),$$

which means that program A should be offered twice, and program B should be offered four times to obtain a maximum audience of 100,000 viewers. ∎

At this point we remark that the optimum of the objective function will always occur at a vertex of the feasible set. For the case where the graph of the objective equation is parallel to a certain edge of the feasible set, then the whole edge, including both vertices, produces an optimum. Also, an optimum always exists if the feasible set is bounded, as it was in Example 2; but an optimum may or may not exist if the feasible set is not bounded, as was the case in Example 1.

APPLICATIONS

In order that the reader may obtain some idea of how linear programming can be applied to various fields, the basic mathematical model of Example 1 will be given five different interpretations.

Example 3

Two factories can manufacture three different grades of paper at known rates. The company that owns the factories has contracts to supply certain quantities of each grade of paper. How many days should each factory be operated in order to satisfy the demand and yet minimize the cost of production?

	Factory 1	Factory 2	Tons needed
High grade	3 tons/day	4 tons/day	340
Medium grade	1 ton/day	1 ton/day	100
Low grade	1 ton/day	5 tons/day	150
Cost per day (thousands of dollars)	3	5	

Let x_1 be the number of days that factory 1 is to be operated. Let x_2 be the number of days that factory 2 is to be operated. ▮

Example 4

The minimum requirements of the amount of certain chemicals necessary to grow a certain crop successfully have been found. Knowing the content and cost of two special types of commercial fertilizers, packaged in 100-pound bags, a grower wants to know how many bags of each type should be applied to his crop to assure proper growth at minimum cost.

	Fertilizer 1	Fertilizer 2	Units needed
Chemical *A*	3 units/bag	4 units/bag	340
Chemical *B*	1 unit/bag	1 unit/bag	100
Chemical *C*	1 unit/bag	5 units/bag	150
Cost per bag	$3	$5	

Let x_1 be the number of bags of fertilizer 1 that are applied. Let x_2 be the number of bags of fertilizer 2 that are applied. ▮

Example 5

A fruit buyer needs a certain number of units of various varieties of fruit. Two suppliers can supply his needs but will sell only in full truckloads, consisting of a specified number of units of each variety (let 1 unit equal 10 boxes). How many loads should the buyer order from each supplier in order to save time and money by holding the total shipping distance to a minimum?

	Supplier 1	Supplier 2	Units needed
Variety *A*	3 units/load	4 units/load	340
Variety *B*	1 unit/load	1 unit/load	100
Variety *C*	1 unit/load	5 units/load	150
Distance (hundreds of miles)	3	5	

Let x_1 be the number of truckloads ordered from supplier 1. Let x_2 be the number of truckloads ordered from supplier 2. ▮

Example 6

An office manager must assign his 2 groups of employees in such a way that each of three tasks is performed a certain number of times and the time to do them is held to a minimum.

	Group 1	Group 2	Times needed
Task A	3 times/assignment	4 times/assignment	340
Task B	1 time/assignment	1 time/assignment	100
Task C	1 time/assignment	5 times/assignment	150
Hours taken per assignment	3	5	

Let x_1 be the number of assignments given to group 1. Let x_2 be the number of assignments given to group 2. ▮

Example 7

A nutritionist working for a spacecraft designer is faced with the problem of minimizing weight subject to certain nutritional requirements. He is considering 2 foods which are packaged in tubes.

	Food 1	Food 2	Units needed
Carbohydrates	3 units/tube	4 units/tube	340
Fats	1 unit/tube	1 unit/tube	100
Proteins	1 unit/tube	5 units/tube	150
Wt. per tube (ounces)	3	5	

Let x_1 be the number of tubes of food 1 sent on the spacecraft. Let x_2 be the number of tubes of food 2 sent on the spacecraft. ▮

The reader should observe that in Examples 5 and 7 the solution vector had to have integral components; when applied problems of this type are encountered, and if such a solution vector is not obtained by our methods, then techniques beyond the scope of this text must be employed. The study of such problems is a very interesting branch of linear programming known as *integer programming*.

EXERCISES

Suggested minimum assignment: Exercises 1, 3, 5, 7, and 9.

In Exercises 1–4 solve the given linear programming problem (the feasible set of each problem was graphed in the corresponding exercises of the last section).

1. Maximize $(x_1 + 2x_2)$ subject to $\begin{aligned} x_1 + x_2 &\leq 6, \\ 3x_1 - x_2 &\leq 10, \end{aligned}$ and $\begin{aligned} x_1 &\geq 0, \\ x_2 &\geq 0. \end{aligned}$

2. Maximize $(x_1 + x_2)$ subject to $\begin{aligned} x_1 - x_2 &\le 4, \\ x_2 &\le 3, \end{aligned}$ and $\begin{aligned} x_1 &\ge 0, \\ x_2 &\ge 0. \end{aligned}$

3. Minimize $(2x_1 + 3x_2)$ subject to $\begin{aligned} 2x_1 + x_2 &\ge 8, \\ x_1 + x_2 &\ge 6, \\ x_1 + 2x_2 &\ge 9, \end{aligned}$ and $\begin{aligned} x_1 &\ge 0, \\ x_2 &\ge 0. \end{aligned}$

4. Minimize $(2x_1 + x_2)$ subject to $\begin{aligned} x_1 + x_2 &\ge 1, \\ x_1 - x_2 &\ge -1, \\ x_1 + 2x_2 &\ge 4, \end{aligned}$ and $\begin{aligned} x_1 &\ge 0, \\ x_2 &\ge 0. \end{aligned}$

In Exercises 5–6 solve the given linear programming problem by maximizing the negative of the objective function, and thereby illustrate that minimization problems in linear programming can be transformed easily to maximization problems.

5. Minimize $(x_1 + x_2)$ subject to $\begin{aligned} x_1 + 2x_2 &\le 7, \\ x_1 - 3x_2 &\ge 2, \end{aligned}$ and $\begin{aligned} x_1 &\ge 0, \\ x_2 &\ge 0. \end{aligned}$

6. Minimize $(3x_1 + x_2)$ subject to $\begin{aligned} x_1 + 2x_2 &\ge 5, \\ 2x_1 + x_2 &\ge 4, \end{aligned}$ and $\begin{aligned} x_1 &\ge 0, \\ x_2 &\ge 0. \end{aligned}$

In Exercises 7 and 8 show that more than one feasible solution will optimize the objective function (the feasible set of each problem was graphed in the corresponding exercise of the last section).

7. Minimize $(2x_1 + 5x_2)$ subject to $\begin{aligned} 4x_1 + 3x_2 &\ge 18, \\ 2x_1 + 5x_2 &\ge 16, \end{aligned}$ and $\begin{aligned} x_1 &\ge 0, \\ x_2 &\ge 0. \end{aligned}$

8. Maximize $(x_1 + x_2)$ subject to $\begin{aligned} -x_1 + x_2 &\le 1, \\ x_1 + x_2 &\le 3, \end{aligned}$ and $\begin{aligned} x_1 &\ge 0, \\ x_2 &\ge 0. \end{aligned}$

In Exercises 9 and 10 show that the given linear programming problem has no solution.

9. Maximize $(4x_1 - 5x_2)$ subject to the constraints of Exercise 7.
10. Minimize $(-x_1 + x_2)$ subject to the constraints of Exercise 3.
11. There are 2 factories that manufacture three different grades of paper. There is some demand for each grade. The company that controls the factories has contracts to supply 16 tons of low-grade, 5 tons of medium-grade, and 20 tons of high-grade paper. It cost \$1000 per day to operate the first factory and \$2000 per day to operate the second factory. Factory 1 produces 8 tons of low-grade, 1 ton of medium-grade, and 2 tons of high-grade paper in 1 day's operation. Factory 2 produces 2 tons of low-grade, 1 ton of medium-grade, and 7 tons of high-grade paper per day. How many days should each factory be operated in order to fill the orders most economically?
12. Two oil refineries produce three grades of gasoline: *A*, *B*, and *C*. At each refinery the various grades of gasoline are produced in a single operation so that they are in fixed proportions. Assume that 1 operation at refinery 1 produces 1 unit of *A*, 3 units of *B*, and 1 unit of *C*. One

operation at refinery 2 produces 1 unit of A, 4 units of B, and 5 units of C. Refinery 1 charges \$300 for what is produced in 1 operation, and refinery 2 charges \$500 for the production of 1 operation. A consumer needs 100 units of A, 340 units of B, and 150 units of C. How should the orders be placed if the consumer is to meet his needs most economically?

13. Suppose that a certain company has 2 methods, M_1 and M_2, of manufacturing 3 automobile gadgets G_1, G_2, and G_3. The first method will produce 1 of each gadget in 3 hours. The second method will produce 3 G_1's and 1 G_3 in 4 hours. The company has an order for 6 G_1's, 2 G_2's, and 4 G_3's. How many times should we employ each method to fill the order and minimize the time we spend in production?

14. A fruit dealer ships 800 boxes of fruit north on a certain truck. If he must ship at least 200 boxes of oranges at \$0.20 per box profit, at least 100 boxes of grapefruit at \$0.10 per box profit, and at *most* 200 boxes of tangerines at \$0.30 per box profit, how should he load his truck for maximum profit? *Hint:* Use a three-dimensional graph or use the constraint that a total of 800 boxes are shipped, to reduce the number of unknowns in the linear programming model from three to two.

6.3 An Algebraic Method of Solution

In Section 6.2 a geometric method of solution was illustrated. Obviously, if the number of unknowns exceeds 3, such a method is inadequate. In this section we shall introduce an algebraic method of solution of a linear programming problem with any number of unknowns.

Example 1

Consider the linear programming problem:

Maximize $(6x_1 + x_3)$ subject to

$$\begin{aligned} 2x_1 + x_2 + x_3 &= 10, \\ x_1 + 4x_2 &\leq 12, \end{aligned} \qquad \text{and} \qquad \begin{aligned} x_1 &\geq 0, \\ x_2 &\geq 0, \\ x_3 &\geq 0. \end{aligned}$$

First write the structural constraints as a system of linear equations. If there is a structural constraint that is an inequality, then it can be changed to an equation by adding or subtracting a new nonnegative variable that will be called a **slack variable.** For example, in our illustration the left side of the constraint,

$$x_1 + 4x_2 \leq 12,$$

is smaller than or equal to the right side; hence there is some unknown nonnegative quantity x_4 that can be added to the left side so that

$$(x_1 + 4x_2) + x_4 = 12.$$

If the sense of the inequality had been reversed, we simply would have subtracted, rather than added, x_4. The problem can be expressed now as maximize $(6x_1 + 0x_2 + x_3 + 0x_4)$ subject to

$$\begin{aligned} 2x_1 + x_2 + x_3 \qquad &= 10, \\ x_1 + 4x_2 \qquad + x_4 &= 12, \end{aligned} \quad \text{and} \quad \begin{aligned} x_1 &\geq 0, \\ x_2 &\geq 0, \\ x_3 &\geq 0, \\ x_4 &\geq 0. \end{aligned}$$

Notice that the introduction of a slack variable has no actual effect on values of the objective function because the coefficient of the slack variable is zero. ▮

Definition 2

A **slack variable** is a nonnegative variable x_q that changes an inequality

$$a_{i1}x_1 + \cdots + a_{ip}x_p \leq p_{i0},$$

or

$$a_{i1}x_1 + \cdots + a_{ip}x_p \geq p_{i0},$$

to an equation

$$a_{i1}x_1 + \cdots + a_{ip}x_p + x_q = p_{i0},$$

or

$$a_{i1}x_1 + \cdots + a_{ip}x_p - x_q = p_{i0},$$

respectively.

Example 1 (continued)

If we designate values of the objective function as f, the linear programming problem can be restated as: Find nonnegative values of x_1, x_2, x_3, and x_4 of the linear system

$$\begin{aligned} 2x_1 + x_2 + x_3 \qquad &= 10, \\ x_1 + 4x_2 \qquad + x_4 &= 12, \end{aligned} \tag{6.1}$$

so that $f = 6x_1 + x_3$ is a maximum.

Because there may be an infinite number of solutions of system (6.1), the task of finding one satisfying the specified conditions seems at first to be an impossible task. Fortunately, however, it can be proved that the solution of this problem, if a solution exists, is a *nonnegative* particular solution of system (6.1), where the parameters of a complete solution are assigned the value 0; solutions of (6.1) in which the parameters of a complete solution are assigned a value of 0 are known as **basic solutions.** Because there is always a finite number of basic solutions, the task of solving a problem with a solution may be long but not impossible.

The calculation of a couple of basic solutions of (6.1) will now be illustrated. An obvious complete solution of (6.1) is

$$\begin{aligned} x_3 &= 10 - 2x_1 - x_2, \\ x_4 &= 12 - x_1 - 4x_2. \end{aligned}$$

If we let the parameters x_1 and x_2 be 0, we have a basic solution

$$(0, 0, 10, 12),$$

with a corresponding f equal to 10. Suppose we now seek another basic solution by replacing x_2 as a parameter. Using the Gauss–Jordan method, we perform the elementary row operations of $(\frac{1}{4}R_2)$ and $(-R_2 + R_1)$ in that order on the augmented matrix of (6.1); the following equivalent system will be obtained:

$$\begin{aligned} \tfrac{7}{4}x_1 + x_3 - \tfrac{1}{4}x_4 &= 7, \\ \tfrac{1}{4}x_1 + x_2 + \tfrac{1}{4}x_4 &= 3. \end{aligned} \tag{6.2}$$

The complete solution corresponding to (6.2) is

$$\begin{aligned} x_3 &= 7 - \tfrac{7}{4}x_1 + \tfrac{1}{4}x_4, \\ x_2 &= 3 - \tfrac{1}{4}x_1 - \tfrac{1}{4}x_4. \end{aligned}$$

If we let the new parameters x_1 and x_4 be zero, we have another basic solution,

$$(0, 3, 7, 0),$$

with a corresponding f equal to 7. We can continue in this manner to find the other nonnegative basic solutions corresponding to complete solutions of the structural constraints. They are $(5, 0, 0, 7)$ with $f = 30$ and $(4, 2, 0, 0)$ with $f = 24$. The largest value of f corresponding to the four nonnegative basic solutions is 30, and hence if there is a maximum value of f that satisfies (6.1), then that value must be 30 and occur at

$$(5,\ 0,\ 0,\ 7).$$

(There can be as many as $C_2^4 = 6$ ways of obtaining 2 parameters from 4 variables. In this example there are two other basic solutions, but it turns out that they are *not* nonnegative; hence they are not considered because of the nonnegativity constraints.) ▮

We now give formal definitions to some of the concepts that were introduced in the previous illustration, and we state the theorem that assures that the maximum f of Example 1 occurs at $(5, 0, 0, 7)$ rather than at some other vector.

▶ **Definition 3**

Consider a system of m linear equations in n unknowns $(n \geq m)$ having a complete solution with $n - m$ parameters. If the parameters are assigned the value 0, the resulting particular solution is called a **basic solution.** The m unknowns or variables *not* serving as parameters in the complete solution associated with a basic solution are called **basic variables.** The $n - m$ parameters that *are* assigned the value 0 are called **nonbasic variables.**

Those basic solutions of a linear system of structural constraints (equations) that are feasible, and hence have no negative coordinates, are called **basic feasible solutions.**

▶ Definition 4

An **optimal solution** is a feasible solution at which the linear objective function is optimum.

We now state the theorem that is the cornerstone of the algebraic method of solving linear programming problems because it assures us that if there is an optimal solution of a linear programming problem, then that solution can be found among the basic feasible solutions. Because there are a finite number of basic feasible solutions, the task of finding an existing, optimal solution is possible.

▶ Theorem 1

If there is an optimal solution of a linear programming problem, then an optimal solution can be found among the basic feasible solutions. Moreover, there may be more than one optimal solution.

A proof of Theorem 1 is beyond the scope of this text (but an outline may be found in Campbell [8] pp. 372–373). ◀

The final example of this section will again illustrate the use of Theorem 1, under the assumption that an optimal solution does exist. We shall also give a geometric interpretation of the basic feasible solutions.

Example 2

Maximize $(2x_1 + x_2)$ subject to

$$\begin{aligned} x_1 + \ \ x_2 &\le 5, \\ x_1 + 3x_2 &\le 9, \end{aligned} \qquad \text{and} \qquad \begin{aligned} x_1 &\ge 0, \\ x_2 &\ge 0. \end{aligned}$$

Solution: Introduce slack variables x_3 and x_4 to change the structural constraints from inequalities to equations, and let $f = 2x_1 + x_2$, where f is to be maximized. Form the system

$$\begin{aligned} x_1 + \ \ x_2 + x_3 \qquad &= 5, \\ x_1 + 3x_2 \qquad + x_4 &= 9. \end{aligned} \tag{6.3}$$

By Theorem 1, maximum f, if it exists, will occur at a basic feasible solution of (6.3). In Exercise 7 the reader will be asked to verify that

$$\begin{aligned} &(0, 0, 5, 9) \text{ with } f = 0, \\ &(5, 0, 0, 4) \text{ with } f = 10, \\ &(0, 3, 2, 0) \text{ with } f = 3, \\ &(3, 2, 0, 0) \text{ with } f = 8, \end{aligned} \tag{6.4}$$

are the basic feasible solutions of (6.3). If we know that an optimal solution exists, then by Theorem 1 that basic feasible solution must be

$$(5, 0, 0, 4)$$

because 10 is the largest value of f from among the 4 candidates. Of course, in the original statement of the problem we were not concerned with the slack variables; hence the solution of the original problem is: Maximum $(2x_1 + x_2) = 10$ at $(5, 0)$. In Figure 6.5, notice that each basic feasible solution of (6.3) corresponds to a vertex of the feasible set (colored region). Each vertex is an intersection of the bounding lines of the solution sets of two of the constraints of the original problem. Notice, also, that the point $(9, 0)$ that lies on the intersection of the bounding lines of the solution sets of the constraints

$$x_1 + 3x_2 \leq 9 \qquad \text{and} \qquad x_2 \geq 0,$$

corresponds to the basic solution $(9, 0, -4, 0)$ of (6.3). The point $(9, 0)$, as well as its corresponding basic solution, is *not* feasible, however, since it fails to satisfy *all* the constraints. ■

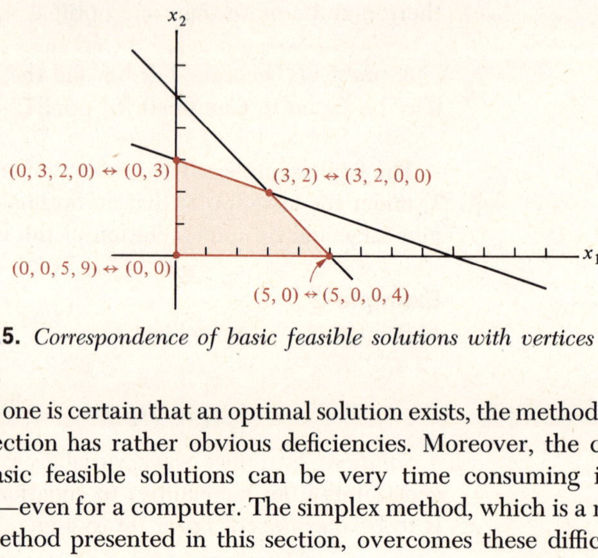

Figure 6.5. *Correspondence of basic feasible solutions with vertices of feasible set.*

Unless one is certain that an optimal solution exists, the method presented in this section has rather obvious deficiencies. Moreover, the calculation of all basic feasible solutions can be very time consuming in a large problem—even for a computer. The simplex method, which is a refinement of the method presented in this section, overcomes these difficulties and is the subject of Sections 6.5 and 6.6.

APPLICATIONS

Example 3

A survey of applications of linear programming may be found in Gass [16], pp. 11–18. This survey includes the following areas: agriculture, contract awards, economics, military practice, personnel assignment, production scheduling, inventory control, structural design, traffic analysis, transportation, and network theory. Industrial applications from the following kinds of industries are also discussed: chemical, coal, commercial aviation, communication, iron and steel, paper, petroleum, and railroad. On pages 325–337 of [16] a bibliography is given for applications in each of the categories listed above. The reader is strongly encouraged to at least glance through the material in [16] in order to appreciate properly the role of linear programming in applications of mathematics in our society; such an appreciation should help to motivate the student to study the simplex method, which is to be presented in Sections 6.5 and 6.6, and which is the primary method of solution of linear programming problems. ▮

Example 4

The usefulness of linear programming in management accounting is made quite evident by Ijiri in [19]. Any serious student of business should at least glance through [19] to observe first-hand the degree of application of linear programming in accounting; other references that illustrate the use of linear programming in accounting are [10], [13], and [20]. ▮

EXERCISES

Suggested minimum assignment: Exercises 1, 3, 5, and 9.

In Exercises 1 and 2 use the method of Example 1 to find all the basic solutions of the given system of equations. If the unknowns in each system of equations are required to be nonnegative, which of the basic solutions would be basic feasible solutions?

1. $$\begin{aligned} x_1 + 2x_2 + x_3 &= 14, \\ 3x_1 + x_2 + x_4 &= 12. \end{aligned}$$

2. $$\begin{aligned} 2x_1 + 4x_2 + x_3 &= 10, \\ 4x_1 + x_2 + x_4 &= 6. \end{aligned}$$

In Exercises 3 and 4 express the given structural constraints of a linear programming problem as a system of linear equations. What are the new nonnegativity constraints?

3. $$\begin{aligned} x_1 + 2x_2 &\leq 7, \\ x_1 - 3x_2 + x_3 &\geq 2, \\ 2x_1 + x_2 + x_3 &= 4. \end{aligned}$$

4. $$\begin{aligned} x_1 - x_2 - x_3 &= 2, \\ 2x_1 - x_2 + x_4 &\leq 1, \\ x_1 + 2x_2 + x_4 &\geq 3. \end{aligned}$$

5. (*a*) On a two-dimensional graph illustrate the feasible set determined by the constraints

$$\begin{aligned} x_1 + 3x_2 &\leq 9, \\ 2x_1 + x_2 &\leq 8, \end{aligned} \quad \text{and} \quad \begin{aligned} x_1 &\geq 0, \\ x_2 &\geq 0. \end{aligned}$$

(*b*) The feasible set in part (*a*) has four bounding lines; find the six points, each of which is common to some two of these bounding lines.

(*c*) Introduce slack variables to make the structural constraints a system of linear equations. Find the six basic solutions and show that they correspond to the six points of intersection found in part (*b*).

(*d*) Find the four basic *feasible* solutions from among the six basic solutions of part (*c*).

(e) Which basic feasible solution is the optimal solution if we seek maximum $(x_1 + x_2)$?

6. Repeat Exercise 5 if the constraints are

$$\begin{aligned} x_1 + 4x_2 &\leq 13, \\ 3x_1 + x_2 &\leq 6, \end{aligned} \qquad \text{and} \qquad \begin{aligned} x_1 &\geq 0, \\ x_2 &\geq 0. \end{aligned}$$

7. In Example 2 of this section verify that (6.4) are the basic feasible solutions of (6.3).

In Exercises 8 and 9 solve the given linear programming problem by the procedure followed in Example 1. Be careful about the sense of each inequality and the consequent addition or subtraction of the slack variable. (The reader was asked to solve each of the exercises by the geometric method in Exercises 5 and 6 of Section 6.2.)

8. Minimize $(x_1 + x_2)$ subject to

$$\begin{aligned} x_1 + 2x_2 &\leq 7, \\ x_1 - 3x_2 &\geq 2, \end{aligned} \qquad \text{and} \qquad \begin{aligned} x_1 &\geq 0, \\ x_2 &\geq 0. \end{aligned}$$

9. Minimize $(3x_1 + x_2)$ subject to

$$\begin{aligned} x_1 + 2x_2 &\geq 5, \\ 2x_1 + x_2 &\geq 4, \end{aligned} \qquad \text{and} \qquad \begin{aligned} x_1 &\geq 0, \\ x_2 &\geq 0. \end{aligned}$$

10. A trucking company owns two types of trucks. Type A has 20 cubic yards of refrigerated space and 30 cubic yards of nonrefrigerated space. Type B has 20 cubic yards of refrigerated space and 10 cubic yards of nonrefrigerated space. A customer wants to transport some produce a certain distance; he will require 160 cubic yards of refrigerated space and 120 cubic yards of nonrefrigerated space. The trucking company figures it will take 300 gallons of gas for the type A truck to make the trip and 200 gallons of gas for the type B truck. Find the number of trucks of each type that the company should allow for the job in order to minimize gas consumption.

11. An insurance consultant must provide the necessary life and disability insurance for a client. Suppose that policies P_1 and P_2 provide the following monthly income to the beneficiary per thousand dollars of insurance.

	P_1	P_2
Life	\$20	\$40
Disability	\$50	\$20

At least \$800 of life insurance income and \$600 of disability insurance income must be provided. If the costs of P_1 and P_2 are \$1 and \$.80 per thousand, respectively, how many thousands of each type of policy should be bought to meet the desired coverage while minimizing the total cost of the insurance purchased?

6.4 Matrix Notation

In the literature that deals with linear programming, it is customary to express linear programming problems using matrix notation; to do this, the following definition is needed.

▶ Definition 5

A matrix A is **"greater than or equal to"** ($\geq$) matrix B of the same order when each entry of A is greater than, or equal to, the corresponding entry of B ($<$, $>$, or $\leq$ can be substituted for $\geq$ with corresponding changes in meaning).

Example 1

By Definition 5,

$$\begin{bmatrix} 2 & 3 \\ 4 & 5 \end{bmatrix} \geq \begin{bmatrix} 1 & 3 \\ 0 & 2 \end{bmatrix}$$

because each entry of the matrix on the left is larger than or equal to the corresponding entry on the right. ▮

Example 2

Consider the diet problem of Section 6.2: Minimize $(3x_1 + 5x_2)$ subject to

$$\begin{aligned} 3x_1 + 4x_2 &\geq 340, \\ x_1 + x_2 &\geq 100, \\ x_1 + 5x_2 &\geq 150, \end{aligned} \qquad \text{and} \qquad \begin{aligned} x_1 &\geq 0, \\ x_2 &\geq 0. \end{aligned}$$

Using matrix multiplication and Definition 5, we can rewrite the problem as

$$\text{Minimize } [3 \quad 5]\begin{bmatrix} x_1 \\ x_2 \end{bmatrix} \text{ subject to}$$

$$\begin{bmatrix} 3 & 4 \\ 1 & 1 \\ 1 & 5 \end{bmatrix}\begin{bmatrix} x_1 \\ x_2 \end{bmatrix} \geq \begin{bmatrix} 340 \\ 100 \\ 150 \end{bmatrix} \qquad \text{and} \qquad \begin{bmatrix} x_1 \\ x_2 \end{bmatrix} \geq \begin{bmatrix} 0 \\ 0 \end{bmatrix},$$

or as

Minimize CX subject to $AX \geq P_0$ and $X \geq \mathbf{0}$, where

$$C = [3 \quad 5], \; X = \begin{bmatrix} x_1 \\ x_2 \end{bmatrix}, \; A = \begin{bmatrix} 3 & 4 \\ 1 & 1 \\ 1 & 5 \end{bmatrix}, \text{ and } P_0 = \begin{bmatrix} 340 \\ 100 \\ 150 \end{bmatrix}.$$

Although technically CX is a 1 by 1 matrix, it is permissible here to use the matrix and its single entry interchangeably. ▮

Example 3

If nonnegative slack variables are subtracted from each of the constraints of Example 2 in order to make equations, then the resulting problem can be written:

Minimize CX subject to $AX = P_0$ and $X \geq \mathbf{0}$, where

$$C = [3 \quad 5 \quad 0 \quad 0 \quad 0],$$

$$X = \begin{bmatrix} x_1 \\ x_2 \\ x_3 \\ x_4 \\ x_5 \end{bmatrix}, \qquad A = \begin{bmatrix} 3 & 4 & -1 & 0 & 0 \\ 1 & 1 & 0 & -1 & 0 \\ 1 & 5 & 0 & 0 & -1 \end{bmatrix}, \qquad \text{and} \qquad P_0 = \begin{bmatrix} 340 \\ 100 \\ 150 \end{bmatrix}. \blacksquare$$

In expressing constraints in matrix notation, one must be careful that the constraints are all of the same type; that is, they must all be ($\leq$) inequalities or ($\geq$) inequalities or equations. The constraints

$$\begin{aligned} x_1 + x_2 &\geq 1, \\ x_1 - x_3 &\leq 4, \end{aligned}$$

are not all of the same type but can be made so by multiplying the last constraint by (-1). The result will then be

$$\begin{aligned} x_1 + x_2 &\geq 1, \\ -x_1 + x_3 &\geq -4, \end{aligned}$$

which is suitable for matrix expression.

EXERCISES

Suggested minimum assignment: Exercises 1, 3, 5, and 7.

In each of Exercises 1 and 2 express the given problem using matrices and *without* the introduction of slack variables. Do *not* solve the problem.

1. Maximize $(x_1 + x_2)$ subject to $\begin{aligned} 2x_1 + x_2 &\leq 4, \\ x_1 - x_2 &\geq -5, \end{aligned}$ and $\begin{aligned} x_1 &\geq 0, \\ x_2 &\geq 0. \end{aligned}$

2. Maximize $(3x_1 - x_2)$ subject to $\begin{aligned} x_1 - x_2 &\geq 5, \\ x_1 + 4x_2 &\leq -3, \end{aligned}$ and $\begin{aligned} x_1 &\geq 0, \\ x_2 &\geq 0. \end{aligned}$

In each of Exercises 3 and 4 express the given problem using matrices and *with* the introduction of slack variables. Do *not* solve the problem.

3. Minimize $(5x_1 + x_2)$ subject to $\begin{aligned} 2x_1 + x_2 &= 3, \\ x_1 + x_2 &\leq 4, \\ x_1 - x_2 &\geq 2, \end{aligned}$ and $\begin{aligned} x_1 &\geq 0, \\ x_2 &\geq 0. \end{aligned}$

4. Minimize $(x_1 - x_2)$ subject to $\begin{aligned} x_1 + x_2 &\geq 2, \\ 2x_1 - x_2 &\leq 4, \\ x_1 - 3x_2 &= 2, \end{aligned}$ and $\begin{aligned} x_1 &\geq 0, \\ x_2 &\geq 0. \end{aligned}$

In each of Exercises 5 and 6 find the optimal solution, if one exists, by the geometric method.

5. Maximize CX subject to $AX \leq P_0$ and $X \geq \mathbf{0}$, where $C = [2 \quad 1]$, $X = \begin{bmatrix} x_1 \\ x_2 \end{bmatrix}$, $A = \begin{bmatrix} 1 & 1 \\ 3 & 2 \end{bmatrix}$, and $P_0 = \begin{bmatrix} 5 \\ 12 \end{bmatrix}$.
6. Minimize CX subject to $AX \geq P_0$ and $X \geq \mathbf{0}$, where $C = [2 \quad -1]$, $X = \begin{bmatrix} x_1 \\ x_2 \end{bmatrix}$, $A = \begin{bmatrix} 1 & 2 \\ -1 & 1 \end{bmatrix}$, and $P_0 = \begin{bmatrix} 2 \\ -1 \end{bmatrix}$.

In each of Exercises 7 and 8, find the optimal solution, given that one exists, by use of the algebraic method.

7. Minimize CX subject to $AX = P_0$ and $X \geq \mathbf{0}$, where

$$C = [1 \quad 2 \quad 0 \quad 0],\ X = \begin{bmatrix} x_1 \\ x_2 \\ x_3 \\ x_4 \end{bmatrix},\ A = \begin{bmatrix} 1 & 1 & 1 & 0 \\ 3 & 1 & 0 & 1 \end{bmatrix},\ \text{and } P_0 = \begin{bmatrix} 6 \\ 12 \end{bmatrix}.$$

8. Maximize CX subject to $AX = P_0$ and $X \geq \mathbf{0}$, where

$$C = [-3 \quad 1 \quad 0 \quad 0],\ X = \begin{bmatrix} x_1 \\ x_2 \\ x_3 \\ x_4 \end{bmatrix},\ A = \begin{bmatrix} 2 & 1 & 1 & 0 \\ -1 & 1 & 0 & 1 \end{bmatrix},\ \text{and } P_0 = \begin{bmatrix} 8 \\ 5 \end{bmatrix}.$$

6.5 An Introduction to the Simplex Method

In Section 6.3 we learned that an optimal solution, if one exists, will be among the basic feasible solutions of the structural constraints. Some difficulties remain, however. First, there must be some way to determine whether or not an optimal solution exists, and second, an efficient procedure must be developed for finding the optimal solution without having to find all the basic feasible solutions of the structural constraints; in a large problem, the job of calculating all the basic feasible solutions can be very time consuming—even for a computer. The **simplex method,** developed by George Dantzig in 1947, will overcome these difficulties. A brief history of the development of the simplex method may be found in Dantzig [11], pp. 12–31.

Although the execution of the simplex method is rather involved, the basic idea is reasonably easy. That basic idea is to use the objective function to identify, one at a time, only those basic feasible solutions that can enlarge (or maintain) the value of the objective function of a maximum problem. This avoids having to find all the basic feasible solutions and hence effects considerable efficiency. Moreover, the objective function can be used to determine when an optimal solution is reached or to determine that no solution exists. We shall illustrate these basic ideas of the simplex method in the next example.

Example 1

In Example 1 of Section 6.3 we found maximum $(6x_1 + x_3)$ by evaluating $(6x_1 + x_3)$ at *all* the basic feasible solutions of the structural constraints

$$\begin{aligned} 2x_1 + x_2 + x_3 &= 10, \\ x_1 + 4x_2 + x_4 &= 12. \end{aligned}$$

The simplex method begins with only one basic feasible solution as determined by the obvious complete solution

$$\begin{aligned} x_3 &= 10 - 2x_1 - x_2, \\ x_4 &= 12 - x_1 - 4x_2. \end{aligned}$$

In this initial complete solution, the basic variables are x_3 and x_4 and the nonbasic variables (parameters) are x_1 and x_2. In terms of only the nonbasic variables the objective equation

$$f = 6x_1 + x_3$$

becomes

$$f = 6x_1 + (10 - 2x_1 - x_2),$$

or when simplified

$$f = 10 + 4x_1 - x_2.$$

When the nonbasic variables (parameters) x_1 and x_2 are set equal to 0, then $f = 10$ at the initial basic feasible solution (0, 0, 10, 12). The question then arises: Would f be any larger than 10 if one of the nonbasic variables x_1 or x_2 became a basic variable? If, for instance, x_1 replaced x_3 as a basic variable and if x_1 assumed a positive value, then certainly

$$f = 10 + 4x_1 - x_2$$

would be larger because the coefficient 4 of x_1 is positive. We then proceed to make x_1 a basic variable by finding a new complete solution. To accomplish this and to find the corresponding new expression for f, we apply the Gauss–Jordan method to the augmented matrix of the system

$$\begin{aligned} 2x_1 + x_2 + x_3 &= 10, \\ x_1 + 4x_2 + x_4 &= 12, \\ \hline 4x_1 - x_2 &= f - 10, \end{aligned}$$

in such a way as to maintain the nonnegative numbers on the right side of the structural constraints (this assures that the new basic solution will be feasible). The augmented matrix is

$$M_1 = \left[\begin{array}{cccc|c} 2 & 1 & 1 & 0 & 10 \\ 1 & 4 & 0 & 1 & 12 \\ \hline 4 & -1 & 0 & 0 & f - 10 \end{array}\right].$$

If x_1 is to become a basic variable and if the numbers on the right side of the first equations are to remain nonnegative, then the successive elementary row operations $(-\frac{1}{2}R_1 + R_2)$, $(-2R_1 + R_3)$, and $\frac{1}{2}R_1$ are performed on M_1 (methods for obtaining these operations will be explained in the next section). The resulting matrix is called M_2.

$$M_2 = \left[\begin{array}{cccc|c} 1 & \frac{1}{2} & \frac{1}{2} & 0 & 5 \\ 0 & \frac{7}{2} & -\frac{1}{2} & 1 & 7 \\ \hline 0 & -3 & -2 & 0 & f-30 \end{array}\right].$$

The latter matrix is the augmented matrix of the system

$$\begin{array}{rcl} x_1 + \frac{1}{2}x_2 + \frac{1}{2}x_3 \qquad & = & 5, \\ \frac{7}{2}x_2 - \frac{1}{2}x_3 + x_4 & = & 7, \\ \hline -3x_2 - 2x_3 \qquad & = & f - 30. \end{array}$$

From this system we see that the new complete solution of the structural constraints is

$$\begin{aligned} x_1 &= 5 - \tfrac{1}{2}x_2 - \tfrac{1}{2}x_3, \\ x_4 &= 7 - \tfrac{7}{2}x_2 + \tfrac{1}{2}x_3, \end{aligned}$$

and the new expression for f in terms of only the nonbasic variables x_2 and x_3 is

$$f = 30 - 3x_2 - 2x_3.$$

When the nonbasic variables are set equal to 0, then $f = 30$ at the second basic feasible solution $(5, 0, 0, 7)$. Again the question is raised: Would f be any larger than 30 if one of the nonbasic variables x_2 or x_3 became basic? Notice that if either x_2 or x_3 could assume positive values, then f would not be larger than 30 because the coefficients of x_2 and x_3 are negative. It can be proved that no further improvement in f is possible and hence maximum $f = 30$ at $(5, 0, 0, 7)$. Notice that we arrived at the optimal solution by calculating only two of the basic feasible solutions; moreover, we had a method of determining that the final result was indeed the maximum. ▮

In summary, the simplex method is an iterative procedure that begins with one basic feasible solution of the structural constraints (equations) of the maximum problem and then, by means of the Gauss–Jordan method, replaces the old basic feasible solution by a new basic feasible solution that will increase (or in some cases only maintain) the value of the objective function. The procedure is repeated until an optimal solution is found or until it is determined that no optimal solution exists.

In this section we have been concerned primarily with the basic rationale behind the simplex method. Some of the finer points in the execution of the simplex method will be discussed in the next section.

EXERCISES

Suggested minimum assignment: Exercise 1.

1. Consider the problem: Maximize $(2x_1 + x_3)$ subject to

$$\begin{aligned} -x_1 + x_2 + x_3 \quad &= 1, \\ x_1 + 2x_2 \quad + x_4 &= 5, \end{aligned} \qquad \text{and} \qquad X \geq \mathbf{0}.$$

   (*a*) Find an initial complete solution and the associated basic feasible solution.

   (*b*) Express the value f of the objective function in terms of only the nonbasic variables.

   (*c*) Can the initial value of f be enlarged by making one of the nonbasic variables basic? In other words, can the value of the objective function be enlarged if one of the nonbasic variables could take on positive values? If your answer is yes, name such a variable.

   (*d*) Write the augmented matrix of the system consisting of the structural constraints and your answer to part (*b*). Call the result M_1.

   (*e*) Use the Gauss–Jordan method to find a new complete solution (of the structural constraints) that introduces your answer of part (*c*) as a basic variable. Remember that the first two entries of the last column of M_1 must remain nonnegative.

   (*f*) Express the value f of the objective function in terms of the new nonbasic variables.

   (*g*) Can the value of the objective function be enlarged by making one of the current nonbasic variables basic?

   (*h*) If your answer to part (g) was yes, what nonbasic variable should be made basic? If your answer to part (g) was no, what is maximum f and at what basic feasible solution does maximum f occur?

2. Repeat Exercise 1 for the problem: Maximize $(-2x_1 - x_3)$ subject to

$$\begin{aligned} 3x_1 + x_2 + x_3 \quad &= 9, \\ x_1 + 2x_2 \quad + x_4 &= 8, \end{aligned} \qquad \text{and} \qquad X \geq \mathbf{0}.$$

6.6 The Simplex Method

In this section, formal statements of the steps of the simplex method will be given and illustrated. The plausibility of some of these steps will be discussed, but the reader should be aware of the fact that most of the assertions are not proved. Such justifications may be found in more advanced books on linear programming.

The steps formally listed below apply to the linear programming problem: Maximize CX subject to $AX = P_0$, where $X \geq \mathbf{0}$ and $P_0 \geq \mathbf{0}$. If a minimum problem is encountered, it can be changed to a maximum problem by simply maximizing $(-CX)$ rather than by minimizing CX; the constraints remain the same. Each step will be illustrated by the same example used in Section 6.5.

Step A. Write the augmented matrix $M_0 = \left[\begin{array}{c|c} A & P_0 \\ \hline C & f \end{array}\right]$ of the system

$$AX = P_0,$$
$$CX = f.$$

For the linear programming problem: Maximize $(6x_1 + x_3)$ subject to

$$\begin{aligned} 2x_1 + \ x_2 + x_3 \qquad &= 10, \\ x_1 + 4x_2 \qquad + x_4 &= 12, \end{aligned} \qquad \text{and} \qquad X \geq \mathbf{0},$$

we have

$$M_0 = \left[\begin{array}{cccc|c} 2 & 1 & 1 & 0 & 10 \\ 1 & 4 & 0 & 1 & 12 \\ \hline 6 & 0 & 1 & 0 & f \end{array}\right].$$

Step B. Perform, if possible,† the elementary row operations on M_0 that (1) will produce a complete solution of the structural constraints $AX = P_0$ while maintaining nonnegative values for the transformed P_0, and (2) will simultaneously express f in terms of only nonbasic variables. Call the result M_1.

For the matrix M_0 stated above, no operations are needed to obtain a complete solution of the structural constraints. The basic variables of the existing complete solution are x_3 and x_4. In order that f be expressed in terms of only *non*basic variables x_1 and x_2, the operation $(-R_1 + R_3)$ is performed on M_0 to make the coefficient of x_3 become 0. The result is

$$M_1 = \left[\begin{array}{cccc|c} 2 & 1 & 1 & 0 & 10 \\ 1 & 4 & 0 & 1 & 12 \\ \hline 4 & -1 & 0 & 0 & f - 10 \end{array}\right],$$

and from the last row of M_1 we have $4x_1 - x_2 = f - 10$ or $f = 10 + 4x_1 - x_2$.

Step C. Locate any positive number among all but the final entry of the last row of M_1 and designate the column in which that number appears as the **pivot column.** If there are no such positive numbers, then the simplex method terminates and the current basic feasible solution is an optimal solution.

For our example the only such positive number in the last row of M_1 is 4 and that appears in the first column. Hence the first column is the pivot column.

Step D. Divide each *positive* entry of the pivot column (except the final entry) into the corresponding entry of the last column of M_1 and select

† There will be some problems (see Exercise 13) where step B is either impossible or very difficult. This difficulty can be overcome easily by creating an auxiliary problem for which step B is possible (see Campbell [8], pp. 382–387). The auxiliary problem will include what are called artificial variables.

the smallest of these ratios. The row of M_1 that produces this smallest ratio is designated the **pivot row.**† That entry that belongs to *both* the pivot row and the pivot column is called the **pivot.** If there are no positive entries in the pivot column above the last row, then the simplex method terminates and there is no optimal solution of the problem.

For our example, the ratios are formed in Figure 6.6, and since $\frac{10}{2}$ is the smaller of the ratios, and since this ratio was produced by the first row of M_1, then the first row is the pivot row, and the entry in the first row and first column of M_1 is the *pivot*.

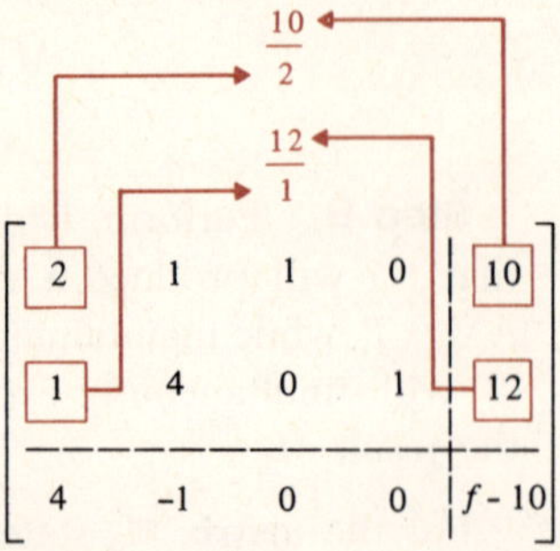

Figure 6.6. *Formation of ratios of step D.*

The purpose of step D is to assure that the nonnegative entries of the last column P_0 are maintained when the elementary row operations of the next step are performed.

Step E. Perform on M_1 the elementary row operations necessary to change the pivot to a 1 and to change the other entries of the pivot column to a 0. Call the result M_2.

In our example the operation $(-\frac{1}{2}R_1 + R_2)$ will produce a 0 in the second entry of the pivot column. The operation $(-2R_1 + R_3)$ will produce a 0 in the third entry of the pivot column. Finally, the operation $\frac{1}{2}R_1$ will make the pivot a 1.

$$\overbrace{\left[\begin{array}{cccc|c} 2^{\circ} & 1 & 1 & 0 & 10 \\ 1 & 4 & 0 & 1 & 12 \\ \hline 4 & -1 & 0 & 0 & f-10 \end{array}\right]}^{M_1} \overset{\text{row}}{\sim} \overbrace{\left[\begin{array}{cccc|c} 1 & \frac{1}{2} & \frac{1}{2} & 0 & 5 \\ 0 & \frac{7}{2} & -\frac{1}{2} & 1 & 7 \\ \hline 0 & -3 & -2 & 0 & f-30 \end{array}\right]}^{M_2}.$$

The purpose of step E is to generate a new complete solution of the structural constraints, together with a new expression for f in terms of the new nonbasic variables.

† It is possible to have a "tie" for the smallest ratio; this situation is symptomatic of what is known as degeneracy. Although degeneracy causes considerable theoretical difficulties, the practical difficulties are few. Arbitrarily pick any one of the candidates as the pivot row and continue.

Step F. Return to step C for M_2, and continue until the method terminates.

When step C is reconsidered for M_2, we see that all the entries of the last row (excluding the final entry) are nonpositive, and hence, according to step C, the simplex method terminates. Thus the maximum $f = 30$ is obtained from $f = 30 - 3x_2 - 2x_3$ at the basic feasible solution $(5, 0, 0, 7)$ as found from the last complete solution,

$$x_1 = 5 - \tfrac{1}{2}x_2 - \tfrac{1}{2}x_3,$$
$$x_4 = 7 - \tfrac{7}{2}x_2 + \tfrac{1}{2}x_3.$$

Example 1

Minimize $(2x_1 + x_2)$ subject to

$$\begin{aligned} -x_1 - x_2 &\le -1, \\ 2x_1 + 3x_2 &\le 6, \end{aligned} \qquad \text{and} \qquad X \ge \mathbf{0}.$$

First, we must write the structural constraints in the form $AX = P_0$, where $P_0 \ge \mathbf{0}$. To do this, add slack variables to the left side and then multiply the first equation by -1. We get

$$\begin{aligned} x_1 + x_2 - x_3 \phantom{{}+x_4} &= 1, \\ 2x_1 + 3x_2 \phantom{{}-x_3} + x_4 &= 6. \end{aligned}$$

Also, to minimize $(2x_1 + x_2)$ is to maximize $(-2x_1 - x_2)$.

Step A

$$M_0 = \left[\begin{array}{rrrr|c} 1 & 1 & -1 & 0 & 1 \\ 2 & 3 & 0 & 1 & 6 \\ \hline -2 & -1 & 0 & 0 & f \end{array}\right].$$

Step B. We do not have an obvious complete solution of the structural constraints if we require $P_0 \ge \mathbf{0}$; hence we apply the Gauss–Jordan method to try to obtain one. We choose to try to make x_1 a basic variable. To do so requires the operation $(-2R_1 + R_2)$, and if f is to be expressed in terms of only nonbasic variables, we also apply $(2R_1 + R_3)$. The result is

$$M_1 = \left[\begin{array}{rrrr|c} 1 & 1 & -1 & 0 & 1 \\ 0 & 1 & 2 & 1 & 4 \\ \hline 0 & 1 & -2 & 0 & f+2 \end{array}\right].$$

Step C. The second column of M_1 is the pivot column.

Step D. Form the ratios shown in Figure 6.7, and since the smallest ratio is produced by the first row, then that row becomes the pivot row.

Step E. Perform the operations $(-R_1 + R_2)$ and $(-R_1 + R_3)$ to make the nonpivot entries of the second column 0.

$$
\overbrace{\left[\begin{array}{cccc|c} 1 & 1^{*} & -1 & 0 & 1 \\ 0 & 1 & 2 & 1 & 4 \\ \hline 0 & 1 & -2 & 0 & f+2 \end{array}\right]}^{M_1} \underset{\sim}{\text{row}} \overbrace{\left[\begin{array}{cccc|c} 1 & 1 & -1 & 0 & 1 \\ -1 & 0 & 3 & 1 & 3 \\ \hline -1 & 0 & -1 & 0 & f+1 \end{array}\right]}^{M_2}.
$$

Step F. (Return to step C). There are no positive entries in the first four entries of the last row, hence maximum $f = -1$ at $(0, 1, 0, 3)$. Since maximum $(-2x_1 - x_2) = -1$, then minimum $(2x_1 + x_2) = +1$. ∎

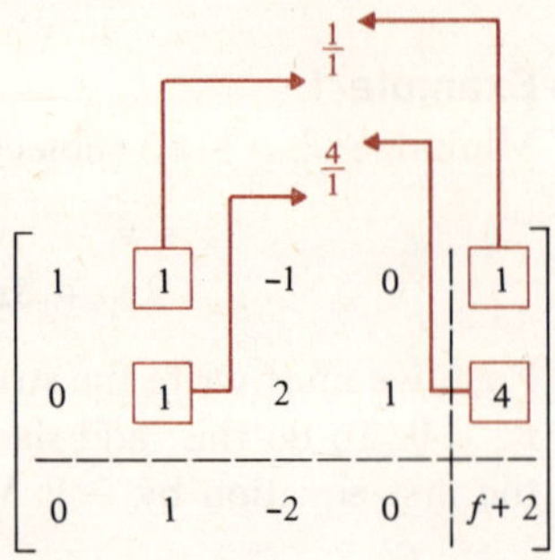

Figure 6.7. *Formation of ratios of step D.*

APPLICATIONS

Example 2

An application of linear programming to a probability problem is described in this example. Suppose that two types of missiles (say A and B) are available to fire at two types of targets, say, target 1 and target 2. Let

x_1 = number of A missiles fired at target 1,
x_2 = number of B missiles fired at target 1,
x_3 = number of A missiles fired at target 2,
x_4 = number of B missiles fired at target 2,
$Q_1 = .10$ = probability that a single A missile will fail to destroy target 1,
$Q_2 = .20$ = probability that a single B missile will fail to destroy target 1,
$Q_3 = .15$ = probability that a single A missile will fail to destroy target 2,
$Q_4 = .25$ = probability that a single B missile will fail to destroy target 2,
c_1 = \$20,000 = cost of a single A missile,
c_2 = \$15,000 = cost of a single B missile,
E = \$1,000,000 = maximum amount that can be spent on a given engagement.

If no more than 20 missiles of each type, and no more than 30 missiles altogether can be expended on a given engagement, determine the number of missiles (quantities x_1, x_2, x_3, and x_4) that will minimize the over-all probability of failure, which is

$$Q = Q_1^{x_1}Q_2^{x_2}Q_3^{x_3}Q_4^{x_4}.$$

Because of certain properties of logarithms, the last equation can be transformed to

$$\ln Q = x_1(\ln Q_1) + x_2(\ln Q_2) + x_3(\ln Q_3) + x_4(\ln Q_4),$$

which for constant values of Q produces a family of linear equations. Since Q decreases if and only if $\ln Q$ decreases, minimum Q occurs when $\ln Q$ is a minimum. Thus the original mathematical model has been converted into a linear programming model, namely,

$$\begin{array}{l}\text{Minimize}\dagger(-2.3x_1 - 1.6x_2 - 1.9x_3 - 1.4x_4), \text{ subject to} \\ 20{,}000x_1 + 15{,}000x_2 + 20{,}000x_3 + 15{,}000x_4 \leq 1{,}000{,}000, \\ x_1 + x_2 + x_3 + x_4 \leq 30, \\ x_1 + x_3 \leq 20, \\ x_2 + x_4 \leq 20,\end{array}$$

and $X \geq \mathbf{0}$.

After dividing the first inequality by 1000, adding four slack variables, and changing to a maximum problem, we obtain the initial augmented matrix M_0:

$$\left[\begin{array}{cccccccc|c} 20 & 15 & 20 & 15 & 1 & 0 & 0 & 0 & 1000 \\ 1 & 1 & 1 & 1 & 0 & 1 & 0 & 0 & 30 \\ 1 & 0 & 1 & 0 & 0 & 0 & 1 & 0 & 20 \\ 0 & 1 & 0 & 1 & 0 & 0 & 0 & 1 & 20 \\ \hline +2.3 & +1.6 & +1.9 & +1.4 & 0 & 0 & 0 & 0 & f-0 \end{array}\right].$$

After three iterations it is possible to determine that the number of missiles that will minimize the over-all probability of failure is

$$x_1 = 20, \quad x_2 = 10, \quad x_3 = 0, \quad \text{and} \quad x_4 = 0.$$

The conversion of models illustrated in this example illustrates one of the strengths of mathematical modeling. ▮

Example 3

The last example need not be restricted to missiles fired at targets. Actually, the model will serve any similar situation in which m "programs" are available for use to solve n "problems," where the probabilities of failure are known and where the constraints are linear. For example, one may have m chemicals available to attack n diseases or m plans available to apply to n social problems. ▮

† The coefficients of the objective function are obtained from a table of natural logarithms: $\ln(.10) \approx -2.3$; $\ln(.20) \approx -1.6$; $\ln(.15) \approx -1.9$; $\ln(.25) \approx -1.4$.

EXERCISES

Suggested minimum assignment: Exercises 1, 5, 7, 9, and 13.

In each of Exercises 1–8 apply the simplex method to solve the problem.

1. Maximize $(x_1 + 2x_2)$ subject to

$$\begin{aligned} x_1 + x_2 + x_3 &= 4, \\ x_1 + 4x_2 + x_4 &= 7, \end{aligned} \quad \text{and} \quad X \geq \mathbf{0}.$$

2. Maximize $(x_1 - 3x_2)$ subject to

$$\begin{aligned} -x_1 + x_2 + x_3 &= 1, \\ x_1 + x_2 + x_4 &= 1, \end{aligned} \quad \text{and} \quad X \geq \mathbf{0}.$$

3. Maximize $(2x_1 + x_2 + 6x_3 + x_4)$ subject to

$$\begin{aligned} x_1 + 3x_2 + x_3 + x_4 &\leq 4, \\ x_1 + x_3 + 2x_4 &\leq 5, \\ x_2 + x_3 &\leq 2, \end{aligned} \quad \text{and} \quad X \geq \mathbf{0}.$$

4. Maximize $(4x_1 - 6x_2 + 5x_3)$ subject to

$$\begin{aligned} -x_1 + x_2 &\leq 1, \\ x_2 + 2x_3 &\leq 4, \\ 2x_1 + x_3 &\leq 6, \\ 2x_2 + x_3 &\leq 1, \end{aligned} \quad \text{and} \quad X \geq \mathbf{0}.$$

5. Maximize $(2x_1 + x_4)$ subject to

$$\begin{aligned} x_1 + x_2 + 2x_3 + x_4 &= 2, \\ x_1 + 2x_2 + x_3 &\leq 5, \end{aligned} \quad \text{and} \quad X \geq \mathbf{0}.$$

6. Minimize $(3x_1 + x_3)$ subject to

$$\begin{aligned} x_1 + x_2 + x_3 &= 3, \\ x_1 + 2x_2 + x_4 &= 7, \end{aligned} \quad \text{and} \quad X \geq \mathbf{0}.$$

7. Minimize $(2x_1 + x_2)$ subject to

$$\begin{aligned} x_1 + x_2 &\geq 1, \\ -x_1 + x_2 &\leq 1, \end{aligned} \quad \text{and} \quad X \geq \mathbf{0}.$$

8. Minimize $(x_1 + x_2 + 4x_3 + x_4)$ subject to

$$\begin{aligned} -x_1 - x_2 + x_3 + x_4 &= 2, \\ -x_1 + x_2 + 2x_3 &\geq 1, \end{aligned} \quad \text{and} \quad X \geq \mathbf{0}.$$

Hint: Subtract slack variable from the left side of the second constraint.

In each of Exercises 9–12 show that the simplex method determines that there is no optimal solution. In each case demonstrate that this determination is correct by means of a graph.

Maximize $(2x_1 + x_2)$ subject to

$$-x_1 + x_2 \leq 1, \quad x_1 - 2x_2 \leq 2, \qquad \text{and} \qquad X \geq \mathbf{0}.$$

10. Maximize $(x_1 + x_2)$ subject to

$$-2x_1 + x_2 \leq 4, \quad x_1 - 3x_2 \leq 6, \qquad \text{and} \qquad X \geq \mathbf{0}.$$

11. Minimize $(x_1 - 2x_2)$ subject to

$$x_1 - x_2 \leq 2, \quad 3x_1 - x_2 \geq -3, \qquad \text{and} \qquad X \geq \mathbf{0}.$$

12. Minimize $(x_1 - x_2)$ subject to

$$2x_1 - 5x_2 \leq 0, \quad -3x_1 + x_2 \leq 6, \qquad \text{and} \qquad X \geq \mathbf{0}.$$

13. (*a*) Show that step B is impossible for the problem:
Maximize $(-x_1 + x_2)$ subject to

$$x_1 + x_2 \leq 3, \quad x_1 - x_2 \geq 5, \qquad \text{and} \qquad X \geq \mathbf{0}.$$

(*b*) Illustrate graphically that the feasible set is empty.

NEW VOCABULARY

feasible set of a system of linear constraints 6.1
feasible solution 6.1
nonnegativity constraints 6.1
structural constraints 6.1
objective function 6.2
objective equation 6.2
slack variable 6.3
basic solution 6.3
basic variables 6.3
nonbasic variables 6.3
basic feasible solution 6.3
optimal solution 6.3
inequality of matrices 6.4
simplex method 6.5
pivot column 6.6
pivot row 6.6
pivot 6.6

7 Game Theory

Game theory is a relatively new area of mathematics that is of considerable interest in some disciplines, particularly the social and managerial sciences. One reason for this interest is that game theory furnishes methods for quantification and analysis of certain conflict situations in which two or more adversaries compete. Examples of such conflict situations include political, economic, and personal competition, plus warfare and competition against nature. The analysis of the behavior of the competitors themselves can be of some interest in psychology and sociology. Many of the problems to which game theory may apply, however, are very complex, and much work remains to be done, both in developing new theory and in applying present theory to applied problems.

Prerequisites: 4.1, 4.10, 5.1–5.3, 6.1, 6.2; some Exercises and Applications in 7.3 and 7.4 require 6.3–6.6, but those Exercises and Applications can be omitted.
Suggested Sections for Moderate Emphasis: 7.1–7.3
Suggested Sections for Minimum Emphasis: 7.1–7.2

7.1 Games and Strategies

In this introduction to the subject of game theory, we shall limit our consideration to a specific class of games known as two-person, zero-sum games; Example 1 is an example of such a game.

Example 1

One of the simplest of all games is that of matching pennies. The game consists of two players, who we will call player R (for row) and player C (for column), and a set of rules: (1) Each player has a penny which he must lay on a flat surface so that one side of the coin is exposed; (2) if both players produce a head, or if both produce a tail, on their respective coins, then R wins a penny from C; (3) if either player turns up a head and the other player turns up a tail, then R will lose a penny to C. *In order to analyze the game just described, we shall construct a mathematical model of the game and analyze it.* The model is called a **2 by 2 matrix game** which includes two players and a **payoff matrix** (or **game matrix**). The payoffs of the penny matching game are exhibited as entries in the payoff matrix

$$A = \overset{\text{Player } C}{\overbrace{\begin{matrix} \text{Heads} & \text{Tails} \end{matrix}}} \atop \begin{bmatrix} +1 & -1 \\ -1 & +1 \end{bmatrix} \begin{matrix} \text{Heads} \\ \text{Tails} \end{matrix} \Big\} \text{ Player } R.$$

A **play** of the matrix game is a choice of a row by R (the row player) and the choice of a column by C (the column player). Notice that the entries of the payoff matrix represent **payoffs** from the *row player's point of view;* therefore, the entry -1 represents a loss to R and a gain for C. It is customary in game theory to always express the payoffs from the row player's point of view. After each play, R receives a payoff from C of an amount equal to the value of the entry that is common to the chosen row and column. If that entry is negative, then one can think of the payoff as a bill to R from C of money owed to C. As an illustration suppose that R chooses row 1 and C chooses column 2 of the payoff matrix; then the payoff to R is -1, which means that R owes C 1 penny. In many games, such as the penny-matching game, it is agreed that R's loss is C's gain and vice versa; such a game is called a **zero-sum game.** ∎

From here on we shall concentrate on matrix games that serve as mathematical models for other zero-sum, two-person games. If in a matrix game the column choice of player C is predictable, then it is obvious that R, if intelligent, will adjust his choice so that he increases his payoff if possible. But if C is also intelligent, we can assume that he too will make a choice to his advantage. The natural question to ask is whether there is some way that a player could determine his play so as to optimize the payoff against an opponent whose responses are unpredictable. The answer to this question is the subject of the remainder of this chapter and is based on the concepts of **expected value** (Section 4.10) and **strategy** (explained and defined next).

Example 2

In the game of matching pennies one common way of determining the play of "heads" or "tails" is to flip the penny and let the choice be determined by chance. A determination of one's play in this manner is an illustration of a **strategy:** For such a strategy there is a probability of $\frac{1}{2}$ of obtaining a "head" and a probability of $\frac{1}{2}$ of obtaining a "tail"; hence this strategy can be represented by the vector $(\frac{1}{2}, \frac{1}{2})$. ∎

▶ Definition 1

Let p_i represent the probability that the ith row of an m by n payoff matrix will be played by the row player. The probability vector $(p_1, p_2, \ldots, p_m)$, where $p_1 + p_2 + \cdots + p_m = 1$, is called the **strategy** of the row player. Let q_j represent the probability that the jth column of an m by n payoff matrix will be played by the column player. The probability vector $(q_1, q_2, \ldots, q_n)$, where $q_1 + q_2 + \cdots + q_n = 1$, is called the **strategy** of the column player.

Example 3

In the penny-matching game of Example 1, suppose that the row player adopts a strategy of $(\frac{1}{4}, \frac{3}{4})$. Such a strategy simply means that R decides to determine his play in such a way that there is $\frac{1}{4}$ probability that he will choose a head and a $\frac{3}{4}$ probability that he will choose a tail. Such a strategy can be executed by the construction of an experiment consisting of the selection of one of two outcomes having respective probabilities of $\frac{1}{4}$ and $\frac{3}{4}$. Experiments having such outcomes are easily designed; for example, a simple spinner similar to that found in children's games and shown in Figure 7.1 can serve as the instrument of such an experiment. An unrehearsed glance at the location of the second hand on a wristwatch will serve as a make-shift spinner (see Figure 7.2). For example, if the spinner (second hand) is on 1–3, R plays heads; if it is on 4–12 he plays tails. ▮

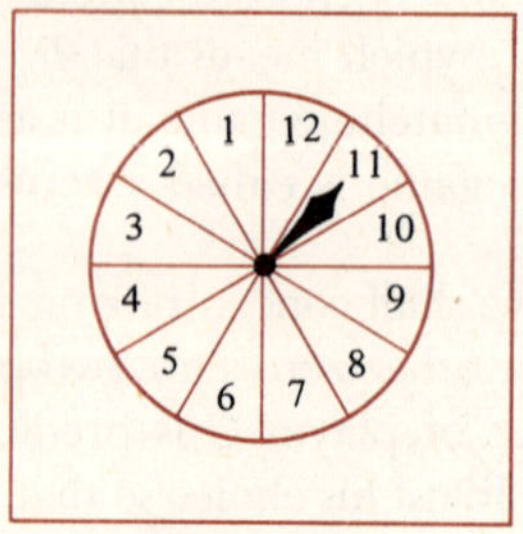

Figure 7.1

Figure 7.2

At first it may seem irresponsible to entrust the selection of a play of a game to a chance device. This is not the case at all, however. As we shall illustrate later, a strategy based on a chance device affords a means of optimizing a player's expectation, regardless of the strategy of the opponent. Furthermore, in most games it is vital that the opponent not be able to anticipate your play; a chance device assures the element of surprise.

By **an expected value of the game to a player,** we mean an expected value for an experiment having outcomes that are the potential payoffs resulting from a single play of the opponent.

Example 4

With a strategy of $(\frac{1}{4}, \frac{3}{4})$ in the penny-matching game, what is an expected value of the game to the row player? The answer to this question depends upon the column choice of the opponent. If the opponent's choice is the first column,

$$\begin{bmatrix} +1 \\ -1 \end{bmatrix},$$

then the possible outcomes are $(+1)$ and (-1) and the expected value to R with strategy $(\frac{1}{4}, \frac{3}{4})$ is

$$\tfrac{1}{4}(+1) + \tfrac{3}{4}(-1) = -\tfrac{1}{2}.$$

If, however, the opponent chooses the second column

$$\begin{bmatrix} -1 \\ +1 \end{bmatrix},$$

then the possible outcomes are (-1) and $(+1)$ and the expected value to R with strategy $(\frac{1}{4}, \frac{3}{4})$ is

$$\tfrac{1}{4}(-1) + \tfrac{3}{4}(+1) = +\tfrac{1}{2}.$$

Although there are two expected values for R's given strategy in this game, it is the smaller of the two expected values that R is concerned about. In Section 7.3 we shall develop a method for determining a strategy that will maximize the smallest expected value to R. ▮

In our discussion of game theory it is assumed that the primary objective of a player is to obtain a strategy that will make his worst expected value as favorable as possible.

▶ **Definition 2**

A strategy of the row player that maximizes his worst (smallest) expected value is called an **optimal strategy** of the row player. A strategy of the column player that minimizes his worst (largest) expected value is called an **optimal strategy** of the column player.

Formulas for calculating optimal strategies for *certain* 2 by 2 matrix games are developed in Exercise 8. General methods for calculating optimal strategies of larger zero-sum matrix games are discussed in the remaining sections of this chapter.

APPLICATIONS

Example 5

The electoral votes of two large states A and B are at stake in a presidential election. Candidate R is leading over candidate C in both states having respective electoral votes of 40 and 30 votes each. As the campaign draws to a close, each candidate has the time and resources to visit exactly one of these states. Candidate R can be expected to successfully defend either state in a face-to-face confrontation, but candidate R can be expected to lose either state that C alone visits. The problem facing candidate R is what strategy will maximize his least expected value against an intelligent opponent. The following matrix game is a mathematical model of the real game situation. The payoff matrix† is

† To assure a zero-sum game, the assumption is made that C originally had all of the electoral votes; thus when a state is carried by R, the payoff represents R's gain and C's loss, but if a state is not carried by R, then there is no movement of electoral votes.

$$\begin{array}{c} \\ \text{Candidate } R \\ \text{selects} \\ \text{state} \end{array} \begin{cases} A \\ B \end{cases} \overset{\begin{array}{c}\text{Candidate } C \\ \text{selects state} \\ \overbrace{A \quad\quad B}\end{array}}{\begin{bmatrix} 70 & 40 \\ 30 & 70 \end{bmatrix}} .$$

The reader will be asked to verify in Exercise 8 that the best strategy for R is $(\frac{4}{7}, \frac{3}{7})$ and that the best strategy for C is $(\frac{3}{7}, \frac{4}{7})$. The reader should recognize that similar models can be constructed to simulate more complicated situations in which any number n of states are visited by the candidates. ▮

Example 6

A special type of game situation is one in which an individual is competing against a set of possible events rather than against another thinking individual with options. For example, suppose that a person wishes to make a long distance telephone call; the night charge for a station-to-station call is \$3 and the night charge for a successful person-to-person call is \$5. If the caller is unsuccessful in getting any answer that night, he knows that he can reach the individual the next day using a station-to-station call at a rate of \$4. If, however, someone other than the desired person answers that night, then the caller can, through the operator, determine a time when a subsequent station-to-station night call will be successful.

The problem facing the caller is what strategy will minimize his expected expenditure. The following matrix game is a mathematical model of the real game situation.

		Individual answers	No one answers	Another person answers
Caller chooses	Person-to-person call	−\$5	−\$4	−\$3
	Station-to-station call	−\$3	−\$4	−\$6

In this particular game, the row player (the caller) is competing against a set of three events over which he has no control; in such games it is customary to refer to such a nonhuman opponent as "Nature." Furthermore, to ensure a zero-sum game, we shall assume that the opponent includes the recipient of the payoff (the phone company).

The reader will be asked to verify in Exercise 10 of Section 7.3 that the best strategy for the caller is $(\frac{3}{5}, \frac{2}{5})$. The latter strategy assumes that the caller has no knowledge of the probabilities of occurrence of each of the three columns. ▮

Example 7

On pages 434–445 of Kemeny et al. [28] the author introduces the model of an expanding economy which was originally proposed by John von

Neumann. The presentation attempts to illustrate how finite mathematics in general, and game theory in particular, can be used in an economic problem. A further reference to this model is Kemeny et al. [26]. ▮

EXERCISES

Suggested minimum assignment: Exercises 1, 2, 3, and 4.

1. Consider a penny-matching game similar to Example 1 except that player R wins 4 cents if both coins show heads and loses 2 cents if both coins show tails; if the coins differ, R loses 1 cent as before.
   (*a*) Construct the payoff matrix of a matrix game that serves as a mathematical model of the matching game.
   (*b*) If on one play of the game R chooses the first row and C chooses the second column, what is the payoff? To whom is the payoff made?
   (*c*) If player R adopts a strategy of $(\frac{1}{2}, \frac{1}{2})$, what are the expected values to R for each column?
   (*d*) If player R adopts a strategy of $(1, 0)$, what are the expected values to R for each column?
   (*e*) If player R adopts a strategy of $(0, 1)$, what are the expected values to R for each column?
   (*f*) Which of the three strategies in parts (*c*)–(*e*) is the "safest" for R to use? (That is, which has the most favorable, least expected value?)
2. Two adversaries are engaged in a competitive situation in which they compete for two positions A and B, with payoffs from loser to winner of 3 and 4 units, respectively. Both adversaries attack exactly one position. If C attacks the position that R does not choose to attack, then C will prevail; R will prevail at every position that R attacks. Positions that are not attacked remain neutral.
   (*a*) Construct the payoff matrix of a matrix game that serves as a mathematical model of the game just described.
   (*b*) If on one play of the matrix game R chooses the second row and C chooses the second column, what is the payoff? To whom is the payoff made?
   (*c*) If R adopts a strategy of $(\frac{1}{3}, \frac{2}{3})$, what are the expected values to R for each column?
   (*d*) If R adopts a strategy of $(1, 0)$, what are the expected values to R for each column?
   (*e*) If R adopts a strategy of $(0, 1)$, what are the expected values to R for each column?
   (*f*) Which of the three strategies in parts (*c*)–(*e*) is the "safest" for R to use? (That is, which has the most favorable, least expected value?)
3. Use Exercise 1 to determine the expected values to C for each row if player C adopts a strategy of
   (*a*) $(\frac{1}{2}, \frac{1}{2})$; (*b*) $(1, 0)$; (*c*) $(0, 1)$.

(*d*) Which of the strategies of parts (*a*)–(*c*) is the "safest" for C to use? (That is, which has the most favorable, worst expected value?)

4. Use Exercise 2 to determine the expected value to C for each row if C adopts a strategy of

   (*a*) $(\frac{1}{2}, \frac{1}{2})$; (*b*) $(1, 0)$; (*c*) $(0, 1)$.

   (*d*) Which of the strategies of parts (*a*)–(*c*) is the "safest" for C to use? (That is, which has the most favorable, worst expected value?)

5. In Example 5 of the text determine the expected value to R for each column if the strategy designated below is adopted by R.

   (*a*) $(\frac{1}{2}, \frac{1}{2})$; (*b*) $(1, 0)$; (*c*) $(0, 1)$; (*d*) $(\frac{4}{7}, \frac{3}{7})$.

6. In Example 6 of the text determine the expected value to R for each column if the strategy designated below is adopted by R.

   (*a*) $(\frac{1}{3}, \frac{2}{3})$; (*b*) $(1, 0)$; (*c*) $(0, 1)$; (*d*) $(\frac{3}{5}, \frac{2}{5})$.

7. In Example 5 of the text determine the expected value to C for each row if the strategy designated below is adopted by C.

   (*a*) $(\frac{1}{4}, \frac{3}{4})$; (*b*) $(1, 0)$; (*c*) $(0, 1)$; (*d*) $(\frac{3}{7}, \frac{4}{7})$.

8. Consider a 2 by 2 zero-sum matrix game with payoff matrix

$$\begin{bmatrix} a_{11} & a_{12} \\ a_{21} & a_{22} \end{bmatrix}.$$

If an unspecified strategy of the row player is $(p, 1 - p)$, then R's two expected values (depending upon C's play) are

$$a_{11}p + a_{21}(1 - p) \qquad \text{and} \qquad a_{12}p + a_{22}(1 - p).$$

Provided the optimal strategy is not $(1, 0)$ or $(0, 1)$, one can show (after studying Section 7.3) that an optimal strategy occurs when these two expected values are equal. Thus we have

$$a_{11}p + a_{21}(1 - p) = a_{12}p + a_{22}(1 - p).$$

Solve this equation for p and for $1 - p$, and call the results p^* and $1 - p^*$. Thus we have

$$\begin{aligned}&(\text{optimal strategy for the row player}) = (p^*, 1 - p^*)\\ &\quad = \left(\frac{a_{22} - a_{21}}{a_{11} - a_{12} - a_{21} + a_{22}}, \frac{a_{11} - a_{12}}{a_{11} - a_{12} - a_{21} + a_{22}}\right).\end{aligned}$$

We can find the optimal strategy for the column player in the same manner [provided it is not $(1, 0)$ or $(0, 1)$].

$$\begin{aligned}&(\text{optimal strategy for the column player}) = (q^*, 1 - q^*)\\ &\quad = \left(\frac{a_{22} - a_{12}}{a_{11} - a_{12} - a_{21} + a_{22}}, \frac{a_{11} - a_{21}}{a_{11} - a_{12} - a_{21} + a_{22}}\right).\end{aligned}$$

Use the formulas just developed to determine the optimal strategy for the row player and the optimal strategy for the column player for the matrix game of Example 5.

9. Use the formulas of Exercise 8 to verify that the optimal strategy of

both players in Example 1 is $(\frac{1}{2}, \frac{1}{2})$. In other words, verify that the best way to play the game of matching pennies with payoff matrix $\begin{bmatrix} 1 & -1 \\ -1 & 1 \end{bmatrix}$ is to determine one's play by flipping the coin.

10. Two adversaries are engaged in a competitive situation in which they compete for two positions A and B worth 3 and 4 units, respectively.† Assume that C holds both positions initially, but can defend only one of them. Assume that R can attack only one position. If R attacks the position that C does not choose to defend, then R will prevail; otherwise C will maintain control. (a) Construct the payoff matrix of a matrix game that serves as a mathematical model of the game just described. (b) Use Exercise 8 to determine the optimal strategy of the row player. (c) Use Exercise 8 to determine the optimal strategy of the column player.

7.2 Pure Strategies

A strategy having 1 as one component and 0 for all the other components is called a **pure strategy.** Strategies that are not pure are called **mixed strategies.** Obviously a pure strategy dictates the choice of a specific row (or column) on a single play. There are some matrix games in which a pure strategy is an optimal strategy. By that we mean that the pure strategy will maximize the smallest expected value to the row player (or minimize the largest expected value to the column player); the primary purpose of this section is to investigate games in which a pure strategy is optimal.

Example 1

From the point of view of the row player R, consider the matrix game

$$A = \begin{bmatrix} 3 & 2 & 4 \\ 5 & 1 & 0 \\ 7 & 1 & -4 \end{bmatrix}.$$

If R selects the pure strategy $(1, 0, 0)$, then the expected values, depending upon his opponent's choice of columns, are

$$1(3) + 0(5) + 0(7) = 3$$

and

$$1(2) + 0(1) + 0(1) = 2$$

and

$$1(4) + 0(0) + 0(-4) = 4.$$

Notice that the expected values are equal to the respective entries of the first row. Thus if R adopts a pure strategy of $(1, 0, 0)$, the smallest payoff

† There is a distinction between this exercise and Exercise 2, in that here the position itself is the payoff and hence must be assumed to belong to one of the players in order that one player's gain is the other's loss (zero-sum requirement).

would be 2, and could be larger if his opponent chooses a column other than the second column. If, however, R selects a pure strategy of $(0, 1, 0)$, then the expected values to R similarly can be shown to be equal to the respective entries of the second row, and the smallest of these values is 0. Finally, if R selects the remaining pure strategy $(0, 0, 1)$, then the expected values are equal to the respective entries of the third row and the smallest of these values is -4. These smallest expected values corresponding to each of the three pure strategies are illustrated in Figure 7.3.

Notice in Figure 7.3 that the smallest or worst expected value of each pure strategy is simply the smallest entry in the respective row, and each represents the worst that can happen to R for the respective pure strategy. Obviously from R's standpoint, the best of the pure strategies is the one that produces the largest of these worst expected values, namely, 2. In other words, R can assure himself a minimum expected value of 2 if he will choose that pure strategy that produces the largest of the worst expected values shown in Figure 7.3. Such a strategy will be called **R's best pure strategy.**

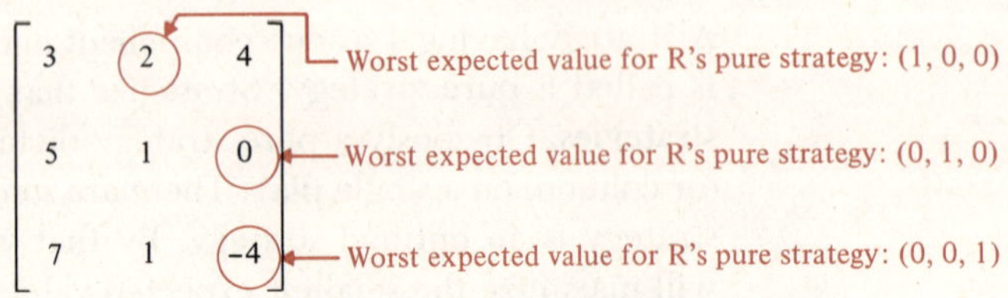

Figure 7.3. *Identification of worst expected value for each pure strategy of* R.

Next consider the same game from C's standpoint, remembering that positive numbers represent losses to C. The worst expected value that C would suffer if he chose the pure strategy $(1, 0, 0)$ is 7 (actually a loss of 7 from C's point of view), which is the largest entry in the first column. The worst expected values to C corresponding to all of the pure strategies of C are shown in Figure 7.4.

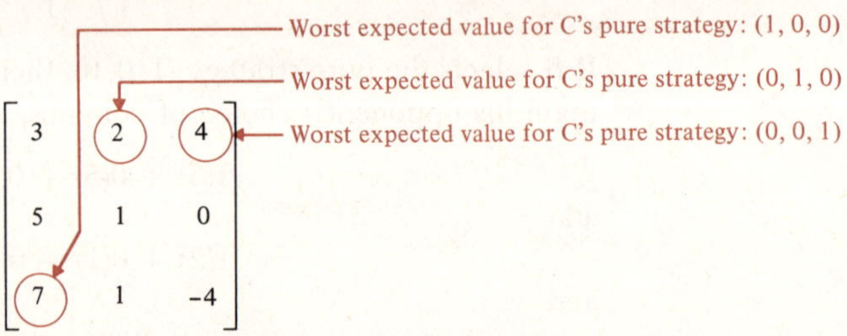

Figure 7.4. *Identification of worst expected value for each pure strategy of* C.

In Figure 7.4 notice that for each pure strategy, the largest or worst expected value to C is simply the largest entry in the respective column and represents the worst that can happen to C for the respective pure strategy chosen. Obviously, from C's standpoint the best of the pure strategies is the one that produces the smallest of these worst expected values, namely, 2. In other words, C can assure himself of an expected value of no worse than 2 (minimum loss to C) if he will choose that pure strategy that produces the least of the worst expected values shown in Figure 7.4. Such a strategy will be called **C's best pure strategy.** ∎

Notice that for the best pure strategies of Example 1, the corresponding expected values for both R and C were the same, namely, 2; as we shall see in the next example, it is not always the case that the expected values corresponding to the best pure strategies of R and C are the same, but when it *is* the case we say that the game is **strictly determined.**

▶ **Definition 3**

If a matrix has an entry that is the common expected value for the best pure strategies of both players, then the game is said to be **strictly determined.**

Strictly determined games are important because it can be proved that in such games the best pure strategies of R and C are optimal strategies. In other words, in a strictly determined game, the common expected value corresponding to the best pure strategy of each player is as good as the worst expected value produced by any pure or mixed strategy of each player. In Example 1 the game was strictly determined because the best expected value for the best pure strategies of both players was the same, namely, $a_{12} = 2$. Because the game in Example 1 is strictly determined, then we know that the optimal strategies are the pure strategies $(1, 0, 0)$ for R and $(0, 1, 0)$ for C.

In the next example we illustrate a matrix game that is not strictly determined.

Example 2

Consider the matrix game having a payoff matrix

$$\begin{bmatrix} 3 & 2 & 1 \\ 0 & -2 & 4 \\ 7 & -3 & 5 \end{bmatrix}.$$

The worst (smallest) expected value to the row player R for each pure strategy is circled in Figure 7.5; the best (largest) of these circled expected

			Row minima
3	2	(1)	1
0	(-2)	4	-2
7	(-3)	5	-3

Figure 7.5. *Worst expected value for each pure strategy of* R.

values is underlined. For the column player C, the worst (largest) expected value for each pure strategy is boxed in Figure 7.6; the best (smallest) of these boxed values is underlined. For the pure strategies of both players, the best of the worst of the expected values are *not* the same; hence the game is not strictly determined, and it can be shown that the optimal strategy is a mixed strategy rather than a pure strategy. ▮

$$\begin{bmatrix} 3 & \underline{\boxed{2}} & 1 \\ 0 & -2 & 4 \\ \boxed{7} & -3 & \boxed{5} \end{bmatrix}$$

Column maxima $\quad 7 \quad 2 \quad 5$

Figure 7.6. *Worst expected value for each pure strategy of* C.

APPLICATIONS

Example 3

Consider an arrangement of 4 towns and 3 shopping centers as shown in Figure 7.7. Each of two firms is planning to open competitive stores in some one of the 3 shopping centers. Both firms, called R and C, agree that the store built by R will get 90 per cent of the customers who are closer to R's store and 20 per cent of the customers who are closer to C's store. Of the customers who are equidistant from both stores, R will get 60 per cent. The question facing each firm is where to locate each store. A matrix game can be constructed to model the decision facing management. The

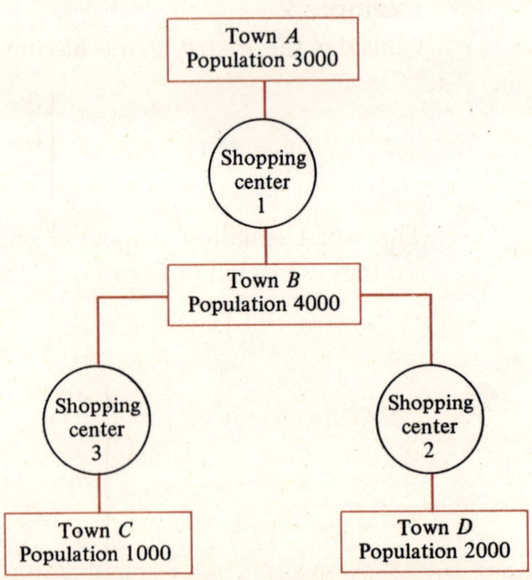

Figure 7.7

payoff matrix of such a game is

$$A = \begin{bmatrix} 6000 & 6100 & 6500 \\ 5400 & 6000 & 6200 \\ 5100 & 5500 & 6000 \end{bmatrix}.$$

To illustrate the calculations of the entries of matrix A, consider the entry a_{12}, which assumes the location of R's store in shopping center 1 and C's store in shopping center 2.

$$\begin{aligned} a_{12} &= (90\% \text{ of } 3000) + (60\% \text{ of } 5000) + (20\% \text{ of } 2000) \\ &= 6100. \end{aligned}$$

In Figure 7.8 the worst expected value corresponding to each pure strategy of R is circled. Remember that for each pure strategy the expected values are simply the entries of the chosen row and the worst expected value in each row is the smallest entry in each row. In Figure 7.9 the worst expected value corresponding to each pure strategy of C is circled.

$$\begin{bmatrix} 6000 & 6100 & 6500 \\ 5400 & 6000 & 6200 \\ 5100 & 5500 & 6000 \end{bmatrix}$$

Figure 7.8. *Worst expected value for each pure strategy of* R.

$$\begin{bmatrix} 6000 & 6100 & 6500 \\ 5400 & 6000 & 6200 \\ 5100 & 5500 & 6000 \end{bmatrix}$$

Figure 7.9. *Worst expected value for each pure strategy of* C.

For the pure strategies of firm R the best of the worst expected values is 6000, which came from the entry a_{11}. For the pure strategies of firm C the best of the worst expected values is also 6000, which again came from the entry a_{11}. Because these two values are the same, the game is strictly determined and the pure strategy $(1, 0, 0)$ is optimal for both firms. Hence both firms should place their stores in shopping center 1. ▮

Example 4

A patient has an undiagnosed disease which is one of 3 types. Two types of medication are available. The probabilities of a cure under the administration of the medications are given in the payoff matrix of the corresponding matrix game. Assume that without medication there is 0 probability of a cure.

$$\begin{array}{c} \\ \text{Medication} \end{array}\begin{cases} 1 \\ 2 \\ \text{Both} \end{cases} \overset{\displaystyle\overbrace{\;1 \quad 2 \quad 3\;}^{\text{Disease}}}{\begin{bmatrix} 1 & \frac{1}{3} & \frac{1}{2} \\ \frac{3}{4} & \frac{3}{4} & \frac{2}{3} \\ 1 & \frac{1}{2} & \frac{1}{2} \end{bmatrix}}$$

The reader will be asked to verify in Exercise 9 that when there is no knowledge of the probability of the diseases, then the best strategy for the doctor is the pure strategy (0, 1, 0). Thus the probability for cure is at least $\frac{2}{3}$, which will be the case if medication 2 is administered. The reader is reminded that similar models can be constructed for similar real problems with known probabilities of success when m various methods are used to attack n types of problems. ▮

Example 5

Two partners in business decide to dissolve their partnership by having one sell out to the other. They agree that the business is worth $60,000. Each partner will produce a sealed bid at a prescribed time stating the amount that he will pay the other for his share of the business. The bids will be made in increments of $10,000 with the understanding that the larger bid will be accepted. The maximum bid allowed will be $40,000. In case the bids are the same, they agree to flip a coin with the winner paying the loser $30,000 for full ownership. In calculating the payoff matrix from R's point of view, a successful bid by R must be subtracted from the $30,000 value of the acquired share of the business; a successful bid by C must be added to the −$30,000 value of the disposed share of the business. In units of $10,000, the payoff matrix is

$$R\text{'s bid}\begin{cases} 0 \\ 1 \\ 2 \\ 3 \\ 4 \end{cases} \overset{\displaystyle\overbrace{\;0 \quad 1 \quad 2 \quad 3 \quad 4\;}^{C\text{'s bid}}}{\begin{bmatrix} 0 & -2 & -1 & 0 & 1 \\ 2 & 0 & -1 & 0 & 1 \\ 1 & 1 & 0 & 0 & 1 \\ 0 & 0 & 0 & 0 & 1 \\ -1 & -1 & -1 & -1 & 0 \end{bmatrix}}.$$

This example illustrates a case in which more than one pure strategy may be optimal. In Exercise 10 the reader is asked to verify that a pure strategy of (0, 0, 1, 0, 0) or (0, 0, 0, 1, 0) is optimal for both bidders. The former, (0, 0, 1, 0, 0), is actually preferable for each bidder if the opponent does not choose an optimal strategy. ▮

EXERCISES

Suggested minimum assignment: Exercises 1, 3, 5, and 7.

In each of the following payoff matrices, determine whether the corresponding game is strictly determined or not. If the game is strictly determined, state the optimal strategy of each player and the corresponding payoff.

1. $\begin{bmatrix} 6 & 3 & 4 \\ 1 & 0 & -2 \\ 2 & -1 & 7 \end{bmatrix}$.

2. $\begin{bmatrix} -1 & 7 & -3 \\ -2 & 3 & 5 \\ 0 & 4 & 1 \end{bmatrix}$.

3. $\begin{bmatrix} 4 & 2 & 3 \\ 0 & -1 & 2 \\ 3 & 1 & 6 \end{bmatrix}$.

4. $\begin{bmatrix} 3 & 4 & 5 \\ 2 & 9 & 6 \\ 1 & 8 & 7 \end{bmatrix}$.

5. $\begin{bmatrix} 3 & 5 & 2 & 7 \\ 6 & 3 & 2 & 2 \\ 1 & 4 & 3 & 5 \\ 6 & 4 & 4 & 1 \end{bmatrix}$.

6. $\begin{bmatrix} 3 & 9 & 1 & 6 \\ 0 & 2 & 1 & 5 \\ 7 & 3 & 2 & 1 \\ 1 & 0 & 8 & 4 \end{bmatrix}$.

7. A manufacturer must use a certain component in a device that he is producing. He can purchase a cheap component for \$6, and if it goes bad, it will cost him an *additional* \$10 to correct the problem. He can, however, purchase a guaranteed component for \$8, and if it goes bad, it will cost him only an *additional* \$2 labor to fix the defect.
   (*a*) Write the payoff matrix for a corresponding matrix game.
   (*b*) Is the game strictly determined?
   (*c*) If the game is strictly determined, what is the optimal strategy of the manufacturer?
8. Suppose that the manufacturer of Exercise 7 considers a third option in which the component can be purchased for \$9, and if found defective, it will be corrected at no cost to the manufacturer; moreover, the \$9 cost will be refunded.
   (*a*) Write the payoff matrix for a corresponding matrix game.
   (*b*) Is the game strictly determined?
9. Verify that the pure strategy (0, 1, 0) is the doctor's optimal strategy for Example 4.
10. In Example 5 verify that the pure strategies (0, 0, 1, 0, 0) or (0, 0, 0, 1, 0) are optimal for each player. Why is the former preferable for each bidder in case the opponent does not use an optimal strategy?

7.3 An Optimal Strategy for the Row Player

The purpose of this section is to establish a general procedure for determining an optimal strategy (mixed or pure) for the row player in an m

by n zero-sum matrix game. We shall introduce the procedure by means of an example.

Example 1

Consider a matrix game with payoff matrix

$$\begin{bmatrix} 5 & 1 \\ 2 & 4 \end{bmatrix}.$$

Suppose that the row player R adopts a mixed strategy of $(\frac{3}{4}, \frac{1}{4})$; that is, there is a $\frac{3}{4}$ probability that R will choose the first row, and a $\frac{1}{4}$ probability that R will choose the second row. The choice for a single play will be determined by a suitable experiment having two outcomes with probabilities of $\frac{3}{4}$ and $\frac{1}{4}$. Thus *on a single play,* if the opponent chooses the first column, then the expected value of the game to R is

$$\tfrac{3}{4}(5) + \tfrac{1}{4}(2) = \tfrac{17}{4}.$$

If, however, the opponent chooses the second column, then the expected value of the game to R is

$$\tfrac{3}{4}(1) + \tfrac{1}{4}(4) = \tfrac{7}{4}.$$

Hence with the strategy $(\frac{3}{4}, \frac{1}{4})$, R anticipates an expected value of at least $\frac{7}{4}$ on a single play no matter what column the column player selects. It is only natural for R to wonder if there is another strategy that will produce a least expected value larger than $\frac{7}{4}$. Suppose that R lets his strategy be an unspecified (p_1, p_2). For a single play, if the opponent chooses the first column, then the expected value of the game to R is

$$p_1(5) + p_2(2).$$

If, however, the opponent chooses the second column, then the expected value of the game to R is

$$p_1(1) + p_2(4).$$

If we let v be the smaller of these two expected values, then with strategy (p_1, p_2), R anticipates an expected value of at least v no matter which column the opponent selects. Obviously, R would like to maximize v. Of course, this maximization is subject to the constraints that each expected value must be no smaller than v; that is,

$$\begin{aligned} p_1(5) + p_2(2) &\geq v, \\ p_1(1) + p_2(4) &\geq v. \end{aligned}$$

Also p_1, p_2, and v must be nonnegative and the sum of the components of the strategy vector must be 1; that is, $p_1 \geq 0$, $p_2 \geq 0$, $v \geq 0$, $p_1 + p_2 = 1$. Thus we have the linear programming problem:

Maximize v subject to

$$\begin{aligned} 5p_1 + 2p_2 - v &\geq 0, & & p_1 \geq 0, \\ p_1 + 4p_2 - v &\geq 0, & \text{and} \quad & p_2 \geq 0, \\ p_1 + p_2 &= 1, & & v \geq 0. \end{aligned} \tag{7.1}$$

The problem can be solved by the methods of Chapter 6. There is a way, however, to reduce both the number of constraints and the number of independent variables. Assume that $v \neq 0$; divide each constraint of (7.1) by v and rearrange the terms to obtain

$$5\left(\frac{p_1}{v}\right) + 2\left(\frac{p_2}{v}\right) \geq 1,$$

$$\left(\frac{p_1}{v}\right) + 4\left(\frac{p_2}{v}\right) \geq 1,$$

$$\left(\frac{p_1}{v}\right) + \left(\frac{p_2}{v}\right) = \frac{1}{v}.$$

Since maximizing positive† v can be accomplished by minimizing $1/v$, and because we can let $x_1 = p_1/v$ and $x_2 = p_2/v$, the original problem can be restated as:

Minimize $(x_1 + x_2)$ subject to

$$\begin{aligned} 5x_1 + 2x_2 &\geq 1, & \text{and} \quad & x_1 \geq 0, \\ x_1 + 4x_2 &\geq 1, & & x_2 \geq 0. \end{aligned}$$

Notice that the coefficient matrix of the constraints is simply the transpose of the payoff matrix. Graphically the solution can be found by Figure 7.10 to be minimum $1/v = \frac{1}{3}$ at $(x_1, x_2) = (\frac{1}{9}, \frac{2}{9})$.

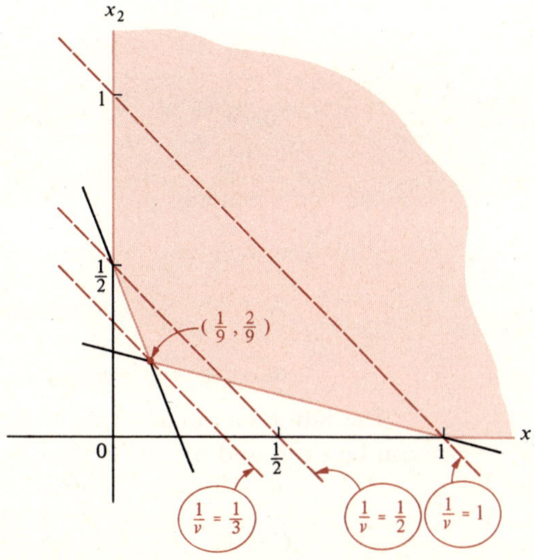

Figure 7.10. *Geometric solution of corresponding linear programming problem.*

† We are now restricting v to be positive rather than nonnegative; if v is zero, then $1/v$ is undefined.

In terms of the original variables p_1, p_2, and v, we find maximum $v = 3$ at $(\frac{1}{3}, \frac{2}{3})$. We designate an optimum v with the symbol v^* and call v^* the **row value** of the game. The strategy at which v^* is obtained is the optimal strategy of the row player and is designated by $(p_1^*, \ldots, p_m^*)$. In summary, if R will use the strategy $(\frac{1}{3}, \frac{2}{3})$, he will always come out with an expected value of at least three per play no matter what column the opponent chooses. ∎

In the procedure introduced in Example 1 it is required that the row value is positive; this can be assured if the payoff matrix has nonnegative entries and does not have a column of zeros. If, however, one encounters a game matrix that does not satisfy these conditions, then the difficulty can be resolved by simply adding to every entry a sufficiently large real number; this resolution is possible because it can be proved (Exercise 13) that if the real number k is added to every entry of the original payoff matrix, then the optimal row strategy will remain unchanged, and the row value of the new game will be the row value of the original game plus k.

Thus in general, for a payoff matrix (of the type described in the last paragraph)

$$A = \begin{bmatrix} a_{11} & \cdots & a_{1n} \\ \vdots & & \vdots \\ a_{m1} & \cdots & a_{mn} \end{bmatrix},$$

the optimal row strategy $(p_1^*, \ldots, p_m^*)$ is the solution of the linear programming problem:

Maximize v subject to

$$\begin{array}{rcl} a_{11}p_1 + \cdots + a_{m1}p_m - v \geq 0, & & p_1 \geq 0, \\ \vdots \qquad\qquad \vdots \qquad \vdots \;\; \vdots & \text{and} & \vdots \quad \vdots \\ a_{1n}p_1 + \cdots + a_{mn}p_m - v \geq 0, & & p_m \geq 0, \\ p_1 + \cdots + \quad p_m \qquad = 1, & & v \geq 0, \end{array}$$

or the corresponding problem where $x_i = p_i/v$:

Minimize $(x_1 + \cdots + x_m)$ subject to

$$\begin{array}{rcl} a_{11}x_1 + \cdots + a_{m1}x_m \geq 1, & & x_1 \geq 0, \\ \vdots \qquad\qquad \vdots \quad \vdots & \text{and} & \vdots \quad \vdots \\ a_{1n}x_1 + \cdots + a_{mn}x_m \geq 1, & & x_m \geq 0. \end{array}$$

The latter formulation of the associated linear programming problem can be expressed in matrix notation as follows. Let

$$C = [1 \;\; 1 \;\; \cdots \;\; 1], \quad X = \begin{bmatrix} x_1 \\ \vdots \\ x_m \end{bmatrix}, \quad \text{and} \quad P_0 = \begin{bmatrix} 1 \\ \vdots \\ 1 \end{bmatrix}.$$

Minimize CX subject to $A^T X \geq P_0$ and $X \geq \mathbf{0}$. If minimum CX occurs at $X = X^*$, then the row value v^* of the associated matrix game with payoff matrix $A \geq \mathbf{0}$ (without a column of zeros) is given by

$$v^* = \frac{1}{\min CX} \text{ at } (p_1^*, \ldots, p_m^*) = v^*(x_1^*, \ldots, x_m^*).$$

APPLICATIONS

Example 2

Consider a conflict situation where each adversary has m objects which can be deployed in n positions. The adversary having the greater number of objects in a position will prevail in that position. In case of a tie suppose that player R prevails. As an illustration consider a presidential election campaign in which each candidate has 3 units of money to distribute to two states A and B having 40 and 60 electoral votes, respectively. Let the notation (a, b) denote that a units of money were distributed to state A and b units were distributed to state B. Under the assumption that C originally holds the electoral votes of the states, the payoff matrix in terms of electoral votes gained by R is

		Distributed by C			
		(3, 0)	(2, 1)	(1, 2)	(0, 3)
Distributed by R	(3, 0)	100	40	40	40
	(2, 1)	60	100	40	40
	(1, 2)	60	60	100	40
	(0, 3)	60	60	60	100

In Exercise 11 the reader can verify that for candidate R the row value is $64\frac{12}{13}$ at the optimal strategy $(\frac{8}{65}, \frac{12}{65}, \frac{18}{65}, \frac{27}{65})$. Notice that the optimal strategy for R is preferable to the pure strategy of $(0, 0, 0, 1)$ because an intelligent opponent might select one of the first three columns and thereby reduce the least expected value to R to 60. The optimal strategy of C is left as Exercise 10 of Section 7.4. ▮

Example 3

On page 101 of Williams [46] an application is given in which a man has 3 options for investing a certain amount of money. The man has estimated the payoffs for each investment for 3 different future economic and political situations. An optimal strategy of the row player is given with the suggestion that the row player could play the odds or invest a mixture of $\frac{5}{17}$ of his principal in one investment and $\frac{12}{17}$ of his principal in another. The author then gives a brief discussion of the ramifications of the latter departure from the established practice of making a clear-cut choice of a single row with a suitable chance device. ▮

Example 4

A hypothetical military situation is developed on page 47 of Williams [46]. In that illustration there are two planes on a bombing mission; one carries the bomb and the other carries various electronic equipment. The planes fly in a formation such that one plane is more protected than the other. The alternatives for the offense are

1. to put the bomb in the most favored position;
2. to put the bomb in the least favored position.

The alternatives for the defense, which can get off only one attack, are

1. to attack the most favored position;
2. to attack the least favored position.

The supposition is made that there is a 60 per cent chance the bomb carrier will survive if attacked in the least favored position, a 80 per cent chance of survival if attacked in the most favored position, and a 100 per cent chance of survival if not attacked. It turns out that the optimal strategy for the offense is $(\frac{2}{3}, \frac{1}{3})$ and the row value is $86\frac{2}{3}$ per cent (Exercise 12). Observe that the chance of survival is $6\frac{2}{3}$ per cent in excess of the intuitive action of placing the bomb in the most favored position. ▮

EXERCISES

Suggested minimum assignment: Exercises 1, 3, 5, and 7.

In Exercises 1–4, for the given payoff matrices, formulate a linear programming problem that will maximize the least expected value to R; that is, formulate the linear programming problem that will determine the row value. Do *not* solve the linear programming problem.

1. $\begin{bmatrix} 2 & 1 \\ 3 & 2 \end{bmatrix}$.

2. $\begin{bmatrix} 8 & 1 \\ 4 & 7 \end{bmatrix}$.

3. $\begin{bmatrix} 3 & 0 & 1 \\ 0 & 4 & 1 \\ 1 & 2 & 6 \end{bmatrix}$.

4. $\begin{bmatrix} 1 & 2 & 3 \\ 0 & 4 & 6 \\ 9 & 2 & 1 \end{bmatrix}$.

For the given payoff matrices in Exercises 5–8 determine the row value of the game and the optimal strategy of the row player. Use the graphical method to solve the associated linear programming problem. The theorem stated in Exercise 13 will be needed to obtain the correct result in Exercises 7 and 8.

5. $\begin{bmatrix} 2 & 1 \\ 1 & 2 \end{bmatrix}$.

6. $\begin{bmatrix} 3 & 0 \\ 1 & 4 \end{bmatrix}$.

7. $\begin{bmatrix} 1 & -2 \\ -1 & 2 \end{bmatrix}$.

8. $\begin{bmatrix} 0 & 3 \\ 3 & -1 \end{bmatrix}$.

9. In Example 5 of Section 7.1 use linear programming to verify that $(\frac{4}{7}, \frac{3}{7})$ is the best strategy for R.
10. In Example 6 of Section 7.1 use linear programming to verify that $(\frac{3}{5}, \frac{2}{5})$ is the best strategy for R.

11. Verify that $(\frac{8}{65}, \frac{12}{65}, \frac{18}{65}, \frac{27}{65})$ is the optimal strategy for candidate R in Example 2. *Hint:* Subtract 40 from every entry of the payoff matrix (see Exercise 13) to shorten the work in this lengthy problem.
12. In Example 4 of this section write the payoff matrix; also verify the optimal strategy of the offense and the row value.
13. Prove that if the real number k is added to every entry of the original payoff matrix, then the optimal strategy of R will remain unchanged and the row value of the new game will be k plus the row value of the old game. (Answer provided.)

7.4 An Optimal Strategy for the Column Player

In this section we shall establish a general procedure for finding an optimal strategy for the column player in an m by n zero-sum matrix game. As an illustration we shall use the same 2 by 2 matrix game that was used in Example 1 of the last section.

Example 1

Consider the matrix game with payoff matrix

$$\begin{bmatrix} 5 & 1 \\ 2 & 4 \end{bmatrix}.$$

Suppose that the column player C adopts a mixed strategy of $(\frac{1}{3}, \frac{2}{3})$; that is, there is a $\frac{1}{3}$ probability that C will choose the first column and a $\frac{2}{3}$ probability that C will choose the second column. The choice for a single play will be determined by a suitable experiment with a pair of outcomes having probabilities $\frac{1}{3}$ and $\frac{2}{3}$. Thus for a *single play,* if the opponent chooses the first row, then the expected value of the game to C is

$$\tfrac{1}{3}(5) + \tfrac{2}{3}(1) = \tfrac{7}{3},$$

or if the opponent selects the second row, then the expected value of the game to C is

$$\tfrac{1}{3}(2) + \tfrac{2}{3}(4) = \tfrac{10}{3}.$$

Hence with strategy $(\frac{1}{3}, \frac{2}{3})$ C has an expected value of at worst $\frac{10}{3}$ on a single play, no matter which row the row player selects. Notice that we said "at worst $\frac{10}{3}$"; we must remember that to C a payoff of $+\frac{10}{3}$ means a $\frac{10}{3}$ loss because C does the paying.

Of course, C is going to wonder if there is another strategy that will produce a smaller "worst expected value" than $\frac{10}{3}$. Suppose that C lets his strategy be an unspecified (q_1, q_2); for a single play, if the opponent chooses the first row, then the expected value of the game to C is

$$q_1(5) + q_2(1),$$

or if the opponent chooses the second row, then the expected value of the game to C is

$$q_1(2) + q_2(4).$$

If we let w be the *larger* of these two expected values, then with strategy (q_1, q_2), C has an expected value of at worst w no matter which row the opponent selects. Because a positive w represents a loss to C, he obviously would like to minimize w. This minimization is subject to the constraints similar to those of the last section, namely, that each expected value be no larger than w; that q_1, q_2, and w be nonnegative; and that $q_1 + q_2 = 1$. Thus we have another linear programming problem:

Minimize w subject to

$$\begin{aligned} 5q_1 + q_2 &\leq w, \\ 2q_1 + 4q_2 &\leq w, \\ q_1 + q_2 &= 1, \end{aligned} \quad \text{and} \quad \begin{aligned} q_1 &\geq 0, \\ q_2 &\geq 0, \\ w &\geq 0. \end{aligned} \tag{7.2}$$

As before, we can reduce both the number of constraints and the number of independent variables by dividing each constraint of (7.2) by positive w, by letting $y_1 = q_1/w$, $y_2 = q_2/w$, and by recognizing that minimizing positive w can be accomplished by maximizing $1/w$.

A corresponding linear programming problem is

Maximize $(y_1 + y_2)$ subject to

$$\begin{aligned} 5y_1 + y_2 &\leq 1, \\ 2y_1 + 4y_2 &\leq 1, \end{aligned} \quad \text{and} \quad \begin{aligned} y_1 &\geq 0, \\ y_2 &\geq 0. \end{aligned}$$

Notice that the coefficient matrix of the constraints is simply the payoff matrix. Graphically, the solution can be found by Figure 7.11 to be

$$\text{Maximum } \frac{1}{w} = \frac{1}{3} \text{ at } (y_1, y_2) = \left(\frac{1}{6}, \frac{1}{6}\right).$$

In terms of the original variables q_1, q_2, and w we find minimum $w = 3$ at $(\frac{1}{2}, \frac{1}{2})$. We designate an optimal w with the symbol w^* and call w^* the **column value** of the game. The strategy at which w^* is obtained is the optimal strategy of the column player and is designated $(q_1^*, \ldots, q_n^*)$. In summary, if C will use the strategy $(\frac{1}{2}, \frac{1}{2})$, he will always come out with an expected value (loss to the column player) of no worse than 3 per play no matter what row the opponent chooses. ▮

In general, the optimal column strategy for the matrix game with payoff matrix $A \geq \mathbf{0}$ (without a column of zeros) can be found as follows: Let

$$C = [1 \ \ 1 \ \ \cdots \ \ 1], \qquad Y = \begin{bmatrix} y_1 \\ y_2 \\ \vdots \\ y_n \end{bmatrix}, \qquad \text{and} \qquad P_0 = \begin{bmatrix} 1 \\ 1 \\ \vdots \\ 1 \end{bmatrix}.$$

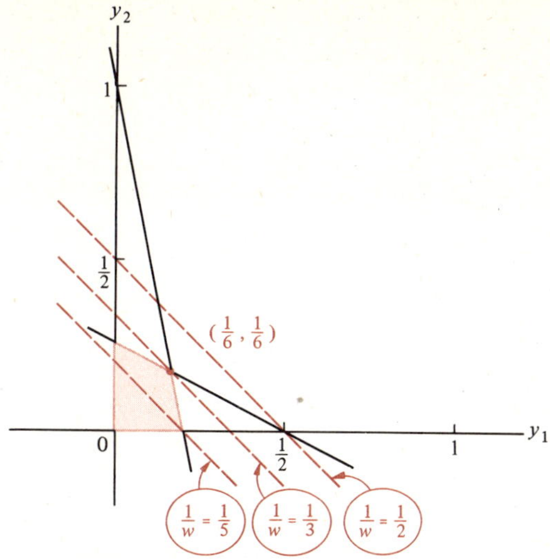

Figure 7.11. *Geometric solution of corresponding linear programming problem.*

Then maximize CY subject to $AY \leq P_0$ and $Y \geq \mathbf{0}$. If maximum CY occurs at $Y = Y^*$, then the column value w^* of the associated matrix game with payoff matrix $A \geq \mathbf{0}$ (without a column of zeros) is given by

$$w^* = \frac{1}{\max CY} \text{ at } (q_1^*, \ldots, q_n^*) = w^*(y_1^*, \ldots, y_n^*).$$

At this point it is natural to wonder whether or not there is any guarantee that every matrix game has an optimal strategy for each player. A famous theorem of game theory states that for every matrix game, there exist optimal strategies for both players. (The proof of this theorem is beyond the scope of this book and may be found in McKinsey [34], pp. 32–37.) Moreover, the corresponding row and column values are the same; this common value is known as the **value** V of the game. In fact, the value V of the matrix game with payoff matrix A can be calculated by the equation†

$$V = P^*AQ^*,$$

where P^* is the row player's optimal strategy in row matrix form and Q^* is the column player's optimal strategy in column matrix form. (A may or may not have all nonnegative entries.) Of course, the value of the game represents the expected value to both players when both use their optimal strategies. When V is 0 the game is called a **fair game.** When V is positive, the row player has the advantage, and when V is negative, the column player has the advantage.

† Although technically P^*AQ^* is a 1 by 1 matrix, it is permissible here to use the matrix and its single entry interchangeably.

The presentation of game theory presented in this chapter has been necessarily elementary and many useful concepts have been left out. For an elementary discussion of some of these concepts the reader is referred to Chapter 5 of Williams [46]. In that chapter the topics, approximations of optimal strategies (particularly important), dominance, simple solutions, multiple solutions, measurements of payoffs, quantitative payoffs, symmetric games, and non-zero-sum games are discussed in a readable style. In fact, Williams's book [46] serves as a very enjoyable primer in game theory.

APPLICATIONS

Example 2

Consider a network of streets shown in Figure 7.12. Suppose that a crime has been committed at point B and the criminals leave the scene of the crime in a car to avoid capture. The police, on the other hand, who are initially at point A, would like to cross paths with the criminals in order to effect their capture. Assume that each car moves from intersection to intersection in the same time and can make $90°$ turns but not $180°$ turns. Further assume that if the criminals can avoid the police for three blocks, they will escape. A meeting on any street or intersection will result in capture. The various paths that each car can travel are identified for reference in Figure 7.13. Using the figure, we calculate the payoff matrix of the associated matrix game by letting a 1 represent a capture and a 0 represent an escape. The resulting payoff matrix follows:

$$P_1 = \overbrace{\begin{matrix} 1 & 2 & 3 & 4 & 5 & 6 & 7 & 8 & 9 & 10 \end{matrix}}^{\text{Path of Criminals}}$$

$$P_1 = \begin{bmatrix} 1 & 0 & 0 & 0 & 0 & 0 & 0 & 0 & 0 & 1 \\ 1 & 0 & 0 & 0 & 0 & 1 & 0 & 0 & 1 & 1 \\ 0 & 1 & 1 & 1 & 0 & 0 & 1 & 1 & 1 & 0 \\ 0 & 1 & 1 & 1 & 0 & 0 & 1 & 1 & 1 & 0 \\ 0 & 1 & 1 & 1 & 0 & 1 & 1 & 1 & 1 & 0 \\ 0 & 0 & 0 & 0 & 1 & 1 & 0 & 0 & 0 & 0 \\ 0 & 1 & 1 & 1 & 0 & 0 & 1 & 1 & 1 & 0 \\ 0 & 1 & 1 & 1 & 0 & 0 & 1 & 1 & 1 & 0 \\ 0 & 1 & 1 & 1 & 0 & 0 & 1 & 1 & 1 & 1 \\ 0 & 0 & 0 & 1 & 1 & 1 & 0 & 0 & 0 & 1 \end{bmatrix} \left.\begin{matrix} 1 \\ 2 \\ 3 \\ 4 \\ 5 \\ 6 \\ 7 \\ 8 \\ 9 \\ 10 \end{matrix}\right\} \text{Path of Police}$$

Because each entry of the second row is greater than or equal to the corresponding entry of the first row, the police would never choose row 1 in any optimal strategy. In cases like this we say that row 2 *dominates* row 1 or that row 1 is a *recessive* row. In any optimal strategy the row

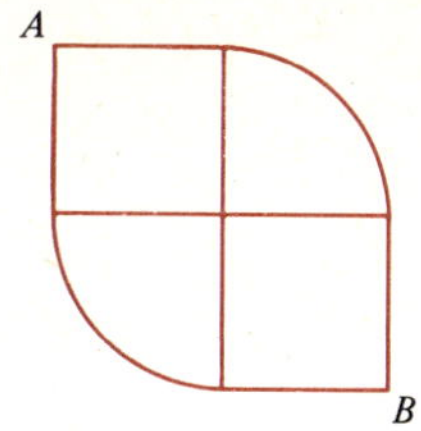

Figure 7.12. *Street arrangement.*

1 component is 0. Hence row 1 can be deleted from consideration. Row 10 dominates row 6 and row 5 dominates rows 3, 4, 7, and 8. Thus if the corresponding strategy vector components of the recessive rows are assigned the value 0 and the rows are deleted from consideration, the size of the payoff matrix can be reduced.

Next consider the criminal's point of view. He observes that each entry of column 5 is less than or equal to each entry of column 6; hence he would always choose column 5 in preference to column 6. In other words, from the criminal's point of view column 5 dominates column 6 and column 1 dominates column 10. Columns 4 and 9 are dominated by columns 2, 3, 7, and 8, all of which are the same. Thus if he chooses column 2 as the representative of these equal columns, the components of the strategy vector that correspond to columns 3, 4, 6, 7, 8, 9, and 10 are 0, and these columns can be deleted. The result of these considerations of dominance of both rows and columns is that the original payoff matrix can be drastically simplified. It is now

$$\begin{array}{cc} & \begin{array}{ccc} 1 & 2 & 5 \end{array} \\ P_2 = & \begin{bmatrix} 1 & 0 & 0 \\ 0 & 1 & 0 \\ 0 & 1 & 0 \\ 0 & 0 & 1 \end{bmatrix} \begin{array}{c} 2 \\ 5 \\ 9 \\ 10 \end{array} \end{array}$$

From the reduced game we see that the middle two rows are the same, and hence we can say that either one dominates the other. Consequently,

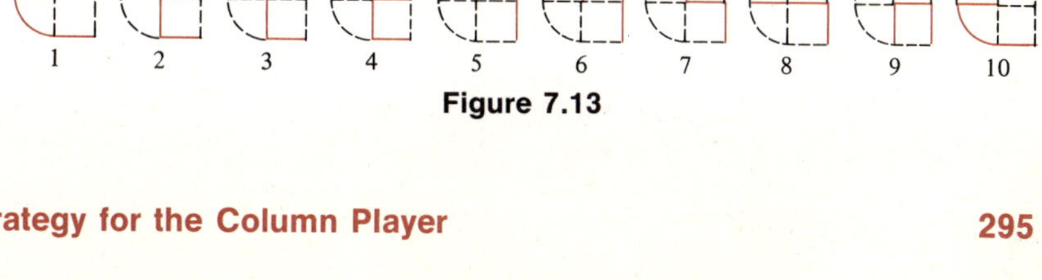

Figure 7.13

the game can be reduced further to the matrix game for which the payoff matrix is

$$P_3 = \begin{array}{c} \begin{matrix} 1 & 2 & 5 \end{matrix} \\ \begin{bmatrix} 1 & 0 & 0 \\ 0 & 1 & 0 \\ 0 & 0 & 1 \end{bmatrix} \end{array} \begin{matrix} 2 \\ 5 \\ 10 \end{matrix} .$$

The methods of this section and the last section can be used to verify that the optimal strategy in this reduced game is $(\frac{1}{3}, \frac{1}{3}, \frac{1}{3})$ for both players (Exercise 11). From this result the original game has an optimal strategy of

$$(0, \tfrac{1}{3}, 0, 0, 0, 0, 0, 0, \tfrac{1}{3}, \tfrac{1}{3}) \qquad \text{or} \qquad (0, \tfrac{1}{3}, 0, 0, \tfrac{1}{3}, 0, 0, 0, 0, \tfrac{1}{3})$$

for the police. Two optimal strategies for the police arise from the fact that two of the rows of P_2 were the same and undominated by the other rows of P_2. One of the optimal strategies for the criminals is

$$(\tfrac{1}{3}, \tfrac{1}{3}, 0, 0, \tfrac{1}{3}, 0, 0, 0, 0, 0).$$

Four optimal strategies for the criminals arise from the fact that four of the columns of P_1 were the same and were undominated by other columns. The value of the game is $\frac{1}{3}$.

The ideas presented in this example can be extended to apply to larger networks of paths of various designs. ▮

Example 3

Although a deeper study of game theory ordinarily is required to develop applications, the titles of the following books will indicate to the reader some areas of applications and furnish references as well.

BELLMAN, R., and BLACKWELL, D., "Red Dog, Blackjack Poker," *Scientific American,* 184, 1951, pp. 44–47.

GALE, DAVID, *The Theory of Linear Economic Models,* McGraw-Hill Book Company, New York, 1960.

LUCE, R. DUNCAN, and RAIFFA, HOWARD, *Games and Decisions: Introduction and Critical Survey,* John Wiley & Sons, Inc., New York, 1957.

MACDONALD, J., *Strategy in Poker, Business, and War,* W. W. Norton & Company, New York, 1950.

RAPOPORT, ANATOL, *Fights, Games, and Debates,* University of Michigan Press, Ann Arbor, Mich., 1960.

RAPOPORT, ANATOL, *Two Person-Game Theory: The Essential Ideas,* University of Michigan Press, Ann Arbor, Mich., 1966.

SHUBIK, MARTIN, *Strategy and Market Structure,* John Wiley & Sons, Inc., New York, 1959.

VON NEUMANN, J., and MORGENSTERN, OSKAR, *Theory of Games and Economic Behavior,* 3rd ed., Princeton University Press, Princeton, N.J., 1953. ▮

EXERCISES

Suggested minimum assignment: Exercises 1, 3, 5, and 7.

In Exercises 1–4, for the given payoff matrices, formulate a linear programming problem that will minimize the greatest expected value to C; that is, formulate a linear programming problem that will determine the column value. Do *not* solve the linear programming problem.

1. $\begin{bmatrix} 2 & 1 \\ 3 & 2 \end{bmatrix}$.

2. $\begin{bmatrix} 8 & 1 \\ 4 & 7 \end{bmatrix}$.

3. $\begin{bmatrix} 3 & 0 & 1 \\ 0 & 4 & 1 \\ 1 & 2 & 6 \end{bmatrix}$.

4. $\begin{bmatrix} 1 & 2 & 3 \\ 0 & 4 & 6 \\ 9 & 2 & 1 \end{bmatrix}$.

In Exercises 5–8, for the given payoff matrices, determine the column value of the game and the optimal strategy of the column player. Using the answers to the same matrix games of Exercises 5–8 of Section 7.3, verify that the column values and row values of each game are the same. A theorem comparable to that given in Exercise 13 of Section 7.3 is needed in Exercises 7 and 8.

5. $\begin{bmatrix} 2 & 1 \\ 1 & 2 \end{bmatrix}$.

6. $\begin{bmatrix} 3 & 0 \\ 1 & 4 \end{bmatrix}$.

7. $\begin{bmatrix} 1 & -2 \\ -1 & 2 \end{bmatrix}$.

8. $\begin{bmatrix} 0 & 3 \\ 3 & -1 \end{bmatrix}$.

9. In Example 5 of Section 7.1 use the method of this section to verify that $(\frac{3}{7}, \frac{4}{7})$ is the optimal strategy for C.
10. In Example 2 of Section 7.3 determine the optimal strategy of C and verify that the row value is the same as the column value of the game. *Hint:* Subtract 40 from every entry of the payoff matrix to shorten the work in this lengthy problem.
11. Verify the optimal strategies given in Example 2 of this section.

NEW VOCABULARY

2 by 2 matrix game 7.1
payoff matrix or game matrix 7.1
play of a game 7.1
zero-sum game 7.1
strategy or strategy vector 7.1
an expected value of the game to a player 7.1
optimal strategy 7.1
pure strategy 7.2
mixed strategy 7.2
R's best pure strategy 7.2
C's best pure strategy 7.2
strictly determined games 7.2
row value of the game 7.3
column value of the game 7.4
value of the game 7.4
fair game 7.4

References

Many of the following books have been referred to in the applied examples of this book. Others have been referred to for proofs of certain theorems. The references marked with an asterisk have many applications of topics in finite mathematics, and it is recommended that those readers who are interested in applications should at least scan the relevant parts of some of these books.

[1] Allen, L. E., and Caldwell, M. C., "Modern Logic and Judicial Decision Making: A Sketch of One View," *Law and Contemporary Problems,* 28, 1963, pp. 213–270.

[2] Alling, D. W., "The After History of Pulmonary TB: A Stochastic Model," *Biometrics,* 14, 1958, pp. 527–547.

[3] Arnold, B. J., *Logic and Boolean Algebra,* Prentice-Hall, Inc., Englewood Cliffs, N.J., 1962.

*[4] Batschelet, Edward, *Introduction to Mathematics for Life Scientists,* Biomathematics, Vol. 2, Springer-Verlag, New York, 1971.

[5] Bellman R., and Blackwell, D., "Red Dog, Blackjack Poker," *Scientific American,* 184, 1951, pp. 44–47.

[6] Bernadelli, H., "Population Waves," *J. Burma Res. Soc.,* 31, 1941, pp. 1–18.

[7] Campbell, Hugh G., *An Introduction to Matrices, Vectors, and Linear Programming,* Appleton-Century-Crofts, New York, 1965.

*[8] Campbell, Hugh G., *Linear Algebra with Applications: Including Linear Programming,* Appleton-Century-Crofts, New York, 1971.

*[9] Campbell, Hugh G., *Matrices with Applications,* Appleton-Century-Crofts, New York, 1968.

[10] Charnes, A., Cooper, W. W., and Millier, M. H., "Application of Linear Programming to Financial Budgeting and the Costing of Funds," *J. Business Univ. Chicago,* 32, 1959, pp. 20–46.

[11] Dantzig, George, *Linear Programming and Extensions,* Princeton University Press, Princeton, N.J., 1963.

[12] Deutsch, K. W., and Madow, W. G., "A Note on the Appearance of Wisdom in Large Bureaucratic Organizations," *Behavioral Sci.,* 6, 1961, pp. 72–78.

[13] Dorfman R., *Application of Linear Programming to the Theory of the Firm, Including an Analysis of Monopolistic Firms by Non-linear Programming,* University of California Press, Berkeley, Calif., 1951.

[14] Freund, John E., *College Mathematics with Business Applications,* Prentice-Hall, Inc., Englewood Cliffs, N.J., 1969.

[15] Gale, David, *The Theory of Linear Economic Models,* McGraw-Hill Book Company, New York, 1960.

[16] Gass, S. I., *Linear Programming,* 3rd ed., McGraw-Hill Book Company, New York, 1969.

[17] Goren, Charles H., *Goren's New Contract Bridge Complete,* Doubleday & Company, Inc., Garden City, N.Y., 1957.

*[18] Hohn, Franz E., *Applied Boolean Algebra,* 2nd ed., Macmillan Publishing Co., Inc., New York, 1966.

*[19] Ijiri, Yuji, *Management Goals and Accounting for Control,* Rand McNally & Company, Skokie, Ill., 1965.

[20] Ijiri, Y., Levy, F. K., and Lyon, R. C., "A Linear Programming Model for Budgeting and Financial Planning," *J. Accounting Res.,* 1, 1963, pp. 198–212.

[21] ITT, Federal Electric Corporation, *Boolean Algebra,* Prentice-Hall, Inc., Englewood Cliffs, N.J., 1966.

[22] Iverson, Kenneth, *A Programming Language,* John Wiley & Sons, Inc., New York, 1962.

*[23] Johnston, J. B., Price, G. B., and Van Vleck, F. S., *Linear Equations and Matrices,* Addison-Wesley Publishing Company, Inc., Reading, Mass., 1966.

[24] Kattsoff, L. O., and Simone, A. J., "A Logical Model for a Problem in Production Management," *APICS* (American Production and Inventory Control Society) *Quart. Bull.,* 6, no. 2, 1965.

[25] Kay, Paul, ed., *Explorations in Mathematical Anthropology,* The MIT Press, Cambridge, Mass., 1971.

[26] Kemeny, John G., Morgenstern, Oskar, and Thompson, Gerald L., "A Generalization of the von Neumann Model of an Expanding Economy," *Econometrica,* 24, 1950, pp. 115–135.

*[27] Kemeny, J. G., Schleifer, A., Snell, J. L., and Thompson, G. L., *Finite Mathematics with Business Applications,* 2nd ed., Prentice-Hall, Inc., Englewood Cliffs, N.J., 1972.

*[28] Kemeny, J. G., Snell, J. L., and Thompson, G. L., *Introduction to Finite Mathematics,* 2nd ed., Prentice-Hall, Inc., Englewood Cliffs, N.J., 1966.

[29] Kort, Fred, "Simultaneous Equations and Boolean Algebra in the Analysis of Judicial Decisions," in Hans Baade, ed., *Jurimetrics,* Basic Books, Inc., New York, 1963.

[30] Lawlor, R. C., "Foundations of Logical Legal Decision Making," *Modern Uses of Logic in Law* (now published as *Jurimetrics Journal*), June 1963, pp. 98–114.

[31] Luce, R. Duncan, and Raiffa, Howard, *Games and Decisions: Introduction and Critical Survey,* John Wiley & Sons, Inc., New York, 1957.

[32] MacDonald, J., *Strategy in Poker, Business, and War,* W. W. Norton & Company, New York, 1950.

[33] McGill, W. J., "Stochastic Latency Mechanisms," in Luce, R. D., Bush, R. R., and Galantes, E. E., eds., *Handbook of Mathematical Psychology,* John Wiley & Sons, Inc., New York, 1963.

[34] McKinsey, J. C. C., *Introduction to the Theory of Games,* McGraw-Hill Book Company, New York, 1952.

[35] Miller, Irwin, and Freund, John E., *Probability and Statistics for Engineers,* Prentice-Hall, Inc., Englewood Cliffs, N.J., 1965.

[36] Rapoport, Anatol, *Fights, Games, and Debates,* University of Michigan Press, Ann Arbor, Mich., 1960.

[37] Riker, William H., *The Theory of Coalitions,* Yale University Press, New Haven, Conn., 1963.

*[38] Searle, S. R., *Matrix Algebra for the Biological Sciences,* John Wiley & Sons, Inc., New York, 1966.

*[39] Searle, S. R., and Hausman, W. H., *Matrix Algebra for Business and Economics,* John Wiley & Sons, Inc. (Interscience Division), New York, 1970.

[40] Shapley, L. S., and Shubik, M., "A Method for Evaluating the Distribution of Power in a Committee System," *Amer. Political Sci. Rev.*, 48, 1954, pp. 787–792.

[41] Shubik, Martin, *Strategy and Market Structure,* John Wiley & Sons, Inc., New York, 1959.

[42] Stoll, R. R., *Introduction to Set Theory and Logic,* W. H. Freeman and Company, San Francisco, 1963.

[43] Ulmer, Sidney, "Leadership in the Michigan Supreme Court," in G. Shubert, ed., *Judicial Decision Making,* The Free Press, New York, 1963.

[44] *United States Life Tables by Causes of Death 1959–1961,* Vol. 1, No. 6, National Health Center for Health Statistics, Public Health Service, Washington, D.C., 1967.

[45] Von Neumann, J., and Morganstern, Oskar, *Theory of Games and Economic Behavior,* 3rd ed., Princeton University Press, Princeton, N.J., 1953.

*[46] Williams, J. D., *The Compleat Strategyst,* rev. ed., McGraw-Hill Book Company, New York, 1966.

Answers to Odd-numbered Exercises

1.1 Page 7

1.

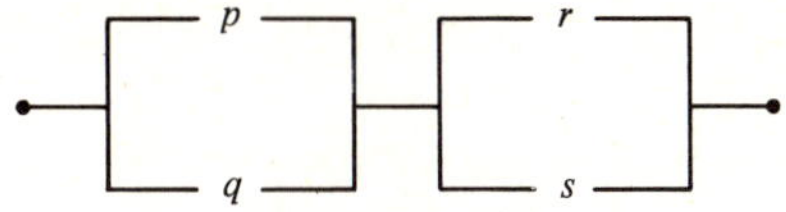

3.

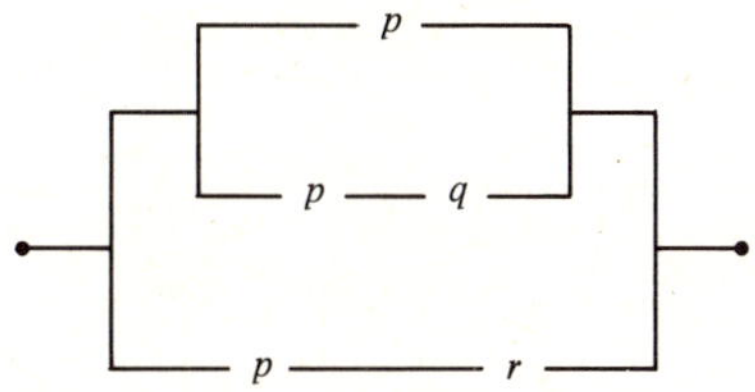

5. $\{[(p \times r) + (q \times p)] + r\} \times r$ is one answer.

7. $\{[p \times (q + r)] + (q \times p)\} + r$ is one answer.

9. p, q, r, $(q + r)$, and $p \times (q + r)$.

1.2 Page 14

1.

p	q	$p \times (p + q)$
1	1	1
1	0	1
0	1	0
0	0	0

3.

p	q	$(p + \sim q) + (q + \sim p)$
1	1	1
1	0	1
0	1	1
0	0	1

5.

p	q	r	$(p + r) \times \sim q$
1	1	1	0
1	1	0	0
1	0	1	1
1	0	0	1
0	1	1	0
0	1	0	0
0	0	1	1
0	0	0	0

7.

p	q	r	$(p \times q) + (r \times \sim q)$
1	1	1	1
1	1	0	1
1	0	1	1
1	0	0	0
0	1	1	0
0	1	0	0
0	0	1	1
0	0	0	0

9. $(p + q) \times (s + t)$.

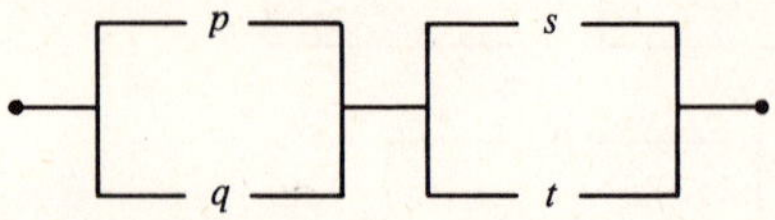

1.3 Page 21

1. Unit switch. **3.** Zero switch. **5.** Equivalent. **7.** Not equivalent.
9. (**a**) Let p be any single compound switch of the set.

p	U	$p \times U$	$U \times p$
1	1	1	1
0	1	0	0

Therefore, $p \equiv p \times U \equiv U \times p$, and hence U is the identity element for $\times$. (**b**) Similar to part (**a**).
11. (**a**) $p \times q$; (**b**) $(\sim p) \times q$; (**c**) $(p \times q) + [(\sim p) \times q)]$; (**d**) equivalent.

1.4 Page 27

1. $\sim p \times q$. **3.** p. **5.** Similar to proof of part (i) in text.
7. Statements:

$$\begin{aligned} p &\equiv p \times U \\ &\equiv p \times (p + \sim p) \\ &\equiv (p \times p) + (p \times \sim p) \\ &\equiv (p \times p) + Z \\ &\equiv p \times p. \end{aligned}$$

9. Given: $p \times x \equiv p$ for every p. Hence if $p \equiv U$ we have $U \times x \equiv U$, but $U \times x \equiv x$ by Theorems 3 and 1. Therefore, $U \equiv U \times x \equiv x$.
11. Statements:

$$\begin{aligned} p \times (p + q) &\equiv (p + Z) \times (p + q) \\ &\equiv p + (Z \times q) \\ &\equiv p + Z \\ &\equiv p. \end{aligned}$$

21. p.

2.1 Page 37

1. (**b**) and (**c**) are statements. **3.** (**a**) and (**b**) are statements.
5. (**a**) $p \rightarrow q$, "if . . . then," conditional. (**b**) $p \rightarrow (q \wedge r)$, "if . . . then," conditional. (**c**) $\sim(p \wedge q)$, "not," negation. **7.** (**a**) $p \wedge \sim q$, "and," conjunction. (**b**) $p \rightarrow q$, "if . . . then," conditional. (**c**) $p \leftrightarrow (q \wedge r)$, "if and only if," biconditional. **9.** (**a**) "or," disjunction. (**b**) "if . . . then," conditional.

(c) "not," negation. **11.** Either the computer is available or I will not get my project finished (but not both).

2.2 Page 43

1.

p	q	$\{p \rightarrow (\sim q)\} \wedge p$
T	T	F
T	F	T
F	T	F
F	F	F

3.

p	q	$[(\sim p) \vee (\sim q)] \leftrightarrow \sim(p \wedge q)$
T	T	T
T	F	T
F	T	T
F	F	T

5.

p	q	r	$[(\sim r) \vee q)] \vee (p \wedge r)$
T	T	T	T
T	T	F	T
T	F	T	T
T	F	F	T
F	T	T	T
F	T	F	T
F	F	T	F
F	F	F	T

7.

p	q	r	$\{(p \vee r) \vee q\} \rightarrow (p \wedge q)$
T	T	T	T
T	T	F	T
T	F	T	F
T	F	F	F
F	T	T	F
F	T	F	F
F	F	T	F
F	F	F	T

9. $q \rightarrow p$ is one answer; $p \vee \sim q$ is another answer. **11.** 16.
13. $[p + (q \times \sim p)] + \sim q$.

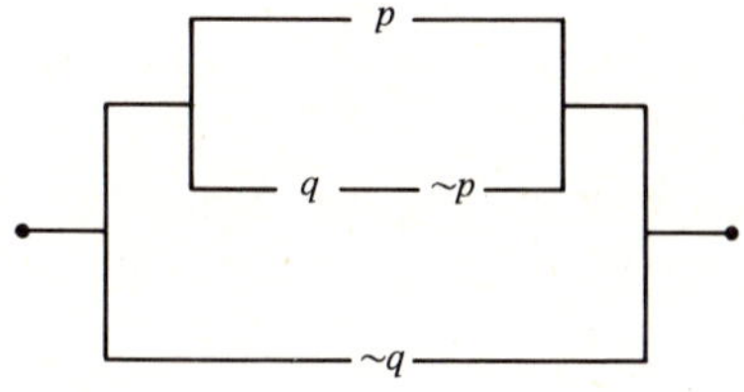

2.3 Page 47

1. Neither. **3.** Tautology; equivalence. **5.** Neither. **7.** Neither. **9.** Show that the evaluations of both are the same. **11.** Show that the evaluations of both are the same. **13.** Yes; the conditional $p \veebar (p \wedge q) \rightarrow \sim q$ is a tautology. **15.** "conjunction," "disjunction," "negation," "tautology," "contradiction," "equivalence."

2.4 Page 52

1. $[(\sim p) \wedge q]$. **3.** p. **5.** $q \vee p$.

7. Statements:

$$\begin{aligned} p &\equiv p \wedge U \\ &\equiv p \wedge (p \vee \sim p) \\ &\equiv (p \wedge p) \vee (p \wedge \sim p) \\ &\equiv (p \wedge p) \vee C \\ &\equiv p \wedge p. \end{aligned}$$

9. Given $p \wedge x \equiv p$ for every p. Hence if $p \equiv U$ we have $U \wedge x \equiv U$, but $U \wedge x \equiv x$ by Theorem 3. Therefore, $U \equiv U \wedge x \equiv x$.

11. Statements:

$$\begin{aligned} p \wedge (p \vee q) &\equiv (p \vee C) \wedge (p \vee q) \\ &\equiv p \vee (C \wedge q) \\ &\equiv p \vee C \\ &\equiv p. \end{aligned}$$

2.5 Page 57

1. Valid argument form because the given conditional is a tautology, as can be proved by an evaluation table or by Theorem 23. Premises are

(1) $p \vee q$,
(2) $\sim p$.

Conclusion is q.

3. Fallacious argument form because the given conditional is *not* a tautology, as can be shown in an evaluation table. Premises are

(1) $p \vee q$,
(2) q.

Conclusion is p.

5. Nonargument form because the given statement form is not a conditional.

7. Valid argument form because the given conditional is a tautology, as can be proved by an evaluation table, or because the conclusion is implied by premises through use of the theorems of this chapter. Premises are

(1) $p \vee (\sim q)$,
(2) $(p \vee q) \wedge q$.

Conclusion is $p \wedge q$.

9. Valid. **11.** Fallacious. **13.** Valid.

23. Argument form is fallacious in Exercise 11, hence argument is unsound. The first premise is false in Exercise 13, hence argument is unsound, even though form is valid.

3.1 Page 63

1. $A = \{$Vt., N.H., Me., Mass., Conn., R.I.$\}$, $A = \{x : x$ is a state in New England$\}$.

3. $C = \{-1, 1\}$, $C = \{x : x^2 - 1 = 0\}$.

5. $E = \{3\}$, E is the set of all positive integers that satisfy the equation $(x - 3)(x + 2) = 0$.

7. One answer is: G is the set of all integers between 1 and 3 inclusive, or $G = \{x : 1 \leq x \leq 3,\ x \text{ is an integer}\}$.
9. One answer is: $\{3, 4\} \subset A$, $3 \in A$.
13. {Chicago, Cleveland, Pittsburgh}, {Chicago, Cleveland}, {Chicago, Pittsburgh}, {Pittsburgh, Cleveland}, {Chicago}, {Cleveland}, {Pittsburgh}, $\varnothing$. **15.** True. **17.** False, $\{2, 3\} \subset U$. **19.** False, $\varnothing \subset U$. **21.** True.
23.

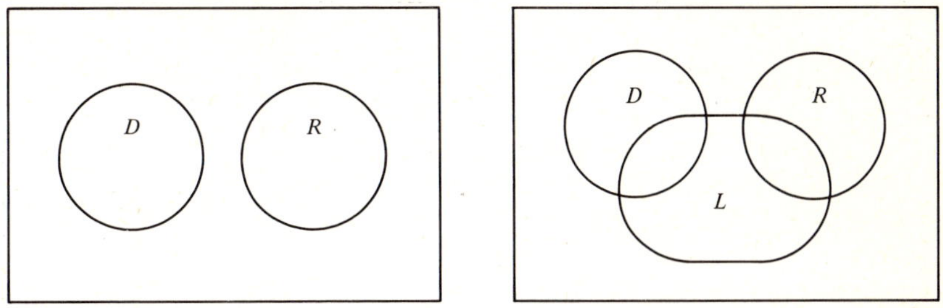

3.2 Page 70

1. **(a)** $A \cup B = \{1, 2, 3, 4, 5, 6\}$; **(b)** $A \cap B = \{4\}$; **(c)** $A' = \{5, 6\}$; **(d)** $A \cup (A' \cap B) = \{1, 2, 3, 4, 5, 6\}$; **(e)** $A \cup \varnothing = \{1, 2, 3, 4\}$; **(f)** $B \cap B' = \varnothing$; **(g)** $U \cup A = \{1, 2, 3, 4, 5, 6\}$; **(h)** $\varnothing \cap B = \varnothing$.
3. **(a)** $P \cup R = \{x : (x \in P) \vee (x \in R)\}$; **(b)** $P \cap R = \{x : (x \in P) \wedge (x \in R)\}$; **(c)** $R' = \{x : (x \notin R)\}$.
5. **(a)** $\varnothing$; **(b)** U; **(c)** U; **(d)** A. **7.** $L \cap D$. **9.** $L \cap D \subseteq D$. **11.** C'.
13. **(a)** U; **(b)** $\varnothing$; **(c)** R; **(d)** N; **(e)** R; **(f)** $\varnothing$; **(g)** No; **(h)** Yes.
15. For Exercise 1, $A \Delta B = \{1, 2, 3, 5, 6\}$; for Exercise 2, $A \Delta B = \{a, d\}$.
17. $B \cap (A \cap B \cap C)'$ is one answer.

3.3 Page 77

1.

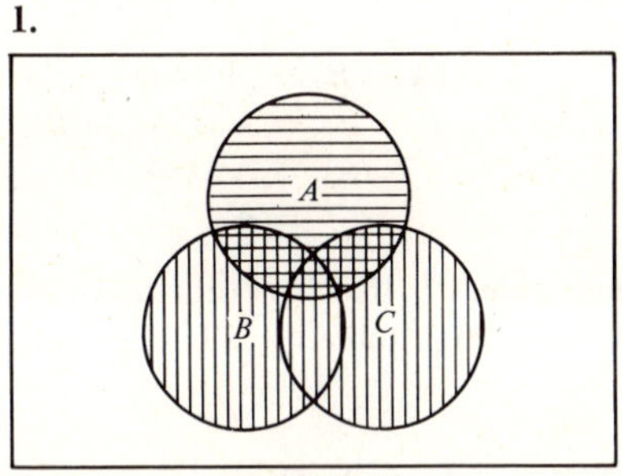

$A \cap (B \cup C)$
(Cross-hatch shading only)

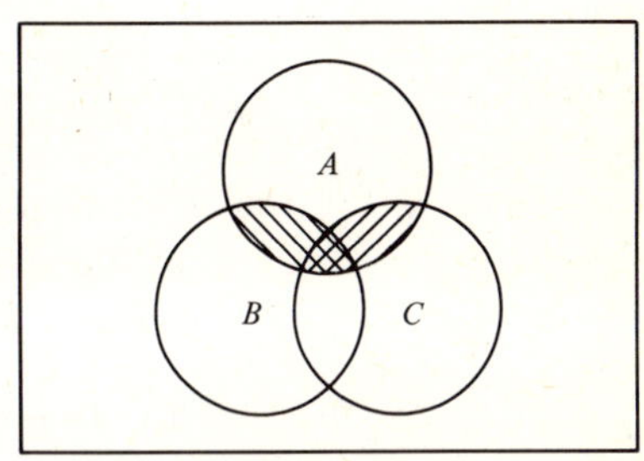

$(A \cap B) \cup (A \cap C)$
(Shaded region)

3. **(i)**

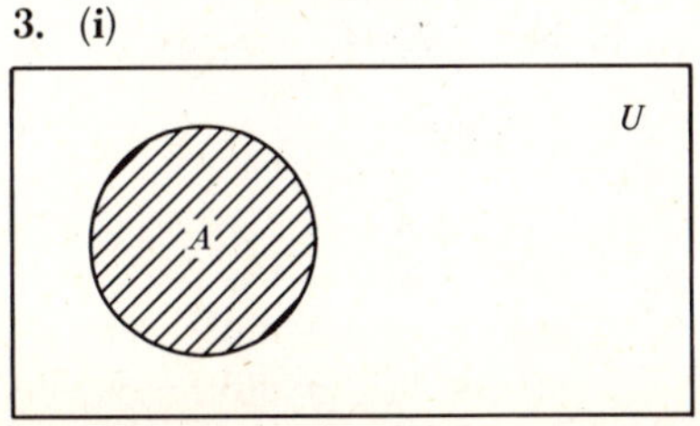

$A \cup \phi$

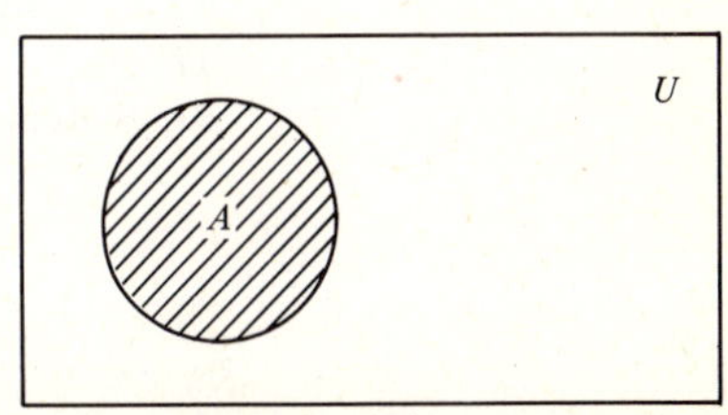

A

(ii)

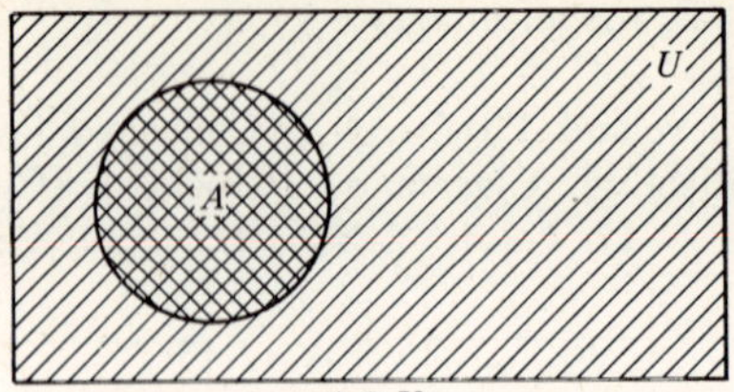

$A \cap U$

(Cross-hatch shading only)

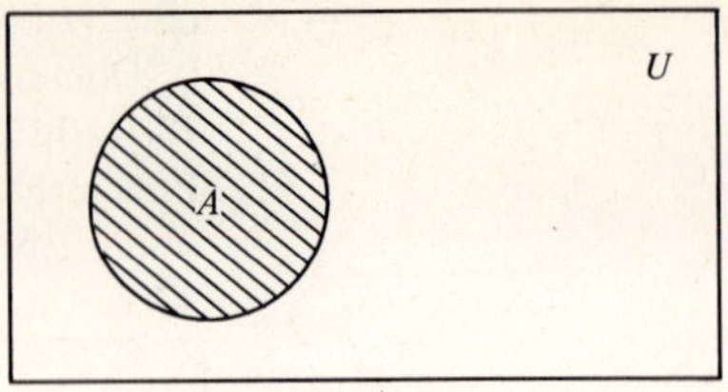

A

(Shaded region)

5. Statements:

$$\begin{aligned} A \cap \varnothing &= (A \cap \varnothing) \cup \varnothing \\ &= \varnothing \cup (A \cap \varnothing) \\ &= (A \cap A') \cup (A \cap \varnothing) \\ &= A \cap (A' \cup \varnothing) \\ &= A \cap A' \\ &= \varnothing . \end{aligned}$$

7. Statements:

$$\begin{aligned} A &= A \cap U \\ &= A \cap (A \cup A') \\ &= (A \cap A) \cup (A \cap A') \\ &= (A \cap A) \cup \varnothing \\ &= A \cap A. \end{aligned}$$

9. Given $A \cap X = A$ for every A. Hence if $A = U$, we have $U \cap X = U$. But $U \cap X = X$ by Theorem 3. Therefore, $U = U \cap X = X$.

11. Statements:

$$\begin{aligned} A \cap (A \cup B) &= (A \cup \varnothing) \cap (A \cup B) \\ &= A \cup (\varnothing \cap B) \\ &= A \cup \varnothing \\ &= A. \end{aligned}$$

13. Use evaluation table. **15.** Use evaluation table. **17.** Use evaluation table. **21.** Use evaluation table. **23.** Use evaluation table.

3.4 Page 81

1. Theorems 1, 2, 3, and 4. **3.** Theorems 1, 2, 3, and 4.

3.5 Page 85

1. 22. **3.** 41. **5.** 14. **7.** 44. **9.** 34. **11.** 3. **13.** 66. **15.** 4. **17.** 19. **19.** 75.

21. Statements:

$$\begin{aligned} U &= A \cup A'; \\ n(U) &= n(A \cup A'); \\ n(U) &= n(A) + n(A') - n(A \cap A'); \\ n(U) &= n(A) + n(A') - n(\varnothing); \\ n(U) - n(A) &= n(A'). \end{aligned}$$

3.6 Page 90

1. $A \times B = \{(a, x), (a, y), (a, z), (b, x), (b, y), (b, z), (c, x), (c, y), (c, z)\}$.

3.

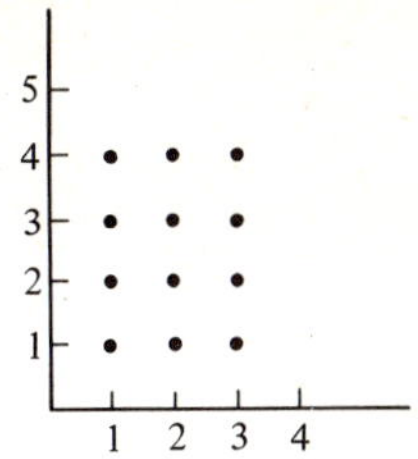

5. $P = \{\{x, y\}, \{x\}, \{y\}, \varnothing\}$. 7. Let C = set of coalitions. Then $C = P \cap \{\varnothing\}'$, where P is the power set of the set of all legislators. $n(C) = 2^{20} - 1$. 9. Diagram is similar to Figure 3.18 without the last column of paths. 11. $A \times M \times E \times D$; 72.

4.1 Page 99

1. $\frac{3}{10}$. 3. $\frac{1}{2}$. 5. $\frac{4}{9}$. 7. $U = \{mmm, mmf, mfm, mff, fmm, fmf, ffm, fff\}$; $E = \{mmf, mfm, fmm\}$. $P(E) = \frac{3}{8}$. 9. $\frac{5}{12}$. 11. $U = \{w, y, o\}$; $E = \{y, o\}$; $w(y) = \frac{3}{11}$; $w(o) = \frac{2}{11}$; $P(E) = \frac{5}{11}$. 13. $\frac{1}{4}$.

4.2 Page 104

1. (a) $U = \{1, 2, 3, 4, 5, 6\}$; (b) $E_1 = \{2, 4, 6\}$, $E_2 = \{3, 4, 5, 6\}$, $E_1 \cap E_2 = \{4, 6\}$, $E_1 \cup E_2 = \{2, 3, 4, 5, 6\}$, $E_2' = \{1, 2\}$; (c) $P(E_1) = \frac{1}{2}$, $P(E_2) = \frac{2}{3}$; (d) $P(E_1 \cap E_2) = \frac{1}{3}$; (e) $P(E_1 \cup E_2) = \frac{5}{6}$; (f) $P(E_2') = \frac{1}{3}$.
3. (a) $U = \{mmm, mmf, mfm, mff, fmm, fmf, ffm, fff\}$; (b) $E_1 = \{mmm, mmf, mfm, mff\}$, $E_2 = \{mmm, fff\}$, $E_1 \cap E_2 = \{mmm\}$, $E_1 \cup E_2 = \{mmm, mmf, mfm, mff, fff\}$, $E_2' = \{mmf, mfm, mff, fmm, fmf, ffm\}$; (c) $P(E_1) = \frac{1}{2}$, $P(E_2) = \frac{1}{4}$; (d) $P(E_1 \cap E_2) = \frac{1}{8}$; (e) $P(E_1 \cup E_2) = \frac{5}{8}$; (f) $P(E_2') = \frac{3}{4}$.
5. (a) $U = \{O_1, O_2, O_3\}$; (b) $E_1 = \{O_1, O_2\}$, $E_2 = \{O_2, O_3\}$, $E_1 \cap E_2 = \{O_2\}$, $E_1 \cup E_2 = \{O_1, O_2, O_3\}$, $E_2' = \{O_1\}$; (c) $P(E_1) = \frac{7}{12}$; $P(E_2) = \frac{3}{4}$; (d) $P(E_1 \cap E_2) = \frac{1}{3}$; (e) $P(E_1 \cup E_2) = 1$; (f) $P(E_2') = \frac{1}{4}$.
7. (a) $w(2) = \frac{1}{36}$, $w(3) = \frac{2}{36}$, $w(4) = \frac{3}{36}$, $w(5) = \frac{4}{36}$, $w(6) = \frac{5}{36}$, $w(7) = \frac{6}{36}$, $w(8) = \frac{5}{36}$, $w(9) = \frac{4}{36}$, $w(10) = \frac{3}{36}$, $w(11) = \frac{2}{36}$, $w(12) = \frac{1}{36}$; (b) $E_1 = \{2, 4, 6, 8, 10, 12\}$, $E_2 = \{7, 11\}$, $E_1 \cap E_2 = \varnothing$, $E_1 \cup E_2 = \{2, 4, 6, 7, 8, 10, 11, 12\}$, $E_2' = \{2, 3, 4, 5, 6, 8, 9, 10, 12\}$; (c) $P(E_1) = \frac{1}{2}$, $P(E_2) = \frac{2}{9}$; (d) $P(E_1 \cap E_2) = 0$; (e) $P(E_1 \cup E_2) = \frac{13}{18}$; (f) $P(E_2') = \frac{7}{9}$.

4.3 Page 111

1. (a) $U = \{JC, JD, JH, JS, QC, QD, QH, QS, KC, KD, KH, KS\}$; (b) $E_1 = \{KC, KD, KH, KS\}$; $E_2 = \{JD, QD, KD\}$, $E_1 \cap E_2 = \{KD\}$; (c) $P(E_1) = \frac{1}{3}$, $P(E_2) = \frac{1}{4}$, $P(E_1 \cap E_2) = \frac{1}{12}$; (d) $P(E_1 | E_2) = \frac{1}{3}$, $P(E_2 | E_1) = \frac{1}{4}$.
3. $\frac{1}{2}$. 5. $\frac{1}{3}$. 7. $\frac{2}{3}$. 9. $\frac{1}{12}$. 11. $\frac{6}{7}$.

4.4 Page 118

1. Not independent. 3. $\frac{3}{100}$. 5. $\frac{72}{100}$. 7. $\frac{1}{12}$. 9. $\frac{6}{100}$; $\frac{94}{100}$.
11. (a) No; (b) yes; (c) yes.
13. Statements:
(1) E_2 is independent of E_1 means $P(E_2 | E_1) = P(E_2)$.

(2) $\dfrac{P(E_1 \cap E_2)}{P(E_1)} = P(E_2 | E_1) = P(E_2).$

(3) $\dfrac{P(E_2 \cap E_1)}{P(E_1)} = P(E_2).$

(4) $\dfrac{P(E_2 \cap E_1)}{P(E_2)} = P(E_1).$

(5) $P(E_1 | E_2) = P(E_1)$.

(6) E_1 is independent of E_2.

4.5 Page 126

1. **(a)** 20; **(b)** $\frac{3}{10}$. **3.** $P_2^4 = 12$; $P_5^7 = 2520$; $P_3^{22} = 9240$; $P_8^8 = 40{,}320$.
5. 120. **7.** 5. **9.** 9999. **11.** **(a)** 360; **(b)** $\frac{1}{15}$; **(c)** $\frac{14}{15}$. **13.** $\frac{551}{1111}$.
15. $\dfrac{P_2^8 + P_2^8 + P_2^8}{P_4^{10}} = \dfrac{1}{30}$.

4.6 Page 133

1. $C_2^6 = 15$; $C_3^5 = 10$; $C_{18}^{20} = 190$. **3.** 1140. **5.** 16. **7.** $\dfrac{C_5^{26}}{C_5^{52}}$.

9. $\dfrac{C_2^{40}C_3^{60}}{C_5^{100}}$. **11.** **(a)** 60; **(b)** 6; **(c)** 0; **(d)** 252; **(e)** $\dfrac{11}{42}$.

13. $\dfrac{C_4^4C_2^8 + C_1^1C_5^{11} - C_4^4C_1^1C_1^7}{C_6^{12}}$.

4.7 Page 141

1. The 6 ordered partitions are $\{\{a, b\}, \{c, d\}\}$, $\{\{a, c\}, \{b, d\}\}$, $\{\{a, d\}, \{b, c\}\}$, $\{\{b, c\}, \{a, d\}\}$, $\{\{b, d\}, \{a, c\}\}$, and $\{\{c, d\}, \{a, b\}\}$.

3. 2520. **5.** $\dfrac{52!}{5!5!5!5!32!}$. **7.** **(a)** $\dfrac{48!5!}{52!}$; **(b)** $\dfrac{4 \cdot 48!5!}{52!}$. **9.** $\dfrac{5}{22}$.

11. **(a)** $\dfrac{52!}{13!13!13!13!}$; **(b)** $\dfrac{48!13!}{52!9!}$; **(c)** $\dfrac{4 \cdot 48!13!}{52!9!}$.

4.8 Page 149

1. $\frac{1}{4}$; yes; because the two trial outcomes are equally likely. **3.** $\frac{432}{2401}$.
5. $\frac{513}{2401}$. **7.** **(a)** .354294; **(b)** .885735. **9.** .31744 (3, 4, or 5 people constitute a majority of the sample). **11.** $\frac{27}{32}$. **13.** .1372.

4.9 Page 155

1. $E = \{(ND_2D_3)\}$; $P(E) = \frac{3}{4} \cdot \frac{13}{51} \cdot \frac{12}{50} = \frac{39}{850}$.
3. $E = \{WWL, WLW, LWW\}$; $P(E) = \frac{1}{4} + \frac{1}{24} + \frac{1}{8} = \frac{5}{12}$.
5. $P(E) = \frac{3}{16} + \frac{1}{24} + \frac{1}{8} = \frac{17}{48}$. **7.** $E = \{RW, DW, IW\}$, $P(E) = \frac{39}{100}$.

9. $\frac{1}{12} \cdot \frac{1}{3} + \frac{1}{9} \cdot \frac{4}{10} + \frac{5}{36} \cdot \frac{5}{11} + \frac{1}{6} + \frac{5}{36} \cdot \frac{5}{11} + \frac{1}{9} \cdot \frac{4}{10} + \frac{1}{12} \cdot \frac{1}{3} + \frac{1}{18} = \frac{244}{495}$.

4.10 Page 161

1. $-\$\frac{1}{19}$. **3.** $\$\frac{1}{2}$. **5.** $-\$\frac{1}{2}$. **7.** $-\$0.40$; odds are 9989 to 11 against winning a prize. **9.** 2.2 hours. **11.** Odds are 5 to 1 against winning; $-\$\frac{1}{3}$.
13. City B because the expected value (crowd) is 6400 compared to 4400 for city A.

5.1 Page 170

1. $\begin{bmatrix} 1 & 5 \\ 6 & 4 \end{bmatrix}$. **3.** $[9 \quad 12 \quad 5]$. **5.** Impossible. **7.** (a) $\begin{bmatrix} 2 & -5 & 12 \\ 2 & 21 & 5 \end{bmatrix}$;

(b) 2 by 3; (c) 5. **9.** $[a_{ij}]_{(2,3)}$. **11.** No.

13. $A + B = \begin{bmatrix} 1 & 1 \\ 7 & 6 \end{bmatrix}$; $B + A = \begin{bmatrix} 1 & 1 \\ 7 & 6 \end{bmatrix}$.

15. Reasons: (1) change in notation;
(2) definition of addition (Definition 3);
(3) addition of real numbers is commutative;
(4) definition of addition (Definition 3);
(5) change in notation.

17. $A + (B + C) = (A + B) + C$ by Associative Property (Theorem 2);
$= (B + A) + C$ by Commutative Property (Theorem 1).

5.2 Page 179

1. (7, 4, 0). **3.** (−3, 12, −8). **5.** 2. **7.** $\sqrt{21}$.

9.

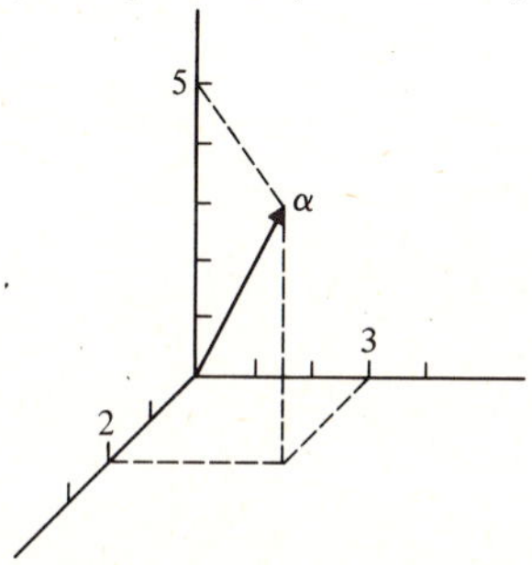

$\|\alpha\|$ represents the length of the arrow shown in the graph.

11. (6, 3, 4); 0.

13. $\left\|\left[\frac{1}{\sqrt{35}} \quad \frac{3}{\sqrt{35}} \quad \frac{5}{\sqrt{35}}\right]\right\| = 1.$

15. $\alpha + \beta = (a_1, \cdots, a_n) + (b_1, \cdots, b_n)$ by change of notation;
$= (a_1 + b_1, \cdots, a_n + b_n)$ by definition of addition;
$= (b_1 + a_1, \cdots, b_n + a_n)$ addition of real numbers is commutative;
$= (b_1, \cdots, b_n) + (a_1, \cdots, a_n)$ by definition of addition;
$= \beta + \alpha$ by change of notation.

17. No, because $\alpha - \beta = \alpha + (-1)\beta$, whereas $\beta - \alpha = \beta + (-1)\alpha$, and these two results may not be the same as can be demonstrated by an example.

19. Clothing industry, $15 profit; construction industry, $26 profit.

5.3 Page 187

1. $AB = \begin{bmatrix} 3 & -2 \\ 0 & 3 \end{bmatrix}$; $BA = \begin{bmatrix} 3 & 0 \\ -6 & 3 \end{bmatrix}$.

3. AB does not exist; because the number of columns of A is not the same as the number of rows of B; $BA = \begin{bmatrix} 2 & 9 & 7 \\ 2 & 21 & 13 \end{bmatrix}$.

5. $AB = [17]$; $BA = \begin{bmatrix} 8 & 24 \\ 3 & 9 \end{bmatrix}$.

7. $AB = \begin{bmatrix} (2x + 3y) \\ (4x + 5y) \end{bmatrix}$; BA does not exist because the number of columns of B is not the same as the number of rows of A.

9. $A^2 = \begin{bmatrix} 16 & 24 \\ 32 & 48 \end{bmatrix}$.

11. $A(BC) = \begin{bmatrix} 2 & 3 \\ 4 & 5 \end{bmatrix}\begin{bmatrix} 0 & 1 \\ 8 & 7 \end{bmatrix} = \begin{bmatrix} 24 & 23 \\ 40 & 39 \end{bmatrix}$;

$$(AB)C = \begin{bmatrix} 6 & 11 \\ 10 & 19 \end{bmatrix}\begin{bmatrix} 4 & 2 \\ 0 & 1 \end{bmatrix} = \begin{bmatrix} 24 & 23 \\ 40 & 39 \end{bmatrix}.$$

13. **(a)** $n = p$; **(b)** m by q; **(c)** $m = q$; **(d)** p by n; **(e)** $m = n$; **(f)** m by m.

15. Statements:

$A(B + C)$

$$= \begin{bmatrix} a_{11} & a_{12} \\ a_{21} & a_{22} \end{bmatrix}\begin{bmatrix} (b_{11} + c_{11}) & (b_{12} + c_{12}) \\ (b_{21} + c_{21}) & (b_{22} + c_{22}) \end{bmatrix}$$

$$= \begin{bmatrix} a_{11}(b_{11} + c_{11}) + a_{12}(b_{21} + c_{21}) & a_{11}(b_{12} + c_{12}) + a_{12}(b_{22} + c_{22}) \\ a_{21}(b_{11} + c_{11}) + a_{22}(b_{21} + c_{21}) & a_{21}(b_{12} + c_{12}) + a_{22}(b_{22} + c_{22}) \end{bmatrix}$$

$$= \begin{bmatrix} (a_{11}b_{11} + a_{12}b_{21}) + (a_{11}c_{11} + a_{12}c_{21}) & (a_{11}b_{12} + a_{12}b_{22}) + (a_{11}c_{12} + a_{12}c_{22}) \\ (a_{21}b_{11} + a_{22}b_{21}) + (a_{21}c_{11} + a_{22}c_{21}) & (a_{21}b_{12} + a_{22}b_{22}) + (a_{21}c_{12} + a_{22}c_{22}) \end{bmatrix}$$

$$= \begin{bmatrix} (a_{11}b_{11} + a_{12}b_{21}) & (a_{11}b_{12} + a_{12}b_{22}) \\ (a_{21}b_{11} + a_{22}b_{21}) & (a_{21}b_{12} + a_{22}b_{22}) \end{bmatrix} + \begin{bmatrix} (a_{11}c_{11} + a_{12}c_{21}) & (a_{11}c_{12} + a_{12}c_{22}) \\ (a_{21}c_{11} + a_{22}c_{21}) & (a_{21}c_{12} + a_{22}c_{22}) \end{bmatrix}$$

$$= \begin{bmatrix} a_{11} & a_{12} \\ a_{21} & a_{22} \end{bmatrix}\begin{bmatrix} b_{11} & b_{12} \\ b_{21} & b_{22} \end{bmatrix} + \begin{bmatrix} a_{11} & a_{12} \\ a_{21} & a_{22} \end{bmatrix}\begin{bmatrix} c_{11} & c_{12} \\ c_{21} & c_{22} \end{bmatrix}$$

$= AB + AC$.

17. $\begin{bmatrix} 0 & 0 & 0 & 0 & 0 \\ 0 & 0 & 0 & 0 & 0 \\ 0 & 0 & 0 & 0 & 0 \\ 1 & 0 & 1 & 0 & 0 \\ 1 & 1 & 2 & 0 & 0 \end{bmatrix}$.

5.4 Page 198

1. **(a)** E; **(b)** B; **(c)** A, B; **(d)** B; **(e)** A, B, D; **(f)** B, D, F; **(g)** A, B, C; **(h)** none.

5. $\left[\begin{array}{c|cc} 3 & 5 & 1 \\ \hline 4 & 0 & 6 \\ 9 & 2 & 1 \end{array}\right]$.

7. **(ii)** $(A + B)^T = \begin{bmatrix} 7 & 5 \\ 4 & 8 \end{bmatrix}$; $A^T + B^T = \begin{bmatrix} 7 & 5 \\ 4 & 8 \end{bmatrix}$.

(iii) $(AB)^T = \begin{bmatrix} 14 & 28 \\ 12 & 15 \end{bmatrix}$; $B^TA^T = \begin{bmatrix} 14 & 28 \\ 12 & 15 \end{bmatrix}$.

(iv) $(kA)^T = \begin{bmatrix} 3 & 9 \\ 12 & 15 \end{bmatrix}$; $kA^T = \begin{bmatrix} 3 & 9 \\ 12 & 15 \end{bmatrix}$.

9. Statements:
(1) Let $C = A + B$
(2) ijth entry of C^T = jith entry of C
(3) $= a_{ji} + b_{ji}$
(4) = (jith entry of A) + (jith entry of B)
(5) = (ijth entry of A^T) + (ijth entry of B^T)
(6) = ijth entry of $A^T + B^T$.

11. Statements:
(1) ijth entry of $(kA)^T$ = jith entry of (kA)
(2) = k(jith entry of A)
(3) = k(ijth entry of A^T)
(4) = ijth entry of kA^T.

13. A, B, C, D.

15. **(a)** Statements: (1) $A + B = A^T + B^T$
(2) $= (A + B)^T$.
(b) Statements: (1) $kA = kA^T$
(2) $= (kA)^T$.
(c) Statements: (1) $A^2 = AA$
(2) $= A^TA^T$
(3) $= (AA)^T$
(4) $= (A^2)^T$.
(d) Statements: (1) $AA^T = (A^T)^TA^T$
(2) $= (AA^T)^T$.
Also $AA^T = A^TA$ by direct substitution of $A = A^T$.

17. Proofs are similar to those of Exercise 15.

5.5 Page 208

1. $A = \begin{bmatrix} 1 & -1 \\ 2 & 4 \end{bmatrix}$; $[A \mid B] = \left[\begin{array}{cc|c} 1 & -1 & 2 \\ 2 & 4 & 1 \end{array}\right]$.

3. $A = \begin{bmatrix} 1 & 3 & 1 \\ 0 & -1 & -2 \end{bmatrix}$; $[A \mid B] = \begin{bmatrix} 1 & 3 & 1 & 2 \\ 0 & -1 & -2 & -3 \end{bmatrix}$.

5. (3, 0). **7.** (3, −2, 1). **9.** (1, 2). **11.** (1, 3, −2).

13. A single point which is the intersection of two lines.

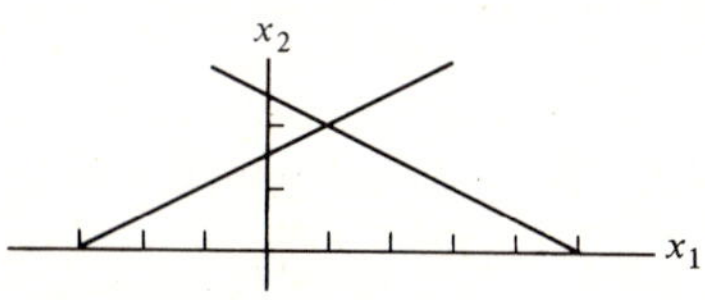

15. The graphs of the first part of the exercise will depend upon the choices of elementary operations and the order in which they are applied. A geometric interpretation of the sequence of algebraic steps is a sequence of rotations of the lines through the solution (1, 2).

5.6 Page 215

1. $\frac{1}{7}, \frac{5}{2}, \frac{1}{\sqrt{2}}$.

3. $\begin{bmatrix} -1 & \frac{1}{2} \\ 3 & -1 \end{bmatrix}$. **5.** No inverse because matrix is not row equivalent to I_2.

7. No inverse because matrix is not square.

9. $\begin{bmatrix} -9 & -11 & 5 \\ -4 & -3 & 2 \\ 2 & 2 & -1 \end{bmatrix}$. **11.** (2, 3). **13.** (12, 9, 2).

15. (a) (ii) $(AB)^{-1} = \begin{bmatrix} 1 & 6 \\ 2 & 13 \end{bmatrix}^{-1} = \begin{bmatrix} 13 & -6 \\ -2 & 1 \end{bmatrix}$;

$$B^{-1}A^{-1} = \begin{bmatrix} 1 & -3 \\ 0 & 1 \end{bmatrix}\begin{bmatrix} 7 & -3 \\ -2 & 1 \end{bmatrix} = \begin{bmatrix} 13 & -6 \\ -2 & 1 \end{bmatrix};$$

(b) No, because $A^{-1}B^{-1} \neq B^{-1}A^{-1}$.

17. Statements: (1) $(AB)(B^{-1}A^{-1}) = A((BB^{-1})A^{-1})$
(2) $= A(I_nA^{-1})$
(3) $= AA^{-1}$
(4) $= I_n$.
(5) Likewise, $(B^{-1}A^{-1})(AB) = I_n$.
(6) $B^{-1}A^{-1} = (AB)^{-1}$.

19. Statements: (1) $B = I_nB$
(2) $= (A^{-1}A)B$
(3) $= A^{-1}(AB)$
(4) $= A^{-1}I_n$
(5) $= A^{-1}$.
(6) Likewise, $B^{-1} = A$.
(7) $AB = AA^{-1} = A^{-1}A = BA$.

21. Statements:

$$\begin{aligned} AX = B &\Rightarrow A^{-1}(AX) = A^{-1}B \\ &\Rightarrow (A^{-1}A)X = A^{-1}B \\ &\Rightarrow I_nX = A^{-1}B \\ &\Rightarrow X = A^{-1}B, \end{aligned}$$

which is unique because A^{-1} is unique.

5.7 Page 221

1. $x_1 = 2 - \frac{1}{2}x_3$,
$x_2 = 1 - \frac{3}{4}x_3$; (2, 1, 0).

An interpretation of the complete solution is the set of all points on the line of intersection of two planes; an interpretation of a particular solution is a point on that line.

3. $x_1 = 2 - \frac{11}{3}x_3$,
$x_2 = 1 + \frac{4}{3}x_3$; $\left(\frac{-5}{3}, \frac{7}{3}, 1\right)$.

5. $x_1 = 3 - \frac{3}{2}x_3$,
$x_2 = -1 - \frac{1}{2}x_3$; $\left(\frac{3}{2}, \frac{-3}{2}, 1\right)$.

7. No solution.

9. $x_1 = \frac{14}{3} - x_3 - \frac{1}{3}x_4,$

$x_2 = \frac{-7}{3} - \frac{1}{3}x_4;$ $\quad \left(\frac{10}{3}, \frac{-8}{3}, 1, 1\right).$

11. If x_1 is the number of barrels of P, x_2 is the number of barrels of Q, and x_3 is the number of barrels of R, then a complete solution is

$$x_1 = 4 + 2x_3,$$
$$x_2 = 8 - 5x_3.$$

5.8 Page 225

1. **(b)** is reduced; **(a)** is not reduced because $I_{(1)}$ is *not* among its columns; **(c)** is not reduced because $I_{(3)}$ is *not* among its columns.

3. (1, 3, 2).

5. $x_1 = x_3,$ $x_2 = -x_3;$ $\quad (1, -1, 1).$

7. $x_1 = 7 - x_3 - 2x_4,$ $x_2 = 5 - x_3 - x_4;$ $\quad (4, 3, 1, 1).$

9. Exercise 5.

5.9 Page 230

1. **(a)** 9; **(b)** 18; **(c)** 0.

3. **(a)** $(1)\det\begin{bmatrix}1 & 1\\ 4 & 3\end{bmatrix} + (-1)(3)\det\begin{bmatrix}0 & 1\\ 2 & 3\end{bmatrix} + (4)\det\begin{bmatrix}0 & 1\\ 2 & 4\end{bmatrix} = -3;$

(b) $(-1)(3)\det\begin{bmatrix}0 & 1\\ 2 & 3\end{bmatrix} + (1)\det\begin{bmatrix}1 & 4\\ 2 & 3\end{bmatrix} + (-1)(4)\det\begin{bmatrix}1 & 4\\ 0 & 1\end{bmatrix} = -3;$

(c) $0 + (1)\det\begin{bmatrix}1 & 4\\ 2 & 3\end{bmatrix} + (-1)(1)\det\begin{bmatrix}1 & 3\\ 2 & 4\end{bmatrix} = -3;$

(d) $(1)\det\begin{bmatrix}1 & 1\\ 4 & 3\end{bmatrix} + 0 + (2)\det\begin{bmatrix}3 & 4\\ 1 & 1\end{bmatrix} = -3.$

5. -54. **7.** 3. **9.** -120.

11. $\det(AB) = -6$; $(\det A)(\det B) = -6$.

13. **(a)** If the two identical rows are interchanged, then by Theorem 8, $\det A = -\det A$. Thus $2\det A = 0$ and hence $\det A = 0$.

(b) If each entry of the ith row of A is k times the corresponding entry of the jth row of A, then by Theorem 9,

$$\det A = k \det B,$$

where B is a matrix having identical ith and jth rows. By part **(a)**, $\det B = 0$; hence $\det A = 0$.

17. $\begin{bmatrix}5 & -2\\ -1 & \frac{1}{2}\end{bmatrix}.$ **19.** $\begin{bmatrix}\frac{3}{2} & 1 & \frac{-3}{2}\\ -\frac{1}{2} & 0 & \frac{1}{2}\\ 1 & 0 & 0\end{bmatrix}.$

5.10 Page 237

1. $\lambda^2 - 5\lambda + 6 = 0$, $\lambda_1 = 3$, $\lambda_2 = 2$; two of many answers for X_1 and X_2 are $X_1 = \begin{bmatrix} 1 \\ 1 \end{bmatrix}$, $X_2 = \begin{bmatrix} 2 \\ 1 \end{bmatrix}$.

3. $\lambda^2 - 9 = 0$, $\lambda_1 = 3$, $\lambda_2 = -3$; two of many answers for X_1 and X_2 are $X_1 = \begin{bmatrix} 1 \\ 1 \end{bmatrix}$, $X_2 = \begin{bmatrix} -1 \\ 1 \end{bmatrix}$.

5. $\lambda_1 = 4$, $\lambda_2 = 2$; two of many answers for X_1 and X_2 are $X_1 = \begin{bmatrix} 1 \\ -3 \end{bmatrix}$, $X_2 = \begin{bmatrix} -1 \\ 1 \end{bmatrix}$.

7. $\lambda_1 = 2$, $\lambda_2 = 1$, $\lambda_3 = -2$; three of many answers for X_1, X_2, and X_3 are $X_1 = \begin{bmatrix} 1 \\ 0 \\ 1 \end{bmatrix}$, $X_2 = \begin{bmatrix} 0 \\ 1 \\ 0 \end{bmatrix}$, and $X_3 = \begin{bmatrix} -1 \\ 0 \\ 1 \end{bmatrix}$.

11. The characteristic equation is $\det(A - \lambda I_2) = 0$, or

$$\det\begin{bmatrix} a - \lambda & b \\ b & c - \lambda \end{bmatrix} = 0,$$

or

$$\lambda^2 - (a + c)\lambda + (ac - b^2) = 0.$$

If this equation is solved by the quadratic formula, we get

$$\lambda = \frac{a + c \pm \sqrt{(a + c)^2 - 4(ac - b^2)}}{2}.$$

Rearranging terms under the radical symbol, we get

$$\lambda = \frac{a + c \pm \sqrt{(a - c)^2 + 4b^2}}{2},$$

which is a real number regardless of the real values of a, b, and c.

6.1 Page 245

1.

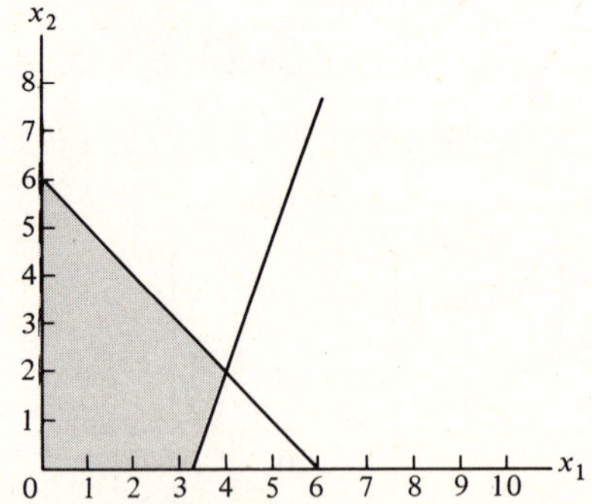

3.

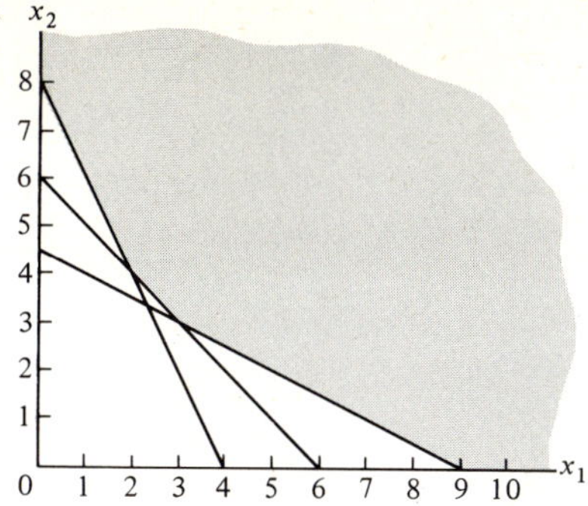

5. Structural constraints of Exercise 1 are
$$\begin{aligned} x_1 + x_2 &\le 6, \\ 3x_1 - x_2 &\le 10. \end{aligned}$$

Structural constraints of Exercise 3 are
$$\begin{aligned} 2x_1 + x_2 &\ge 8, \\ x_1 + x_2 &\ge 6, \\ x_1 + 2x_2 &\ge 9. \end{aligned}$$

Nonnegativity constraints of both exercises are
$$\begin{aligned} x_1 &\ge 0, \\ x_2 &\ge 0. \end{aligned}$$

7.

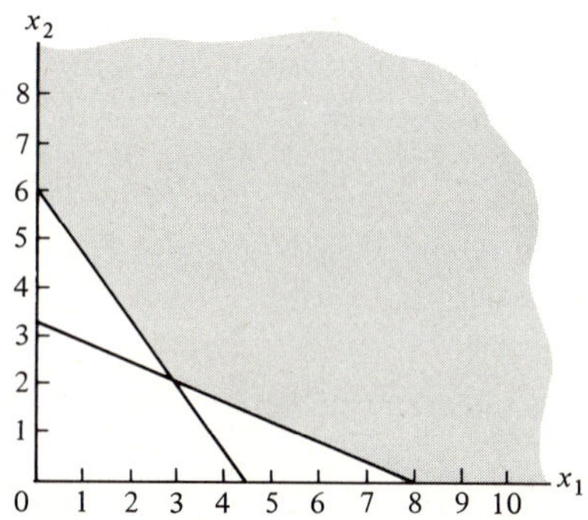

6.2 Page 250

1. Maximum $(x_1 + 2x_2) = 12$ at $(0, 6)$.
3. Minimum $(2x_1 + 3x_2) = 15$ at $(3, 3)$.
5. Minimum $(x_1 + x_2) = 2$ at $(2, 0)$.
7. Minimum $(2x_1 + 5x_2) = 16$ at $(8, 0)$ and at $(3, 2)$ and at all points on the line segment between $(8, 0)$ and $(3, 2)$.
9. Show that the family of lines representing the objective function intersects the feasible set no matter how large values of the objective function are.
11. Factory 1 should operate 3 days, and factory 2 should operate 2 days.
13. M_1 should be employed three times and M_2 should be employed one time.

6.3 Page 257

1. Basic solutions are $(0, 0, 14, 12)$, $(0, 12, -10, 0)$, $(4, 0, 10, 0)$, $(0, 7, 0, 5)$, $(14, 0, 0, -30)$, $(2, 6, 0, 0)$; basic *feasible* solutions are $(0, 0, 14, 12)$, $(4, 0, 10, 0)$, $(0, 7, 0, 5)$, $(2, 6, 0, 0)$.

3. $$\begin{aligned} x_1 + 2x_2 \quad\quad + x_4 \quad\quad &= 7, \\ x_1 - 3x_2 + x_3 \quad\quad - x_5 &= 2, \\ 2x_1 + x_2 + x_3 \quad\quad\quad\quad &= 4, \end{aligned} \qquad \begin{aligned} x_1 &\geq 0, \\ x_2 &\geq 0, \\ x_3 &\geq 0, \\ x_4 &\geq 0, \\ x_5 &\geq 0. \end{aligned}$$

5. **(a)**

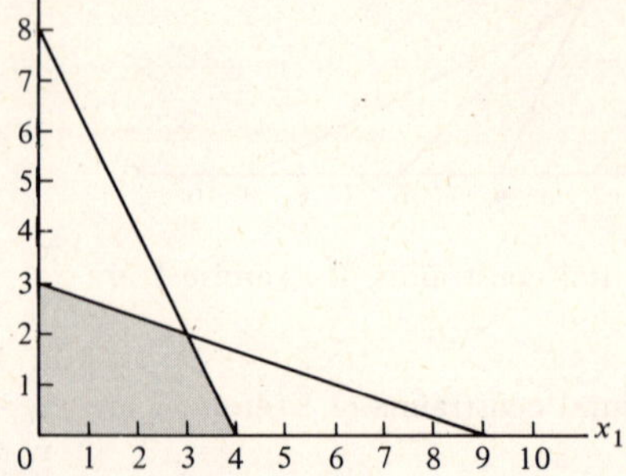

(b) (0, 0), (4, 0), (9, 0), (3, 2), (0, 8), (0, 3);

(c) $$\begin{aligned} x_1 + 3x_2 + x_3 \quad\quad &= 9, \\ 2x_1 + x_2 \quad\quad + x_4 &= 8; \end{aligned}$$

(0, 0, 9, 8), (4, 0, 5, 0), (9, 0, 0, −10), (3, 2, 0, 0), (0, 8, −15, 0), (0, 3, 0, 5);

(d) (0, 0, 9, 8), (4, 0, 5, 0), (3, 2, 0, 0), (0, 3, 0, 5);

(e) (3, 2, 0, 0).

9. Minimum $(3x_1 + x_2) = 4$ at (0, 4).

11. 5 thousand of P_1 and 17.5 thousand of P_2.

6.4 Page 260

1. Maximize CX subject to $AX \leq P_0$ and $X \geq \mathbf{0}$, where

$$A = \begin{bmatrix} 2 & 1 \\ -1 & 1 \end{bmatrix}, \; X = \begin{bmatrix} x_1 \\ x_2 \end{bmatrix}, \; P_0 = \begin{bmatrix} 4 \\ 5 \end{bmatrix}, \text{ and } C = [1 \quad 1].$$

3. Minimize CX subject to $AX = P_0$ and $X \geq \mathbf{0}$, where

$$A = \begin{bmatrix} 2 & 1 & 0 & 0 \\ 1 & 1 & 1 & 0 \\ 1 & -1 & 0 & -1 \end{bmatrix}, \; X = \begin{bmatrix} x_1 \\ x_2 \\ x_3 \\ x_4 \end{bmatrix}, \; P_0 = \begin{bmatrix} 3 \\ 4 \\ 2 \end{bmatrix}, \text{ and } C = [5 \quad 1 \quad 0 \quad 0].$$

5. Maximum $CX = 8$ at (4, 0).

7. Minimum $CX = 0$ at (0, 0).

6.5 Page 264

1. **(a)** $\begin{aligned} x_3 &= 1 + x_1 - x_2, \\ x_4 &= 5 - x_1 - 2x_2; \end{aligned}$ (0, 0, 1, 5);

(b) $f = 1 + 3x_1 - x_2$; **(c)** yes; x_1;

(d) $$M_1 = \left[\begin{array}{cccc|c} -1 & 1 & 1 & 0 & 1 \\ 1 & 2 & 0 & 1 & 5 \\ \hline 3 & -1 & 0 & 0 & f - 1 \end{array}\right];$$

(e) $x_3 = 6 - 3x_2 - x_4$,
$x_1 = 5 - 2x_2 - x_4$;
(f) $f = 16 - 7x_2 - 3x_4$; (g) no;
(h) maximum $f = 16$ at $(5, 0, 6, 0)$.

6.6 Page 270

1. Maximum $(x_1 + 2x_2) = 5$ at $(3, 1, 0, 0)$.
3. Maximum $(2x_1 + x_2 + 6x_3 + x_4) = 16$ at $(2, 0, 2, 0)$.
5. Maximum $(2x_1 + x_4) = 4$ at $(2, 0, 0, 0)$.
7. Minimum $(2x_1 + x_2) = 1$ at $(0, 1)$.
9. Show that eventually in one of the applications of step D, there are no positive entries in the pivot column above the last row.
11. Show that eventually in one of the applications of step D, there are no positive entries in the pivot column above the last row.
13. (b) The intersection of the four sets is empty.

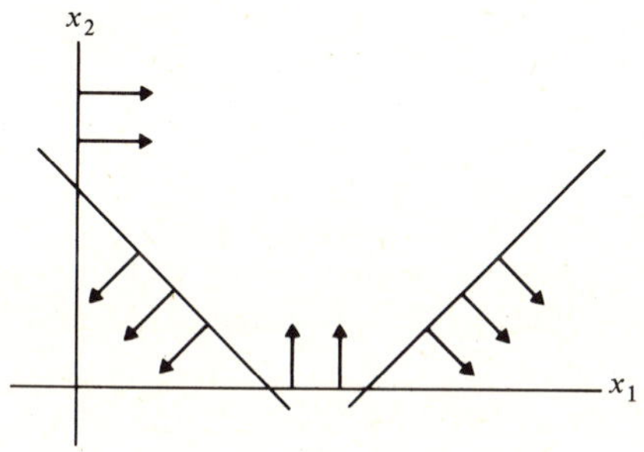

7.1 Page 277

1. (a) C plays
$$\begin{array}{c} H \quad T \\ \begin{bmatrix} 4 & -1 \\ -1 & -2 \end{bmatrix} \begin{matrix} H \\ T \end{matrix} \end{array} \Big\} R \text{ plays};$$
(b) one cent to C; (c) $\frac{3}{2}$ and $\frac{-3}{2}$; (d) 4 and -1; (e) -1 and -2; (f) $(1, 0)$.

3. (a) $\frac{3}{2}$ and $\frac{-3}{2}$ (loss); (b) 4 and -1 (loss); (c) -1 and -2 (loss); (d) $(0, 1)$.
5. (a) 50 and 55; (b) 70 and 40; (c) 30 and 70; (d) $\frac{370}{7}$ and $\frac{370}{7}$.

7. (a) $\frac{95}{2}$ and 60; (b) 70 and 30; (c) 40 and 70; (d) $\frac{370}{7}$ and $\frac{370}{7}$.

7.2 Page 285

1. Strictly determined; $(1, 0, 0)$ for R and $(0, 1, 0)$ for C; payoff is 3.
3. Strictly determined; $(1, 0, 0)$ for R and $(0, 1, 0)$ for C; payoff is 2.
5. Not strictly determined.

7. (a)

	Goes bad	Good part
Cheap	-16	-6
Expensive	-10	-8

$$\begin{array}{l} \\ \text{Cheap} \\ \text{Expensive} \end{array} \begin{array}{c}\text{Goes bad} \quad \text{Good part} \\ \begin{bmatrix} -16 & -6 \\ -10 & -8 \end{bmatrix}\end{array};$$

(b) yes; (c) (0, 1).

7.3 Page 290

1. Minimize $(x_1 + x_2)$ subject to $\begin{aligned} 2x_1 + 3x_2 &\geq 1, \\ x_1 + 2x_2 &\geq 1, \end{aligned}$ and $\begin{aligned} x_1 &\geq 0, \\ x_2 &\geq 0, \end{aligned}$

where $x_1 = \dfrac{p_1}{v}$ and $x_2 = \dfrac{p_2}{v}$.

3. Minimize $(x_1 + x_2 + x_3)$ subject to

$$\begin{aligned} 3x_1 \qquad\quad + \ x_3 &\geq 1, \\ 4x_2 + 2x_3 &\geq 1, \\ x_1 + x_2 + 6x_3 &\geq 1, \end{aligned} \qquad \text{and} \qquad \begin{aligned} x_1 &\geq 0, \\ x_2 &\geq 0, \\ x_3 &\geq 0, \end{aligned}$$

where $x_1 = \dfrac{p_1}{v}$, $x_2 = \dfrac{p_2}{v}$, and $x_3 = \dfrac{p_3}{v}$.

5. $v = \frac{3}{2}$ for $(\frac{1}{2}, \frac{1}{2})$. **7.** $v = 0$ for $(\frac{1}{2}, \frac{1}{2})$.

13. Start with the constraints on the middle of page 288 (but with k added to each entry of the payoff matrix) and v_1 the new value for v.

$$\begin{aligned} (a_{11} + k)p_1 + \cdots + (a_{m1} + k)p_m - v_1 &\geq 0, \\ \vdots \qquad\qquad\qquad \vdots \qquad\quad & \ \vdots \\ (a_{1n} + k)p_1 + \cdots + (a_{mn} + k)p_m - v_1 &\geq 0, \\ p_1 + \cdots + \qquad\qquad p_m \qquad &= 1. \end{aligned}$$

Since $p_1 + \cdots + p_m = 1$, these constraints can be arranged to be

$$\begin{aligned} a_{11}p_1 + \cdots + a_{m1}p_m - (v_1 - k) &\geq 0, \\ \vdots \qquad\qquad \vdots \qquad\quad & \ \vdots \\ a_{1n}p_1 + \cdots + a_{mn}p_m - (v_1 - k) &\geq 0, \\ p_1 + \cdots + \quad p_m \qquad &= 1. \end{aligned}$$

Now let $v_1 - k = v$. The solution of this problem is the same as that shown following the constraints on the middle of page 288, where v^* is maximum v. Since $v_1 = k + v$, max $v_1 = k + v^*$ and so new row value is old row value plus k.

7.4 Page 297

1. Maximize $(y_1 + y_2)$ subject to $\begin{aligned} 2y_1 + y_2 &\leq 1, \\ 3y_1 + 2y_2 &\leq 1, \end{aligned}$ and $\begin{aligned} y_1 &\geq 0, \\ y_2 &\geq 0, \end{aligned}$

where $y_1 = \dfrac{q_1}{w}$ and $y_2 = \dfrac{q_2}{w}$.

3. Maximize $(y_1 + y_2 + y_3)$ subject to

$$\begin{aligned} 3y_1 \qquad\quad + \ y_3 &\le 1, \\ 4y_2 + \ y_3 &\le 1, \\ y_1 + 2y_2 + 6y_3 &\le 1, \end{aligned} \qquad \text{and} \qquad \begin{aligned} y_1 &\ge 0, \\ y_2 &\ge 0, \\ y_3 &\ge 0, \end{aligned}$$

where $y_1 = \dfrac{q_1}{w}$, $y_2 = \dfrac{q_2}{w}$, and $y_3 = \dfrac{q_3}{w}$.

5. $w = \frac{3}{2}$ for $(\frac{1}{2}, \frac{1}{2})$. 7. $w = 0$ for $(\frac{2}{3}, \frac{1}{3})$.

Index